Chemical Kinetics
and Reaction Dynamics

Chemical Kinetics and Reaction Dynamics

Paul L. Houston

Cornell University

McGraw Hill

Boston   Burr Ridge, IL   Dubuque, IA   Madison, WI   New York   San Francisco   St. Louis
Bangkok   Bogotá   Caracas   Kuala Lumpur   Lisbon   London   Madrid   Mexico City
Milan   Montreal   New Delhi   Santiago   Seoul   Singapore   Sydney   Taipei   Toronto

McGraw-Hill Higher Education

*A Division of The **McGraw-Hill** Companies*

CHEMICAL KINETICS AND REACTION DYNAMICS

Publisher: *James M. Smith*
Developmental editor: *Spencer J. Cotkin, Ph.D.*
Marketing manager: *Thomas Timp*
Project manager: *Jane E. Matthews*
Production supervisor: *Laura Fuller*
Designer: *K. Wayne Harms*
Cover designer: *Rokusek Design*
Senior photo research coordinator: *Carrie K. Burger*
Supplement producer: *Jodi K. Banowetz*
Compositor: *Lachina Publishing Services*
Typeface: *10.5/12 Times Roman*
Printer: *Courier Westford*

Cover image: Velocity distribution of $O(^1D)$ produced in the photodissociation of ozone at 265 nm. The four rings correspond to production of the sibling $O_2(^1\Delta_g)$ fragment in $v = 0$ (outer right), 1, 2, or 3 (inner ring). The image was taken by Scott Dylewski in the laboratory of Paul L. Houston using the technique described in Section 7.5.5.

Library of Congress Cataloging-in-Publication Data

Houston, Paul L.
    Chemical kinetics and reaction dynamics / Paul L. Houston. — 1st ed.
      p.   cm.
    Includes index.
    ISBN 0–07–243537–2
    1. Chemical kinetics. 2. Chemical reactions. I. Title.

QD502 .H7   2001
541.3'94—dc21

00–064712
CIP

www.mhhe.com

This work is dedicated to Barbara Lynch, without whose support it would not have been completed.

Contents

Preface

Chemical Kinetics and Reaction Dynamics is a textbook in modern chemical kinetics. There are two operative words here, *textbook* and *modern*. It is a textbook, not a reference book. While the principal aim of a reference book is to cover as many topics as possible, the principal aim of a textbook is to teach. In my view, a serious problem with modern "textbooks" is that they have lost the distinction. As a consequence of incorporating too many topics, these books confuse their audience; students have a difficult time seeing the forest through the trees. This textbook first aims to teach, and to teach as well as possible, the underlying principles of kinetics and dynamics. Encyclopedic completeness is sacrificed for an emphasis on these principles. I aim to present them in as clear a fashion as possible, using several examples to enhance basic understanding rather than racing immediately to more specialized applications. The more technical applications are not totally neglected; many are included as separate sections or appendices, and many are covered in sets of problems that follow each chapter. But the emphasis is on making this a textbook.

The second operative word is modern. Even recently written texts often use quite dated examples. Important aims of this textbook are first to demonstrate that the basic kinetic principles are essential to the solution of modern chemical problems and second to show how the underlying question, "how do chemical reactions occur," leads to exciting, vibrant fields of modern research. The first aim is achieved by using relevant examples in presenting the basic material, while the second is attained by inclusion of chapters on surface processes, photochemistry, and reaction dynamics.

Chemical Kinetics and Reaction Dynamics provides, then, a modern textbook. In addition to teaching and showing modern relevance, any textbook should be flexible enough so that individual instructors may choose their own sequence of topics. In as much as possible, the chapters of this text are self-contained; when needed, material from other sections is clearly referenced. An introduction to each chapter identifies the basic goals, their importance, and the general plan for achieving those goals. The text is designed for several possible formats. Chapters 1, 2, and 3 form a basic package for a partial semester introduction to kinetics. The basic material can be expanded by inclusion of Chapter 4. Chapters 5 through 8 can be included for a full semester course. Taken in its entirety, the text is suitable for a one-semester course at the third-year undergraduate level or above. I have used it for many years in a first-year graduate course.

While rigorous mathematical treatment of the topic cannot and should not be avoided if we are to give precision to the basic principles, the greatest problem students have with physical chemistry is keeping sight of the chemistry while wading through the mathematics. This text endeavors to emphasize the chemistry by two techniques. First, the chemical objectives and the reasons for undertaking the mathematical routes to those objectives are clearly stated; the mathematics is treated as a means to an end rather than an end in itself. Second, the text includes several "conceptual" problems in addition to the traditional "method" problems. Recent research on the teaching of physics has shown that, while students can frequently memorize the recipe for solving particular types of problems, they often fail to develop conceptual intuition.* The first few problems at the end of each chapter are designed as a conceptual self-test for the student.

---

*I. A. Halloun and D. Hestenes, *Am. J. Phys.* **53**, 1043 (1985); **53**, 1056 (1985); **55**, 455 (1987); D. Hestenes, *Am. J. Phys.* **55**, 440 (1987); E. Mazur, *Opt. Photon. News* **2**, 38 (1992).

The text assumes some familiarity with elementary kinetics at the level of high-school or freshman chemistry, physics at the freshman level, and mathematics through calculus. Each chapter then builds upon this basis using observations, derivations, examples, and instructive figures to reach clearly identified objectives.

I am grateful to Professor T. Michael Duncan for providing some of the problems used in Chapters 2 and 3, to Brian Bocknack and Julie Mueller for assistance with the problems and solutions, to Jeffrey Steinfeld and Joseph Francisco for helpful suggestions, to many outside reviewers of the text, especially Laurie Butler, for good suggestions, and to my wife, Barbara Lynch, for support and tolerance during the long periods when I disappeared to work on the text.

Paul Houston
Ithaca, New York

Introduction

A User's Guide to Chemical Kinetics and Reaction Dynamics

Chemistry is the study of the composition, structure, and properties of substances; of the transformation between various substances by reaction; and of the energy changes that accompany reaction. In these broad terms, *physical* chemistry is then the subbranch of the discipline that seeks to understand chemistry in quantitative and theoretical terms; it uses the tools of physics and mathematics to predict and explain macroscopic behavior on a microscopic level.

Physical chemistry can, in turn, be described by its subfields. Thermodynamics deals primarily with *macroscopic* manifestations of chemistry: the transformations between work and heat, the stability of compounds, and the equilibrium properties of reactions. Quantum mechanics and spectroscopy, on the other hand, deal primarily with *microscopic* manifestations of chemistry: the structure of matter, its energy levels, and the transitions between these levels. The subfield of statistical mechanics relates the microscopic properties of matter to the macroscopic observables such as energy, entropy, pressure, and temperature.

At their introductory level, however, all of these fields emphasize properties at equilibrium. Thermodynamics can be used to calculate an equilibrium constant, but it cannot be used to predict the rate at which equilibrium will be approached. For example, a stoichiometric mixture of hydrogen and oxygen is predicted by thermodynamics to react to water, but kinetics can be used to calculate that the reaction will take on the order of 10^{25} years ($\approx 3 \times 10^{32}$ s) at room temperature, though only 10^{-6} s in the presence of a flame. Similarly, quantum mechanics can do a good job at predicting the spacing of energy levels, but it does not do very well, at least at the elementary level, in providing simple reasons why population of some energy levels will be preferred over others following a reaction. Many reactions produce products in a Maxwell-Boltzmann distribution, but some, such as those responsible for chemical lasers, produce an "inverted" distribution that, over a specified energy range, is characterized by a negative temperature. We would like to have an understanding of why the rate for a reaction can be changed by 38 orders of magnitude, or why a reaction yields products in very specific, nonequilibrium distributions over energy levels.

Questions about the rates of processes and about how reactions take place are the purview of chemical kinetics and reaction dynamics. Because this subfield of physical chemistry is the one most concerned with the "how, why, and when" of chemical reaction, it is a central intellectual cornerstone to the discipline of chemistry.

And yet it is of enormous practical importance as well. Chemical reactions control our environment, our life processes, our food production, and our energy utilization. Understanding of and possible influence over the rates of chemical reactions could provide a healthier environment and a better life, with adequate food and more efficient resource management.

Thus, chemical kinetics is both an exciting intellectual frontier and a field that addresses societal needs as well. At the present time both the intellectual and practical forefronts of chemical kinetics are linked to a rapidly developing new set of instrumental techniques, including lasers that can push our time resolution to 10^{-15} s or detect concentrations at sensitivities approaching one part in 10^{16}, microscopes that can see individual atoms, and computers that can calculate some rate constants more accurately than they can be measured. These techniques are being applied to rate processes in all phases of matter, to reactions in solids, liquids, gases, plasmas, and even at the narrow interfaces between such phases. Never before have we been in such a good position to answer the fundamental question "how do molecules react?"

We begin our answer to this question by examining the motions of gas-phase molecules. What are their velocities, and what controls the rate of collisions among them? In Chapter 1, "Kinetic Theory of Gases," we will see that at equilibrium the molecular velocities can be described by the Boltzmann distribution and that factors such as the size, relative velocity, and molecular density influence the number of collisions per unit time. We will also develop an understanding of one of the central tools of physical chemistry, the distribution function.

We then examine the rates of chemical reactions in Chapter 2, first concentrating on the macroscopic observables such as the order of a reaction and its rate constant, but then examining how the overall rate of a reaction can be broken down into a series of elementary, molecular steps. Along the way we will develop some powerful tools for analyzing chemical rates, tools for determining the order of a reaction, tools for making useful approximations (such as the "steady-state" approximation), and tools for analyzing more complex reaction mechanisms.

In Chapter 3, "Theories of Chemical Reactions," we look at reaction rates from a more microscopic point of view, drawing on quantum mechanics, statistical mechanics, and thermodynamics to help us understand the magnitude of chemical rates and how they vary both with macroscopic parameters like temperature and with microscopic parameters like molecular size, structure, and energy spacing.

Chapter 4, "Transport Properties," uses the velocity distribution developed in Chapter 1 to provide a coherent description of thermal conductivity, viscosity, and diffusion, that is, a description of the movement of such properties as energy, momentum, or concentration through a gas. We will see that these properties are passed from one molecule to another upon collision, and that the mean distance between collisions, the "mean free path," is an important parameter governing the rate of such transport.

Armed with the fundamental material of the first four chapters, we move to four exciting areas of modern research: "Reactions in Liquid Solutions" (Chapter 5), "Reactions at Solid Surfaces" (Chapter 6), "Photochemistry" (Chapter 7), and "Molecular Reaction Dynamics" (Chapter 8).

The material of the text can be presented in several different formats depending on the amount of time available. The complete text can be covered in 12–14 weeks assuming 3 hours of lecture per week. In this format, the text might form the basis of an advanced undergraduate or beginning graduate level course. A more likely scenario, given the pressures of current instruction in physical chemistry, is one in which only the very fundamental topics are covered in detail. **Table 1** shows a flow chart giving the order of presentation and the number of lectures required for the fundamental material; the total number of lectures ranges between 11 and 17.

Of course, if more time is available, the instructor can supplement the fundamental material with selected topics from later chapters. Several suggestions, including the number of lectures required, are given in **Table 2** through **Table 5.**

TABLE 1	Fundamental Sections for a Course in Kinetics

Most Important Sections (Lectures)	Supplemental (Lectures)
1.1–1.6 (3)	
1.7 (1)	4.1–4.8 (3)
2.1–2.5 (4)	2.6 (2)
3.1–3.5 (3)	5.1–5.2 (1)
Total Lectures: 11	Total Lectures: 6

TABLE 2	Reactions in Liquid Solutions

Fundamental (Lectures)	Supplemental (Lectures)	Advanced (Lectures)
5.1–5.3 (2)	5.4 (1)	

TABLE 3	An Introduction to Surface Kinetics

Fundamental (Lectures)	Supplemental (Lectures)	Advanced (Lectures)
6.1–6.3, 6.6 (2)	6.4 (1)	6.5 (1)

TABLE 4	Photochemistry and Atmospheric Chemistry

Fundamental (Lectures)	Supplemental (Lectures)	Advanced (Lectures)
7.1, 7.2 (1)		
7.3.1, 7.3.4 (1)	7.3.2, 7.3.3 (1)	
7.4 (1)		7.5 (2)
Total Lectures: 3	Total Lectures: 2	Total Lectures: 2

TABLE 5	Reaction Dynamics

Fundamental (Lectures)	Supplemental (Lectures)	Advanced (Lectures)
8.1, 8.2, 8.3 (2)	8.4 (1)	
8.5 (2)	8.6 (1)	8.7 (1)
Total Lectures: 4	Total Lectures: 2	Total Lectures: 1

Kinetic Theory of Gases

Chapter Outline

1.1 INTRODUCTION

The overall objective of this chapter is to understand macroscopic properties such as pressure and temperature on a microscopic level. We will find that the pressure of an ideal gas can be understood by applying Newton's law to the microscopic motion of the molecules making up the gas and that a comparison between the Newtonian prediction and the ideal gas law can provide a function that describes the distribution of molecular velocities. This distribution function can in turn be used to learn about the frequency of molecular collisions. Since molecules can react only as fast as they collide with one another, the collision frequency provides an upper limit on the reaction rate.

The outline of the discussion is as follows. By applying Newton's laws to the molecular motion we will find that the product of the pressure and the volume is proportional to the average of the square of the molecular velocity, $<v^2>$, or equivalently to the average molecular translational energy ϵ. In order for this result to be consistent with the observed ideal gas law, the temperature T of the gas must also be proportional to $<v^2>$ or $<\epsilon>$. We will then consider in detail how to determine the average of the square of the velocity from a distribution of velocities, and we will use the proportionality of T with $<v^2>$ to determine the Maxwell-Boltzmann distribution of speeds. This distribution, $F(v)\,dv$, tells us the number of molecules with speeds between v and $v + dv$. The speed distribution is closely related to the distribution of molecular energies, $G(\epsilon)\,d\epsilon$. Finally, we will use the velocity distribution to calculate the number of collisions Z that a molecule makes with other molecules in the gas per unit time. Since in later chapters we will argue that a reaction between two molecules requires that they collide, the collision rate Z provides an upper limit to the rate of a reaction. A related quantity λ is the average distance a molecule travels between collisions or the *mean free path*.

The history of the kinetic theory of gases is a checkered one, and serves to dispel the impression that science always proceeds along a straight and logical path.[a] In 1662 Boyle found that for a specified quantity of gas held at a fixed temperature the product of the pressure and the volume was a constant. Daniel Bernoulli derived this law in 1738 by applying Newton's equations of motion to the molecules comprising the gas, but his work appears to have been ignored for more than a century.[b] A school teacher in Bombay, India, named John James Waterston submitted a paper to the Royal Society in 1845 outlining many of the concepts that underlie our current understanding of gases. His paper was rejected as "nothing but nonsense, unfit even for reading before the Society." Bernoulli's contribution was rediscovered in 1859, and several decades later in 1892, after Joule (1848) and Clausius (1857) had put forth similar ideas, Lord Rayleigh found Waterston's manuscript in the Royal Society archives. It was subsequently published in *Philosophical Transactions*. Maxwell (*Illustrations of Dynamical Theory of Gases*, 1859–1860) and Boltzmann (*Vorlesungen über Gastheorie*, 1896–1898) expanded the theory into its current form.

1.2   PRESSURE OF AN IDEAL GAS

We start with the basic premise that the pressure exerted by a gas on the wall of a container is due to collisions of molecules with the wall. Since the number of molecules in the container is large, the number colliding with the wall per unit time is large enough so that fluctuations in the pressure due to the individual collisions are immeasurably small in comparison to the total pressure. The first step in the calculation is to apply Newton's laws to the molecules to show that the product of the pressure and the volume is proportional to the average of the square of the molecular velocity, $<v^2>$.

Consider molecules with a velocity component v_x in the x direction and a mass m. Let the molecules strike a wall of area A located in the z-y plane, as shown in **Figure 1.1.** We would first like to know how many molecules strike the wall in a time Δt, where Δt is short compared to the time between molecular collisions. The distance along the x axis that a molecule travels in the time Δt is simply $v_x \Delta t$, so that all molecules located in the volume $A v_x \Delta t$ and moving toward the wall will strike it. Let n^* be the number of molecules per unit volume. Since one half of the molecules will be moving toward the wall in the $+x$ direction while the other half will be moving in the $-x$ direction, the number of molecules which will strike the wall in the time Δt is $\frac{1}{2} n^* A v_x \Delta t$.

The force on the wall due to the collision of a molecule with the wall is given by Newton's law: $F = ma = m \, dv/dt = d(mv)/dt$, and integration yields $F \Delta t = \Delta(mv)$. If a molecule rebounds elastically (without losing energy) when it hits the wall, its momentum is changed from $+mv_x$ to $-mv_x$, so that the total momentum change is $\Delta(mv) = 2mv_x$. Consequently, $F \Delta t = 2mv_x$ for one molecular collision, and $F \Delta t = (\frac{1}{2} n^* A v_x \Delta t)(2mv_x)$ for the total number of collisions. Canceling Δt from both sides and recognizing that the pressure is the force per unit area, $p = F/A$, we obtain $p = n^* m v_x^2$.

[a]The history of the kinetic theory of gases is outlined by E. Mendoza, *Physics Today* **14**, 36–39 (1961).
[b]A translation of this paper has appeared in *The World of Mathematics*, J. R. Newman, Ed., Vol. 2 (Simon and Schuster, New York, 1956), p. 774.

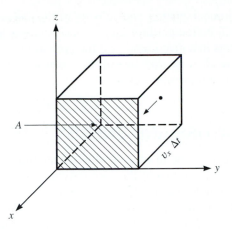

Figure 1.1

All the molecules in the box that are moving toward the z-y plane will strike the wall.

Of course, not all molecules will be traveling with the same velocity v_x. We will learn below how to characterize the distribution of molecular velocities, but for now let us simply assume that the pressure will be proportional to the average of the square of the velocity in the x direction, $p = n^* m <v_x^2>$.[c] The total velocity of an individual molecule most likely contains other components along y and z. Since $\mathbf{v} = \hat{i} v_x + \hat{j} v_y + \hat{k} v_z$,[d] where $\hat{i}$, $\hat{j}$, and $\hat{k}$ are unit vectors in the x, y, and z directions, respectively, $\mathbf{v}^2 = v_x^2 + v_y^2 + v_z^2$ and $<\mathbf{v}^2> = <v_x^2> + <v_y^2> + <v_z^2>$. In an isotropic gas the motion of the molecules is random, so there is no reason for the velocity in one particular direction to differ from that in any other direction. Consequently, $<v_x^2> = <v_y^2> = <v_z^2> = <v^2>/3$. When we combine this result with the calculation above for the pressure, we obtain

$$p = \frac{1}{3} n^* m <v^2>. \qquad (1.1)$$

Of course, n^* in **equation 1.1** is the number of molecules per unit volume and can be rewritten as $n N_A / V$, where N_A is Avogadro's number and n is the number of moles. The result is

$$pV = \frac{1}{3} n N_A m <v^2>. \qquad (1.2)$$

Since the average kinetic energy of the molecules is $<\epsilon> = \frac{1}{2} m <v^2>$, another way to write **equation 1.2** is

$$pV = \frac{2}{3} n N_A <\epsilon>. \qquad (1.3)$$

Equations 1.2 and **1.3** bear a close resemblance to the ideal gas law, $pV = nRT$. The ideal gas law tells us that the product of p and V will be constant if the temperature is constant, while **equations 1.2** and **1.3** tell us that the product will be constant if $<v^2>$ or $<\epsilon>$ is constant. The physical basis for the constancy of pV with $<v^2>$ or $<\epsilon>$ is clear from our previous discussion. If the volume is

---

[c]In this text, as in many others, we will use the notation $<x>$ or $\bar{x}$ to mean "the average value of x."

[d]Throughout the text we will use **boldface** symbols to indicate vector quantities and normal weight symbols to indicate scalar quantities. Thus, $v = |\mathbf{v}|$. Note that $\mathbf{v}^2 = \mathbf{v} \cdot \mathbf{v} = v^2$.

increased while the number, energy, and velocity of the molecules remain constant, then a longer time will be required for the molecules to reach the walls; there will thus be fewer collisions in a given time, and the pressure will decrease. To identify **equation 1.3** with the ideal gas law, we need to consider in more detail the relationship between temperature and energy.

1.3   TEMPERATURE AND ENERGY

Consider two types of molecule in contact with one another. Let the average energy of the first type be $<\epsilon>_1$ and that of the second type be $<\epsilon>_2$. If $<\epsilon>_1$ is greater than $<\epsilon>_2$, then when molecules of type 1 collide with those of type 2, energy will be transferred from the former to the latter. This energy transfer is a form of heat flow. From a macroscopic point of view, as heat flows the temperature of a system of the type 1 molecules will decrease, while that of the type 2 molecules will increase. Only when $<\epsilon>_1 = <\epsilon>_2$ will the temperatures of the two macroscopic systems be the same. In mathematical terms, we see that $T_1 = T_2$ when $<\epsilon>_1 = <\epsilon>_2$ and that $T_1 > T_2$ when $<\epsilon>_1 > <\epsilon>_2$. Consequently, there must be a correspondence between $<\epsilon>$ and T so that the latter is some function of the former: $T = T(<\epsilon>)$.

The functional form of the dependence of T on $<\epsilon>$ cannot be determined solely from kinetic theory, since the temperature scale can be chosen in many possible ways. In fact, one way to define the temperature is through the ideal gas law: $T = pV/(nR)$. Experimentally, this corresponds to measuring the temperature either by measuring the volume of an ideal gas held at constant pressure or by measuring the pressure of an ideal gas held at constant volume. Division of both sides of **equation 1.3** by nR and use of the ideal gas relation gives us the result

$$T = \frac{pV}{nR} = \frac{2}{3}\frac{N_A}{R}<\epsilon>, \tag{1.4}$$

or

$$<\epsilon> = \frac{3}{2}kT, \tag{1.5}$$

where k, known as Boltzmann's constant, is defined as R/N_A. Note that since $<\epsilon> = \frac{1}{2}m<v^2>$,

$$<v^2> = \frac{3kT}{m}. \tag{1.6}$$

example 1.1

Calculation of Average Energies and Squared Velocities

Objective   Calculate the average molecular energy, $<\epsilon>$, and the average squared velocity, $<v^2>$, for a nitrogen molecule at $T = 300$ K.

Method   Use **equations 1.5** and **1.6** with $m =$ (28 g/mole)(1 kg/1000 g)/(N_A molecule/mole) and $k = 1.38 \times 10^{-23}$ J/K.

Solution $<\epsilon> = 3kT/2 = 3(1.38 \times 10^{-23}$ J/K)(300 K)/2 $= 6.21 \times 10^{-21}$ J.

$$<v^2> = \frac{3kT}{m}$$

$$= 3 \frac{(1.38 \times 10^{-23} \text{ J/K})(300 \text{ K})}{[(28/6.02 \times 10^{26})]}$$

$$= 2.67 \times 10^5 \text{ (m/s)}^2 = (516 \text{ m/s})^2.$$

To summarize the discussion so far, we have seen from **equation 1.2** that pV is proportional to $<v^2>$ and that the ideal gas law is obtained if we take the definition of temperature to be that embodied in **equation 1.5.** Since $<\epsilon> = \frac{1}{2}m<v^2>$, both temperature and pV are proportional to the average of the square of the velocity. The use of an average recognizes that not all the molecules will be moving with the same velocity. In the next few sections we consider the *distribution* of molecular speeds. But first we must consider what we mean by a distribution.

1.4   DISTRIBUTIONS, MEAN VALUES, AND DISTRIBUTION FUNCTIONS

Suppose that five students take a chemistry examination for which the possible grades are integers in the range from 0 to 100. Let their scores be $S_1 = 68$, $S_2 = 76$, $S_3 = 83$, $S_4 = 91$, and $S_5 = 97$. The average score for the examination is then

$$<S> = \frac{S_1 + S_2 + S_3 + S_4 + S_5}{N_T} = \frac{1}{N_T} \sum_{i=1}^{N_T} S_i, \tag{1.7}$$

where $N_T = 5$ is the number of students. In this case, the average is easily calculated to be 83.

Now suppose that the class had 500 students rather than 5. Of course, the average grade could be calculated in a manner similar to that in **equation 1.7** with an index i running from 1 to $N_T = 500$. However, another method will be instructive. Clearly, if the examination is still graded to one-point accuracy, it is certain that more than one student will receive the same score. Suppose that, instead of summing over the students, represented by the index i in **equation 1.7,** we form the average by summing over the scores themselves, which range in integer possibilities from $j = 0$ to 100. In this case, to obtain the average, we must weight each score S_j by the number of students who obtained that score, N_j:

$$<S> = \frac{1}{N_T} \sum_{j=0}^{100} S_j N_j. \tag{1.8}$$

Note that the definition of N_j requires that $\Sigma N_j = N_T$. The factor $1/N_T$ in **equation 1.8** is included for normalization, since, for example, if all the students happened to get the same score $S_j = S$ then

$$<S> = \frac{1}{N_T} \sum_j S_j N_j = \frac{S}{N_T} \sum_j N_j = S. \tag{1.9}$$

Now let us define the probability of obtaining score S_j as the fraction of students receiving that score:

$$P_j = \frac{N_j}{N_T}. \tag{1.10}$$

Then another way to write **equation 1.8** is

$$<S> = \sum_j S_j P_j,\qquad(1.11)$$

where $\sum_j P_j = 1$ from normalization.

   Equation 1.11 provides an alternative to **equation 1.7** for finding the average score for the class. Furthermore, we can generalize **equation 1.11** to provide a method for finding the average of *any* quantity,

$$<Q> = \sum_j P_j Q_j,\qquad(1.12)$$

where P_j is the probability of finding the *j*th result.

example 1.2

Calculating Averages from Probabilities

Objective   Find the average throw for a pair of dice.

Method   Each die is independent, so the average of the sum of the throws will be twice the average of the throw for one die. Use **equation 1.12** to find the average throw for one die.

Solution   The probability for each of the six outcomes, 1–6, is the same, namely, 1/6. Factoring this out of the sum gives $<T> = (1/6) \sum T_i$, where $T_i = 1,2,3,4,5,6$ for $i = 1$–6. The sum is 21, so that the average throw for one die is $<T> = 21/6 = 3.5$. For the sum of two dice, the average would thus be 7.

   The method can be extended to calculate more complicated averages. Let $f(Q_j)$ be some arbitrary function of the observation Q_j. Then the average value of the function $f(Q)$ is given by

$$<f(Q)> = \sum_j P_j f(Q_j).\qquad(1.13)$$

For example, if Q were the square of a score, then

$$<S^2> = \sum_j P_j S^2.\qquad(1.14)$$

   Suppose now that the examination is a very good one, indeed, and that the talented instructor can grade it not just to one-point accuracy (a remarkable achievement in itself!) but to an accuracy of dS, where dS is a very small fraction of a point. Let $P(S)$ dS be the probability that a score will fall in the range between S and $S +$ dS, and let dS become infinitesimally small. The fundamental theorems of calculus tell us that we can convert the sum in **equation 1.11** to the integral

$$<S> = \int P(S)S\,\mathrm{d}S,\qquad(1.15)$$

or, more generally for any observable quantity,

$$<Q> = \int P(Q)Q \, dQ. \tag{1.16}$$

Equation 1.16 will form the basis for much of our further work. The probability function $P(Q)$ is sometimes called a *distribution function,* and the range of the integral is over all values of Q where the probability is nonzero. Note that normalization of the probability requires

$$\int P(Q) \, dQ = 1. \tag{1.17}$$

The quantity $|\psi(x)|^2 \, dx$ is simply a specific example of a distribution function. Although knowledge of quantum mechanics is not necessary to solve it, you may recognize a connection to the particle in the box in Problem 1.7, which like **Example 1.3** is an exercise with distribution functions.

example 1.3

Determining Distribution Functions

Objective   Bees like honey. A sphere of radius r_0 is coated with honey and hanging in a tree. Bees are attracted to the honey such that the average number of bees per unit volume is given by Kr^{-5}, where K is a constant and r is the distance from the center of the sphere. Derive the normalized distribution function for the bees. They can be at any distance from the honey, but they cannot be inside the sphere. Using this distribution, calculate the average distance of a bee from the center of the sphere.

Method   First we need to find the normalization constant K by applying **equation 1.17,** recalling that we have a three-dimensional problem and that in spherical coordinates the volume element for a problem that does not depend on the angles is $4\pi r^2 \, dr$. Then, to evaluate the average, we apply **equation 1.16.**

Solution   Recall that, by hypothesis, there is no probability for the bees being at $r < r_0$, so that the range of integration is from r_0 to infinity. To determine K we require

$$\int_{r_0}^{\infty} (Kr^{-5}) \, 4\pi r^2 \, dr = 1, \tag{1.18}$$

or

$$4\pi K \int_{r_0}^{\infty} r^{-3} \, dr = 1 = 4\pi K \left(-\frac{r^{-2}}{2} \right) \Bigg|_{r_0}^{\infty} = \frac{4\pi K}{2r_0^2}, \tag{1.19}$$

so that

$$K = \frac{r_0^2}{2\pi}.$$

Having determined the normalization constant, we now calculate the average distance:

$$\langle r \rangle = \int_{r_0}^{\infty} r \left(\frac{r_0^2}{2\pi} \right) r^{-5} 4\pi r^2 \, dr$$

$$= 2r_0^2 \int_{r_0}^{\infty} r^{-2} \, dr \tag{1.20}$$

$$= 2r_0^2 (-r^{-1}) \Big|_{r_0}^{\infty} = 2r_0^2 \frac{1}{r_0} = 2r_0.$$

1.5   THE MAXWELL DISTRIBUTION OF SPEEDS

We turn now to the distribution of molecular speeds. We will denote the probability of finding v_x in the range from v_x to $v_x + dv_x$, v_y in the range from v_y to $v_y + dv_y$, and v_z in the range from v_z to $v_z + dv_z$ by $F(v_x, v_y, v_z) \, dv_x \, dv_y \, dv_z$. The object of this section is to determine the function $F(v_x, v_y, v_z)$. There are four main points in the derivation:

1.   In each direction, the velocity distribution must be an even function of v.
2.   The velocity distribution in any particular direction is independent from and uncorrelated with the distributions in orthogonal directions.
3.   The average of the square of the velocity $\langle v^2 \rangle$ obtained using the distribution function should agree with the value required by the ideal gas law: $\langle v^2 \rangle = 3kT/m$.
4.   The three-dimensional velocity distribution depends only on the magnitude of v (i.e., the speed) and not on the direction.

We now examine these four points in detail.

1.5.1   The Velocity Distribution Must Be an Even Function of v

Consider the velocities v_x of molecules contained in a box. The number of molecules moving in the positive x direction must be equal to the number of molecules moving in the negative x direction. This conclusion is easily seen by examining the consequences of the contrary assumption. If the number of molecules moving in each direction were not the same, then the pressure on one side of the box would be greater than on the other. Aside from violating experimental evidence that the pressure is the same wherever it is measured in a closed system, our common observation is that the box does not spontaneously move in either the positive or negative x direction, as would be likely if the pressures were substantially different. We conclude that the distribution function for the velocity in the x direction, or more generally in any arbitrary direction, must be symmetric; i.e., $F(v_x) = F(-v_x)$. Functions possessing the property that $f(x) = f(-x)$ are called *even functions,* while those having the property that $f(x) = -f(-x)$ are called *odd functions.* We can ensure that $F(v_x)$ be an even function by requiring that the distribution function depend on the square of the velocity: $F(v_x) = f(v_x^2)$. As shown in Section 1.5.3, this condition is also in accord with the Boltzmann distribution law.[e]

---

[e]Other even functions, for example, $F = f(v_x^4)$ would be mathematically acceptable, but would not satisfy the requirement of Section 1.5.3.

1.5.2   The Velocity Distributions Are Independent and Uncorrelated

We now consider the relationship between the distribution of x-axis velocities and y- or z-axis velocities. In short, there should be no relationship. The three components of the velocity are independent of one another since the velocities are uncorrelated. An analogy might be helpful. Consider the probability of tossing three honest coins and getting "heads" on each. Because the tosses t_i are independent, uncorrelated events, the joint probability for a throw of three heads, $P(t_1 =$ heads, $t_2 =$ heads, $t_3 =$ heads), is simply equal to the product of the probabilities for the three individual events, $P(t_1 =$ heads$) \times P(t_2 =$ heads$) \times P(t_3 =$ heads$) = \frac{1}{2} \times \frac{1}{2} \times \frac{1}{2}$. In a similar way, because the x-, y-, and z-axis velocities are independent and uncorrelated, we can write that

$$F(v_x, v_y, v_z) = F(v_x)F(v_y)F(v_z). \tag{1.21}$$

We can now use the conclusion of the previous section. We can write, for example, that $F(v_x) = f(v_x^2)$ and similarly for the other directions. Consequently,

$$F(v_x, v_y, v_z) = F(v_x)F(v_y)F(v_z) = f(v_x^2)f(v_y^2)f(v_z^2). \tag{1.22}$$

What functional form has the property that $f(a + b + c) = f(a)f(b)f(c)$? A little thought leads to the exponential form, since $\exp(a + b + c) = e^a e^b e^c$. It can be shown, in fact, that the exponential is the *only* form having this property (see Appendix 1.1), so that we can write

$$F(v_x) = f(v_x^2) = K \exp(\pm \kappa v_x^2), \tag{1.23}$$

where K and κ are constants to be determined. Note that although κ can appear mathematically with either a plus or a minus sign, we must require the minus sign on physical grounds because we know from common experience that the probability of very high velocities should be small.

The constant K can be determined from normalization since, using **equation 1.17,** the total probability that v_x lies somewhere in the range from $-\infty$ to $+\infty$ should be unity:

$$\int_{-\infty}^{\infty} F(v_x)\, dv_x = 1. \tag{1.24}$$

Substitution of **equation 1.23** into **equation 1.24** leads to the equation

$$1 = K \int_{-\infty}^{\infty} \exp(-\kappa v_x^2)\, dv_x = K\left(\frac{\pi}{\kappa}\right)^{1/2}, \tag{1.25}$$

where the integral was evaluated using **Table 1.1.** The solution is then $K = (\kappa/\pi)^{1/2}$.

1.5.3 $<v^2>$ Should Agree with the Ideal Gas Law

The constant κ is determined by requiring $<v^2>$ to be equal to $3kT/m$, as in **equation 1.6.** From **equation 1.16** we find

$$<v_x^2> = \int_{-\infty}^{\infty} v_x^2 F(v_x)\, dv_x = \left(\frac{\kappa}{\pi}\right)^{1/2} \int_{-\infty}^{\infty} v_x^2 \exp(-\kappa v_x^2)\, dv_x. \tag{1.26}$$

The integral is a standard one listed in **Table 1.1,** and using its value we find that

$$<v_x^2> = \frac{1}{2}\left(\frac{\kappa}{\pi}\right)^{1/2}\left(\frac{\pi}{\kappa^3}\right)^{1/2} = \frac{1}{2\kappa}. \tag{1.27}$$

TABLE 1.1　Integrals of Use in the Kinetic Theory of Gases

$$\int_{-\infty}^{\infty} x^{2n}e^{-\beta x^2}\,dx = 2\int_{0}^{\infty} x^{2n}e^{-\beta x^2}\,dx \qquad\qquad \int_{-\infty}^{\infty} x^{2n+1}e^{-\beta x^2}\,dx = 0$$

$$\int_{0}^{\infty} e^{-\beta x^2}\,dx = \tfrac{1}{2}\sqrt{\pi}\,\beta^{-1/2} \qquad\qquad \int_{0}^{\infty} xe^{-\beta x^2}\,dx = \tfrac{1}{2}\beta^{-1}$$

$$\int_{0}^{\infty} x^{2}e^{-\beta x^2}\,dx = \tfrac{1}{2}\sqrt{\pi}\,\tfrac{1}{2}\beta^{-3/2} \qquad\qquad \int_{0}^{\infty} x^{3}e^{-\beta x^2}\,dx = \tfrac{1}{2}\beta^{-2}$$

$$\int_{0}^{\infty} x^{4}e^{-\beta x^2}\,dx = \tfrac{1}{2}\sqrt{\pi}\,\tfrac{3}{4}\beta^{-5/2} \qquad\qquad \int_{0}^{\infty} x^{5}e^{-\beta x^2}\,dx = \beta^{-3}$$

$$\int_{0}^{\infty} x^{2n}e^{-\beta x^2}\,dx = \tfrac{1}{2}\sqrt{\pi}\,\frac{(2n)!\,\beta^{-(n+1/2)}}{2^{2n}n!} \qquad\qquad \int_{0}^{\infty} x^{2n+1}e^{-\beta x^2}\,dx = \tfrac{1}{2}(n!)\beta^{-(n+1)}$$

As a consequence, the average of the square of the total speed, $<v^2> = <v_x^2> + <v_y^2> + <v_z^2> = 3<v_x^2>$, is simply

$$<v^2> = \frac{3}{2\kappa}. \tag{1.28}$$

From **equation 1.6** we have that $<v^2> = 3kT/m$ for agreement with the ideal gas law, so that $3kT/m = 3/(2\kappa)$, or $\kappa = m/(2kT)$. The complete one-dimensional distribution function is thus

$$F(v_x)\,dv_x = \left(\frac{m}{2\pi kT}\right)^{1/2} \exp\left(-\frac{1}{2}\frac{mv_x^2}{kT}\right)\,dv_x. \tag{1.29}$$

This equation is known as the *one-dimensional Maxwell-Boltzmann distribution for molecular velocities*. Plots of $F(v_x)$ are shown in **Figure 1.2.**

Note that **equation 1.29** is consistent with the Boltzmann distribution law, which states that the probability of finding a system with energy ϵ is proportional to $\exp(-\epsilon/kT)$. Since $\epsilon_x = \tfrac{1}{2}mv_x^2$ is equal to the translational energy of the molecule in the x direction, the probability of finding a molecule with an energy ϵ_x should be proportional to $\exp(-\epsilon_x/kT)$, as it is in **equation 1.29**. In Section 1.5.1 we ensured $F(v_x)$ to be even by choosing it to depend on the square of the velocity, $F(v_x) = f(v_x^2)$. Had we chosen some other even function, say $F(v_x) = f(v_x^4)$, the final expression for the one-dimensional distribution would not have agreed with the Boltzmann distribution law.

Equation 1.29 provides the distribution of velocities in one dimension. In three dimensions, because $F(v_x,v_y,v_z) = F(v_x)F(v_y)F(v_z)$, and because $v^2 = v_x^2 + v_y^2 + v_z^2$, we find that the probability that the velocity will have components v_x between v_x and $v_x + dv_x$, v_y between v_y and $v_y + dv_y$, and v_z between v_z and $v_z + dv_z$ is given by

$$F(v_x,v_y,v_z)\,dv_x\,dv_y\,dv_z = F(v_x)F(v_y)F(v_z)\,dv_x\,dv_y\,dv_z \tag{1.30}$$

$$= \left(\frac{m}{2\pi kT}\right)^{3/2} \exp\left(-\frac{mv^2}{2kT}\right)\,dv_x\,dv_y\,dv_z.$$

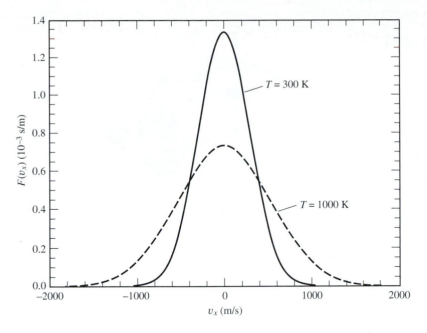

■ Figure 1.2

One-dimensional velocity distribution for a mass of 28 amu and two temperatures.

1.5.4   The Distribution Depends Only on the Speed

Note that the right-hand side of **equation 1.30** depends on v^2 and not on the directional property of **v.** When we have a function that depends only on the length of the velocity vector, $v = |\mathbf{v}|$, and not on its direction, we can be more precise by saying that the function depends on the *speed* and not on the *velocity*. Since $F(v_x, v_y, v_z) = f(v^2)$ depends on the speed, it is often more convenient to know the probability that molecules have a speed in a particular range than to know the probability that their velocity vectors will terminate in a particular volume. As shown in **Figure 1.3,** the probability that the speed will be between v and $v + dv$ is simply the probability that velocity vectors will terminate within the volume of a spherical shell between the radius v and the radius $v + dv$. The volume of this shell is $dv_x\, dv_y\, dv_z = 4\pi v^2\, dv$, so that the probability that speed will be in the desired range is[f]

---

[f]An alternate method for obtaining **equation 1.31** is to note that $dv_x\, dv_y\, dv_z$ can be written as $v^2 \sin\theta\, d\theta\, d\phi\, dv$ in spherical coordinates (see Appendix 1.2) and then to integrate over the angular coordinates. Since the distribution does not depend on the angular coordinates, the integrals over $d\theta$ and $d\phi$ simply give 4π and we are left with the factor $v^2\, dv$.

$$F(v)\,dv = \int_{\phi=0}^{2\pi}\int_{\theta=0}^{\pi} v^2 \left(\frac{m}{2\pi kT}\right)^{3/2}\exp\left(-\frac{mv^2}{2kT}\right)\sin\theta\,dv\,d\theta\,d\phi$$

$$= \int_{\phi=0}^{2\pi}d\phi\int_{\theta=0}^{\pi}\sin\theta\,d\theta\;v^2\left(\frac{m}{2\pi kT}\right)^{3/2}\exp\left(-\frac{mv^2}{2kT}\right)dv$$

$$= 4\pi v^2\left(\frac{m}{2\pi kT}\right)^{3/2}\exp\left(-\frac{mv^2}{2kT}\right)dv.$$

A more complete description of spherical coordinates is found in Appendix 1.2.

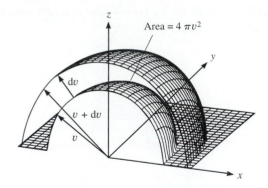

Figure 1.3

The shell between v and $v + dv$ has a volume of $4\pi v^2\, dv$. The thickness of the shell here is exaggerated for clarity.

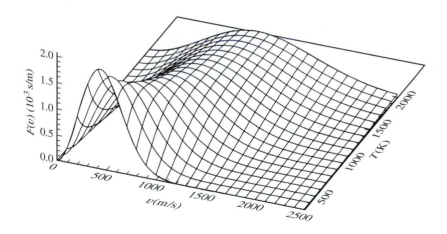

Figure 1.4

Maxwell-Boltzmann speed distribution as a function of temperature for a mass of 28 amu.

$$F(v)\, dv = 4\pi v^2 \left(\frac{m}{2\pi kT} \right)^{3/2} \exp\left(-\frac{mv^2}{2kT} \right) dv. \qquad \textbf{(1.31)}$$

By analogy to **equation 1.29,** we will call **equation 1.31** the *Maxwell-Boltzmann speed distribution.* Speed distributions as a function of temperature are shown in **Figure 1.4.**

We often characterize the speed distribution by a single parameter, for example, the temperature. Equivalently, we could specify one of several types of "average" speed, each of which is related to the temperature. One such average is called the *root-mean-squared* (rms) speed and can be calculated from **equation**

1.6: $c_{rms} \equiv <v^2>^{1/2} = (3kT/m)^{1/2}$. Another speed is the *mean* speed defined by using **equation 1.16** to calculate $<v>$:

$$<v> = \int_0^\infty vF(v)\,dv$$

$$= \int_0^\infty v4\pi v^2 \left(\frac{m}{2\pi kT}\right)^{3/2} \exp\left(-\frac{mv^2}{2kT}\right) dv = \left(\frac{8kT}{\pi m}\right)^{1/2},$$

$$\text{(1.32)}$$

where the integral was evaluated using **Table 1.1** as described in detail in **Example 1.4**. Finally, the distribution might also be characterized by the *most probable* speed, c^*, the speed at which the distribution function has a maximum (Problem 1.8):

$$c^* = \left(\frac{2kT}{m}\right)^{1/2}.$$

$$\text{(1.33)}$$

example 1.4

Using the Speed Distribution

Objective   The speed distribution can be used to determine averages. For example, find the average speed, $<v>$.

Method   Once one has the normalized distribution function, **equation 1.16** gives the method for finding the average of any quantity. Identifying Q as the velocity and $P(Q)\,dQ$ as the velocity distribution function given in **equation 1.31**, we see that we need to integrate $vF(v)\,dv$ from limits $v = 0$ to $v = \infty$.

Solution
$$<v> = \int_0^\infty vF(v)\,dv = \int_0^\infty 4\pi v^3 \left(\frac{a^3}{\pi^{3/2}}\right) \exp(-a^2 v^2)\,dv$$

$$\text{(1.34)}$$

$$= \frac{4}{\sqrt{\pi}} \int_0^\infty a^3 v^3 \exp(-a^2 v^2)\,dv,$$

where $a \equiv (m/2kT)^{1/2}$. We now transform variables by letting $x \equiv av$. The limits will remain unchanged, and $dv = dx/a$. Thus the integral in **equation 1.34** becomes

$$\frac{4}{a\sqrt{\pi}} \int_0^\infty x^3 \exp(-x^2)\,dx = \frac{4}{a\sqrt{\pi}} \frac{1}{2}$$

$$\text{(1.35)}$$

$$= \frac{2}{\sqrt{\pi}} \left(\frac{2kT}{m}\right)^{1/2} = \left(\frac{8kT}{\pi m}\right)^{1/2},$$

where we have used **Table 1.1** to evaluate the integral.

The molecular speed is related to the speed of sound, since sound vibrations cannot travel faster than the molecules causing the pressure waves. For example, in **Example 1.5** we find that the most probable speed for O_2 is 322 m/s, while the

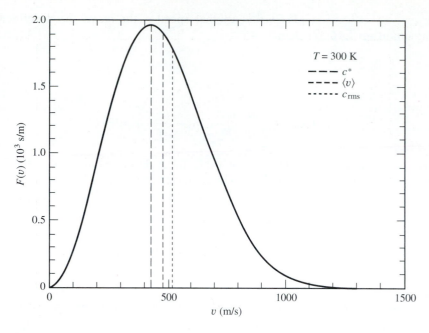

Figure 1.5

Maxwell-Boltzmann speed distribution for a mass of 28 amu and a temperature of 300 K. The vertical lines mark c^*, $<v>$, and c_{rms}.

speed of sound in O_2 is measured to be 330 m/s. For an ideal gas the speed of sound can be shown to be $(\gamma kT/m)^{1/2}$, where γ is the ratio of heat capacities, $\gamma = C_p/C_V$. The *Mach number* is defined as the ratio of the speed of an object in a medium to the speed of sound through the same medium, so that when an aircraft "breaks the sound barrier" (or exceeds "Mach 1") it is actually traveling faster than the speed of the molecules in the medium.

Figure 1.5 shows the shape of the distribution function for $T = 300$ K and the locations of the variously defined speeds.

example 1.5

Comparison of the Most Probable Speeds for Oxygen and Helium

Objective   Compare the most probable speed for O_2 to that for He at 200 K.

Method   Use **equation 1.33** with $T = 200$ K and $m = 2$ amu or $m = 32$ amu. Note that the relative speeds should be proportional to $m^{-1/2}$.

Solution $c^*(He) = (2kT/m)^{1/2} = [2(1.38 \times 10^{-23}$ J K$^{-1})(200$ K$)(6.02 \times 10^{23}$ amu/g$)(1000$ g/kg$)/(2$ amu$)]^{1/2} = 1290$ m/s. A similar calculation substituting 32 amu for 2 amu gives $c^*(O_2) = 322$ m/s.

Comment   The escape velocity from the Earth's gravitational field is roughly $v_e = 1.1 \times 10^4$ m/s, only about 10 times the most probable speed for helium. Because the velocity distribution shifts so strongly toward high velocities as the mass decreases, the fraction of helium

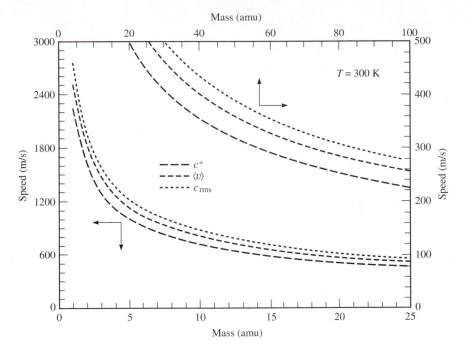

Figure 1.6

Various average speeds as a function of mass for $T = 300$ K.

atoms having speeds in excess of v_e, while minuscule (about 10^{-31}), is still 10^{475} times larger than the fraction of oxygen molecules having speeds in excess of v_e! As a consequence, the composition of the atmosphere is changing; much of the helium released during the lifetime of the planet has already escaped into space. A plot of various speeds as a function of mass for $T = 300$ K is shown in **Figure 1.6.**

1.5.5   Experimental Measurement of the Maxwell Distribution of Speeds

Experimental verification of the Maxwell-Boltzmann speed distribution can be made by direct measurement using the apparatus of **Figure 1.7.** Two versions of the measurement are shown. In **Figure 1.7a,** slits (S) define a beam of molecules moving in a particular direction after effusing from an oven (O). Those that reach the detector (D) must successfully have traversed a slotted, multiwheel chopper by traveling a distance d while the chopper rotated through an angle ϕ. In effect, the chopper selects a small slice from the velocity distribution and passes it to the detector. The speed distribution is then measured by recording the integrated detector signal for each cycle of the chopper as a function of the angular speed of the chopper.

A somewhat more modern technique, illustrated in **Figure 1.7b,** clocks the time it takes for molecules to travel a fixed distance. A very short pulse of molecules leaves the chopper at time $t = 0$. Because these molecules have a distribution of speeds, they spread out in space as they travel toward the detector, which records as a function of time the signal due to molecules arriving a distance L from the chopper.

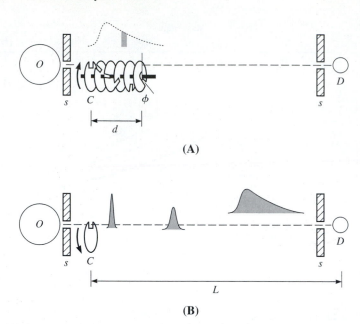

■ **Figure 1.7**

Two methods for measuring the Maxwell-Boltzmann speed distribution.

Analysis of the detector signal from this second experiment is instructive, since it introduces the concept of *flux*. Recall that the distribution $F(v)\,dv$ gives the fraction of molecules with speeds in the range from v to $v + dv$; it is dimensionless. If the number density of molecules is n^*, then $n^*F(v)\,dv$ will be the number of molecules per unit volume with speeds in the specified range. The *flux* of molecules is defined as the number of molecules crossing a unit area per unit time. It is equal to the density of molecules times their velocity: flux (number/m^2/s) = density (number/m^3) $\times$ velocity (m/s).[g] Thus, the flux J of molecules with speeds between v and $v + dv$ is

$$J\,dv = vn^*F(v)\,dv. \qquad (1.36)$$

We will consider the flux in more detail in Section 4.3.2 and make extensive use of it in Chapter 4.

We now return to the speed measurement. Most detectors actually measure the number of molecules in a particular volume during a particular time duration. For example, the detector might measure current after ionizing those molecules that enter a volume defined by a cross-sectional area of A and a length ℓ. Because molecules with high velocity traverse the distance ℓ in less time than molecules with low velocity, the detection sensitivity is proportional to $1/v$. The detector signal $S(t)$ is thus proportional to $JA\ell\,dv/v$, or to $n^*A\ell F(v)\,dv$, where n^* is the number density of molecules in the oven. Assuming that a very narrow pulse of molecules is emitted from the chopper, the speed measured at a particular time t is simply $v = L/t$. We must now transform the velocity distribution from a speed distribution to a time distribution. Note that $dv = d(L/t) = -L\,dt/t^2$, and recall from **equation 1.31** that $F(v)\,dv \propto v^2\exp(-\beta v^2)\,dv \propto (1/t^2)\exp(-\beta L^2/t^2)(L/t^2)$. We thus find that $S(t) \propto t^{-4}\exp(-\beta L^2/t^2)$. **Figure 1.8** displays an arrival time distribution of helium measured

---

[g]Strictly speaking, the flux, **J**, is a vector, since the magnitude of the flux may be different in different directions. Here, since the direction of the flux is clear, we will use just its magnitude, J.

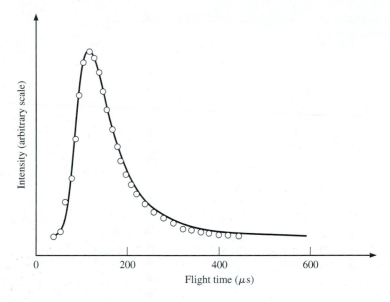

using this "time-of-flight" technique. The open circles are the detector signal, while the smooth line is a fit to the data of a function of the form expected for $S(t)$. The best fit parameter gives a temperature of 300 K.

1.6   ENERGY DISTRIBUTIONS

It is sometimes interesting to know the distribution of molecular energies rather than velocities. Of course, these two distributions must be related since the molecular translational energy ϵ is equal to $\frac{1}{2}mv^2$. Noting that this factor occurs in the exponent of **equation 1.31** and that $d\epsilon = mv\, dv = (2m\epsilon)^{1/2}\, dv$, we can convert velocities to energies in **equation 1.31** to obtain

$$G(\epsilon)\, d\epsilon = 4\pi\left(\frac{2\epsilon}{m}\right)\left(\frac{m}{2\pi kT}\right)^{3/2}\exp\left(-\frac{\epsilon}{kT}\right)\frac{d\epsilon}{\sqrt{2m\epsilon}}$$

$$= 2\pi\left(\frac{1}{\pi kT}\right)^{3/2}\sqrt{\epsilon}\exp\left(-\frac{\epsilon}{kT}\right)d\epsilon.$$

(1.37)

The function $G(\epsilon)\, d\epsilon$ tells us the fraction of molecules which have energies in the range between ϵ and $\epsilon + d\epsilon$. Plots of $G(\epsilon)$ are shown in **Figure 1.9.**

The distribution function $G(\epsilon)$ can be used to calculate the average of any function of ϵ using the relationship of **equation 1.16.** In particular, it can be shown as expected that $<\epsilon> = 3kT/2$ (see Problem 1.9).

Let us pause here to make a connection with thermodynamics. In the case of an ideal monatomic gas, there are no contributions to the energy of the gas from internal degrees of freedom such as rotation or vibration, and there is normally very

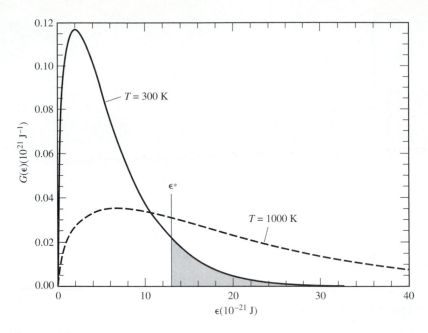

Figure 1.9

Energy distributions for two different temperatures. The fraction of molecules for the 300 K distribution having energy in excess of ϵ^* is shown in the shaded region.

little contribution to the energy from excitation of electronic degrees of freedom. Consequently, the average energy U of n moles of a monatomic gas is simply nN_A times the average energy of one molecule of the gas, or

$$U = nN_A \frac{3}{2} kT = \frac{3}{2} nRT. \qquad (1.38)$$

Note that the heat capacity at constant volume is defined as $C_V = (\partial U/\partial T)_V$, so that for an ideal monatomic gas we find that

$$C_V = \frac{3}{2} nR. \qquad (1.39)$$

This result is an example of the *equipartition principle,* which states that each term in the expression of the molecular energy that is quadratic in a particular coordinate contributes $\frac{1}{2}kT$ to the average kinetic energy and $\frac{1}{2}R$ to the molar heat capacity. Since there are three quadratic terms in the three-dimensional translational energy expression, the molar heat capacity of a monatomic gas should be $3R/2$.

It is sometimes useful to know what fraction of molecules has an energy greater than or equal to a certain value ϵ^*. In principle, the energy distribution $G(\epsilon)$ should be able to provide this information, since the fraction of molecules having energy in the desired range is simply the integral of $G(\epsilon)\,d\epsilon$ from ϵ^* to infinity, as shown by the hatched region in **Figure 1.9.** In practice, the mathematics are somewhat cumbersome, but the result is reasonable. Let $f(\epsilon^*)$ be the fraction of molecules with kinetic energy equal to or greater than ϵ^*. This fraction is given by the integral

$$f(\epsilon^*) = 2\pi \left(\frac{1}{\pi kT} \right)^{3/2} \int_{\epsilon^*}^{\infty} \sqrt{\epsilon}\, \exp\left(-\frac{\epsilon}{kT} \right) d\epsilon. \qquad (1.40)$$

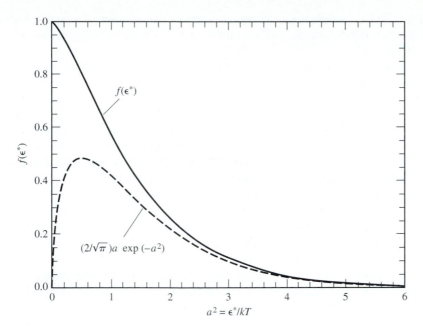

Figure 1.10

The fraction of molecules having energy in excess of ϵ^* as a function of ϵ^*/kT.

Problem 1.10 shows that this integral is given by

$$f(\epsilon^*) = \frac{2}{\sqrt{\pi}}\, ae^{-a^2} + \mathrm{erfc}(a), \qquad (1.41)$$

where $a = (\epsilon^*/kT)^{1/2}$ and $\mathrm{erfc}(a)$ is the co-error function defined in Appendix 1.3. A plot of $f(\epsilon^*)$ as a function of ϵ^*/kT is shown in **Figure 1.10.** Note that for $\epsilon^* > 3kT$, the function $f(\epsilon^*)$ is nearly equal to the first term in **equation 1.41,** $2\sqrt{(\epsilon^*/\pi kT)}\exp(-\epsilon^*/kT)$, shown by the dashed line in the figure. Thus, the fraction of molecules with energy greater than ϵ^* falls off as $\sqrt{\epsilon^*}\exp(-\epsilon^*/kT)$, provided that $\epsilon^* > 3kT$.

1.7   COLLISIONS: MEAN FREE PATH AND COLLISION NUMBER

One of the goals of this chapter is to derive an expression for the number of collisions that molecules of type 1 make with molecules of type 2 in a given time. We will argue later that this collision rate provides an upper limit to the reaction rate, since the two species must have a close encounter to react.

The principal properties of the collision rate can be easily appreciated by anyone who has ice skated at a local rink. Imagine two groups of skaters, some rather sedate adults and some rambunctious 13-year-old kids. If there is only one kid and one adult in the rink, then the likelihood that they will collide is small, but as the number of either adults or kids in the rink increases, so does the rate at which collisions will occur. The collision rate is proportional to the number of possible kid-adult pairs, which is proportional to the number density of adults times the number density of kids.

But the collision rate depends on other factors as well. If all the skaters follow the rules and skate counterclockwise around the rink at the same speed, then there will be no collisions. More often, the kids will skate at much faster or slower speeds, and they will rarely move uniformly. The rate at which they collide with the adults is proportional to the *relative speed* between the adults and kids.

Finally, consider the dependence of the collision rate on the size of the adults and kids. People are typically about 40 cm wide. What would be the effect of increasing or decreasing this diameter by a factor of 10? If the diameter were decreased to 4 cm, the number of collisions would go down dramatically; if the diameter were increased to 4 m, it would be difficult to move around the rink at all. Thus, simple considerations suggest that the collision rate between molecules should be proportional to the relative speed of the molecules, to their size, and to the number of possible collision pairs.

Let us assume that the average of the magnitude of the relative velocity between molecules of types 1 and 2 is $<v_r>$ and that the molecules behave like hard spheres; there are no attractive forces between them, and they bounce off one another like billiard balls when they collide.[h] Let the quantity b, shown in **Figure 1.11,** be defined as the distance of a line perpendicular to the each of the initial velocities of two colliding molecules, one of type 1 and the other of type 2. This distance is often referred to as the *impact parameter.* If the radii of the two molecules are r_1 and r_2, then, as shown in **Figure 1.11,** a "collision" will occur if the two molecules approach one another so that their centers are within the distance $b_{max} \equiv r_1 + r_2$. Thus, b_{max} is the maximum value of the impact parameter for which a collision can occur. From the point of view of one type of molecule striking a molecule of the other type, the target area for a collision is then equal to $\pi(r_1 + r_2)^2 = \pi b_{max}^2$.

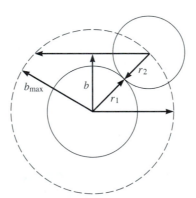

■ **Figure 1.11**

A collision will occur if the impact parameter is less than b_{max}, the sum of the two molecular radii.

---

[h]We consider only the *relative* velocity between the molecules. Appendix 1.4 shows that the *total* velocity of each molecule can be written as a vector sum of the velocity of the center of mass of the pair of molecules and the relative velocity of the molecule with respect to the center of mass. The forces between molecules depend on the relative distance between them and do not change the velocity of their center of mass, which must be conserved during the collision.

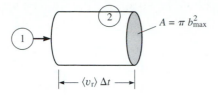

■ **Figure 1.12**

Molecule 1 sweeps out a cylinder of area πb_{max}^2. Any molecule of type 2 whose center is within the cylinder will be struck.

Consider a molecule of type 1 moving through a gas with a speed equal to the average magnitude of the relative velocity $<v_r>$. **Figure 1.12** shows that any molecule of type 2 located in a cylinder of volume $\pi b_{max}^2 <v_r> \Delta t$ will then be struck in the time Δt.[i] If the density of molecules of type 2 is n_2^*, then the number of collisions one molecule of type 1 will experience with molecules of type 2 per unit time is

$$Z_2 = \pi b_{max}^2 <v_r> n_2^*. \tag{1.42}$$

Of course, for a molecule of type 1 moving through other molecules of the same type,

$$Z_1 = \pi b_{max}^2 <v_r> n_1^* = \pi d^2 <v_r> n_1^*, \tag{1.43}$$

where b_{max}^2 has been replaced by d^2 since $r_1 + r_2 = 2r_1 = d$. The quantity πb_{max}^2 is known as the hard-sphere collision cross section. Cross sections are generally given the symbol σ.

Equation 1.42 gives the number of collisions per unit time of one molecule of type 1 with a density n_2^* of molecules of type 2. The *total* number of collisions of molecules of type 1 with those of type 2 per unit time and per unit volume is found simply by multiplying by the density of type 1 molecules:

$$Z_{12} = Z_2 n_1^* = \pi b_{max}^2 <v_r> n_1^* n_2^*. \tag{1.44}$$

Note that the product $n_1^* n_2^*$ is simply proportional to the total number of pairs of collision partners.

By a similar argument, if there were only one type of molecule, the number of collisions per unit time per unit volume is given by

$$Z_{11} = \frac{1}{2} Z_1 n_1^* = \frac{1}{2} \pi b_{max}^2 <v_r> (n_1^*)^2. \tag{1.45}$$

The factor of $\frac{1}{2}$ is introduced for the following reason. The collision rate should be proportional to the number of pairs of collision partners. If there are n molecules, then the number of pairs is $n(n-1)/2$, since each molecule can pair with $n-1$ others and the factor of 2 in the denominator corrects for having counted each pair twice. If n is a large number, then we can approximate $n(n-1)$ as n^2, and since the number of molecules is proportional to the number density, we see that the number of pairs goes as $(n_1^*)^2/2$.

It remains for us to determine the value of the relative speed, averaged over the possible angles of collision and averaged over the speed distribution for each molecule. One way to arrive quickly at the answer for a very specific case is shown in

---

[i]Because of the collisions, the molecule under consideration will actually travel along a zigzag path, but the volume swept out per unit time will be the same.

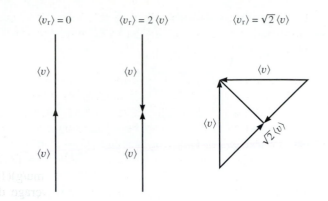

$\langle v_r \rangle = 0$ $\langle v_r \rangle = 2 \langle v \rangle$ $\langle v_r \rangle = \sqrt{2} \langle v \rangle$

$\langle v \rangle$ $\langle v \rangle$ $\langle v \rangle$

$\langle v \rangle$

$\sqrt{2}\langle v \rangle$

$\langle v \rangle$ $\langle v \rangle$

■ Figure 1.13

In a hypothetical collision where two molecules each have a speed equal to the average $\langle v \rangle$, the relative velocity between two molecules, averaged over all collision directions, is $\sqrt{2}\langle v \rangle$.

Figure 1.13. Suppose that the two types of molecules have the same mass, m. Let us assume for the moment that we can accomplish the average of the speed distribution by assuming that the two molecules each have a speed equal to the average of their distribution. Since the two molecules are assumed to have the same mass (and temperature), they will also have the same average speed, $\langle v \rangle$. We now consider the average over collision angles. If the molecules are traveling in the same direction, then the relative velocity between them will have zero magnitude, $v_r = 0$, while if they are traveling in opposite directions along the same line the relative velocity will have a magnitude of $v_r = 2\langle v \rangle$. Suppose that they are traveling at right angles to one another. In that case, which is representative of the average angle of collision, the relative velocity will have a magnitude of $v_r = \langle v_r \rangle = \sqrt{2}\langle v \rangle$. Recalling from **equation 1.32** that $\langle v \rangle = (8kT/\pi m)^{1/2}$, we find that

$$\langle v_r \rangle = \sqrt{2}\langle v \rangle$$
$$= \sqrt{2}\left(\frac{8kT}{\pi m}\right)^{1/2} = \left(\frac{8kT}{\pi(m/2)}\right)^{1/2} = \left(\frac{8kT}{\pi \mu}\right)^{1/2}, \tag{1.46}$$

where we have introduced the *reduced mass, μ*, defined as $\mu = m_1 m_2/(m_1 + m_2)$. When the masses m_1 and m_2 are the same, $\mu = m^2/2m = m/2$. If the masses are different, then the mean velocities will not be the same, and the simple analysis of **Figure 1.13** is not adequate. However, as shown for the general case in Appendix 1.4 and Problem 1.12, the result for $\langle v_r \rangle$ is the same as that given in **equation 1.46.** The appendix also shows why the definition of μ as $m_1 m_2/(m_1 + m_2)$ is a useful one.

example 1.6

The Collision Rate of NO with O_3

Objective Find the collision rate of NO with O_3 at 300 K if the abundances at 1 atm total pressure are each 0.2 ppm and if the molecular diameters are 300 and 375 pm, respectively. Reactive collisions between these two species are important in photochemical smog formation.

Method   Use **equation 1.44,** remembering to convert the abundances to number densities at 300 K and calculating the average relative velocity by use of **equation 1.46.**

Solution   First find the total number density n^* at 1 atm: $n^* = (n/V)N_A = (p/RT)N_A = (1\ \text{atm})(6.02 \times 10^{23}\ \text{molec/mole})/[(0.082\ \text{L atm mol}^{-1}\ \text{K}^{-1})(300\ \text{K})] = 2.45 \times 10^{22}\ \text{molec/L}$. Next determine the number densities of NO and O_3, each being the total density times 0.2×10^{-6}: $n^*(\text{NO}) = n^*(O_3) = (0.2 \times 10^{-6})(2.45 \times 10^{22}) = 4.9 \times 10^{15}$ molec/L. The average relative velocity is $\langle v_r \rangle = (8kT/\pi\mu)^{1/2} = [8(1.38 \times 10^{-23}\ \text{J K}^{-1})(300\ \text{K})(6.02 \times 10^{23}\ \text{amu/g})(1000\ \text{g/kg})/(\pi(48 \times 30/78)\ \text{amu})]^{1/2} = 586$ m/s. The average diameter is $(300 + 375\ \text{pm})/2 = 337.5$ pm. Then $Z_{12} = \pi(337.5 \times 10^{-12}\ \text{m})^2 (586\ \text{m/s})(4.9 \times 10^{15}\ \text{molec/L})^2(1\ \text{L}/10^{-3}\ \text{m}^3)^2 = 5.0 \times 10^{21}$ collisions $\text{s}^{-1}\ \text{m}^{-3}$. If every collision resulted in a reaction, this would be the number of reactions per unit second per cubic meter.

A quantity related to Z_1 is the *mean free path,* λ. This is the average distance a molecule travels before colliding with another molecule. If we divide the average speed $\langle v \rangle$ in meters per second by the collision number Z_1 in collisions per second, we obtain the mean free path in meters per collision:

$$\lambda = \frac{\langle v \rangle}{Z_1} = \frac{\langle v \rangle}{\pi d^2 \sqrt{2} \langle v \rangle n_1^*}$$

$$= \frac{1}{\sqrt{2}\,\pi d^2 n_1^*}. \tag{1.47}$$

Note that the mean free path is inversely proportional to pressure. The mean free path will be important in Chapter 4, where we will see that the transport of heat, momentum, and matter are all proportional to the distance traveled between collisions.

example 1.7

The Mean Free Path of Nitrogen

Objective   Find Z_1 and the mean free path of N_2 at 300 K and 1 atm given that the molecular diameter is 218 pm.

Method   Use **equation 1.46** to calculate $\langle v_r \rangle$, **equation 1.43** to calculate Z_1, and **equation 1.47** to calculate λ.

Solution   We start by calculating $\langle v_r \rangle = (8kT/\pi\mu)^{1/2}$, where $\mu = 28 \times 28/(28 + 28) = 14$ amu.

$$\langle v_r \rangle = \left\{ \frac{8(1.38 \times 10^{-23}\ \text{J K}^{-1})(300\ \text{K})(6.02 \times 10^{23}\ \text{amu/g})(1000\ \text{g/kg})}{(3.1415 \times 14\ \text{amu})} \right\}^{1/2}$$

$$= 673\ \text{m/s}. \tag{1.48}$$

Next, we calculate Z_1 noting that the density

$$n_1^* = \frac{p}{RT} = \frac{(1\text{ atm})(6.02 \times 10^{23}\text{ molec/mole})}{(0.082\text{ L atm mole}^{-1}\text{ K}^{-1})(10^{-3}\text{ m}^3/\text{L})(300\text{ K})}$$

$$= 2.45 \times 10^{25}\text{ molec/m}^3.$$

(1.49)

Then, $Z_1 = \pi(218 \times 10^{-12}\text{ m})^2(673\text{ m/s})(2.45 \times 10^{25}\text{ molec/m}^3) = 2.46 \times 10^9$ collision/s. Finally, $<v_r>/(\sqrt{2}\,Z_1) = (673\text{ m/s})/(\sqrt{2} \times 2.46 \times 10^9$ collision/s$) = 1.93 \times 10^{-7}$ m.

1.8   SUMMARY

By considering the pressure exerted by ideal gas molecules on a wall, we determined that, for agreement with the observed ideal gas law, the average energy of a molecule must be given by

$$<\epsilon> = \frac{3}{2}kT.$$

(1.5)

To learn how to perform averages, we discussed distribution functions of a continuous variable. The average of some observable quantity Q was found to be given by

$$<Q> = \int P(Q)Q\,\mathrm{d}Q,$$

(1.16)

where $P(Q)$ is the distribution function for the quantity Q. We then made the following observations about the molecular speed distribution: (1) the speed distribution must be an even function of v, (2) the speed distribution in any particular direction is independent from and uncorrelated with that in orthogonal directions, (3) the value of $<v^2>$ must be equal to $3kT/m$ to agree with the ideal gas law, and (4) the distribution depends only on the magnitude of v. These four considerations allowed us to determine the Maxwell-Boltzmann distribution of speeds:

$$F(v)\,\mathrm{d}v = 4\pi v^2 \left(\frac{m}{2\pi kT}\right)^{3/2} \exp\left(-\frac{mv^2}{2kT}\right)\mathrm{d}v.$$

(1.31)

Calculations using this distribution gave us an equation for the average speed of a molecule,

$$<v> = \left(\frac{8kT}{\pi m}\right)^{1/2},$$

(1.32)

and the most probable speed,

$$c^*\left(\frac{2kT}{m}\right)^{1/2}.$$

(1.33)

A simple transformation of variables in the speed distribution led to the Maxwell-Boltzmann energy distribution:

$$G(\epsilon)\,\mathrm{d}\epsilon = 2\pi \left(\frac{1}{\pi kT}\right)^{3/2} \sqrt{\epsilon}\, \exp\left(-\frac{\epsilon}{kT}\right)\mathrm{d}\epsilon.$$

(1.37)

Finally, for molecules behaving as hard spheres, we determined the collision rate,

$$Z_1 = \pi b_{max}^2 <v_r> n_1^*,$$

(1.42)

the relative velocity,

$$<v_r> = \sqrt{2}<v> = \left(\frac{8kT}{\pi\mu}\right)^{1/2}, \tag{1.46}$$

and the mean free path,

$$\lambda = \frac{<v>}{Z_1} = \frac{1}{\sqrt{2}\pi d^2 n_1^*}. \tag{1.47}$$

These concepts form the basis for further investigation into transport properties and chemical reaction kinetics.

appendix 1.1

The Functional Form of the Velocity Distribution

We demonstrate in this appendix that the exponential form used in **equation 1.23** is the only function that satisfies the equation $f(a + b + c) = f(a)f(b)f(c)$. Consider first the simpler equation

$$f(z) = f(a)f(b), \tag{1.50}$$

where $z = a + b$. Taking the derivative of both sides of **equation 1.50** with respect to a we obtain

$$\frac{df(z)}{dz}\frac{dz}{da} = f'(a)f(b). \tag{1.51}$$

On the other hand, taking the derivative of both sides of **equation 1.50** with respect to b, we obtain

$$\frac{df(z)}{dz}\frac{dz}{db} = f(a)f'(b). \tag{1.52}$$

Since $z = a + b$, $dz/da = dz/db = 1$. Consequently,

$$\frac{df(z)}{dz} = f'(a)f(b) = f(a)f'(b). \tag{1.53}$$

Division of both sides of the right-hand equality by $f(a)f(b)$ yields

$$\frac{f'(a)}{f(a)} = \frac{f'(b)}{f(b)}. \tag{1.54}$$

Now the left-hand side of **equation 1.54** depends only on a, while the right-hand side depends only on b. Since a and b are independent variables, the only way that **equation 1.54** can be true is if each side of the equation is equal to a constant, $\pm\kappa$, where κ is defined as nonnegative:

$$\frac{f'(a)}{f(a)} = \pm\kappa \qquad \frac{f'(b)}{f(b)} = \pm\kappa. \tag{1.55}$$

Solution of these differential equations using x to represent either a or b leads to

$$\frac{f'(x)}{f(x)} = \pm\kappa \qquad \text{or} \qquad \frac{df(x)}{f(x)} = \pm\kappa\, dx. \tag{1.56}$$

Integration shows that

$$f(x) = Ke^{\pm\kappa x}, \tag{1.57}$$

where K is related to the constant of integration. **Equation 1.23** is obtained by replacing x with v_x^2.

appendix 1.2

Spherical Coordinates

Many problems in physical chemistry can be solved more easily using spherical rather than Cartesian coordinates. In this coordinate system, as shown in **Figure 1.14,** a point P is located by its distance r from the origin, the angle θ between the z axis and the line from the point to the origin, and the angle ϕ between the x axis and the line between the origin and a projection of the point onto the x-y plane. Any point can be described by a value of r between 0 and ∞, a value of θ between 0 and π, and a value of ϕ between 0 and 2π. The Cartesian coordinates are related to the spherical ones by the following relationships: $x = r \sin\theta \cos\phi$, $y = r \sin\theta \sin\phi$, and $z = r \cos\theta$.

The volume element in spherical coordinates can be calculated with the help of **Figure 1.15.** As the variable θ is increased for fixed r, the position of the point described by (r,θ,ϕ) moves along a longitudinal line on the surface of a sphere, while if ϕ is increased at fixed r, the position of the point moves along a latitudinal line. Starting at a point located at (r,θ,ϕ), if r is increased by dr, θ is increased by $d\theta$, and ϕ is increased by $d\phi$, then the volume increase is the surface area on the

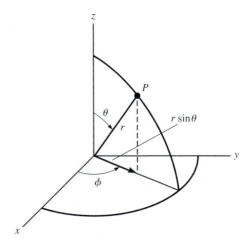

Figure 1.14

Spherical coordinates.

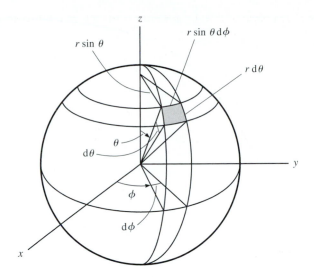

■ Figure 1.15

The volume element in spherical coordinates.

sphere times the thickness dr (for clarity, the thickness dr is not shown in the diagram). The surface area is given by the arc length on the longitude, $r\,d\theta$, times the arc length on the latitude, $r \sin \theta\, d\phi$. Thus, the volume element is $dV = r^2 \sin \theta\, d\theta\, d\phi\, dr$.

appendix 1.3

The Error Function and Co-Error Function

It often occurs that we need to evaluate integrals of the form of those listed in **Table 1.1** but for limits less than the range of 0 to infinity. For such evaluations it is useful to define the *error function*:

$$\text{erf}(x) \equiv \frac{2}{\sqrt{\pi}} \int_0^x e^{-u^2}\, du. \qquad (1.58)$$

From **Table 1.1** we see that for $x = \infty$, the value of the integral is $\sqrt{\pi}/2$, so that $\text{erf}(\infty) = 1$. Note that if we "complement" the error function by $2/\sqrt{\pi}$ times the integral from x to ∞, we should get unity:

$$\frac{2}{\sqrt{\pi}} \int_0^x e^{-u^2}\, du + \frac{2}{\sqrt{\pi}} \int_x^\infty e^{-u^2}\, du = \text{erf}(x) + \frac{2}{\sqrt{\pi}} \int_x^\infty e^{-u^2}\, du$$

$$(1.59)$$

$$= \frac{2}{\sqrt{\pi}} \int_0^\infty e^{-u^2}\, du = 1.$$

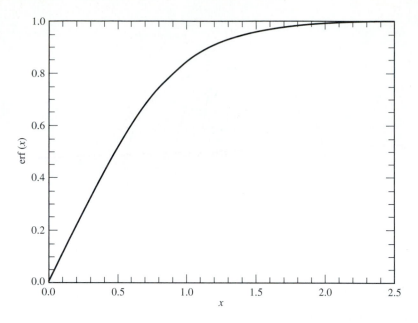

Figure 1.16

Values of the error function.

Consequently, it is also useful to define the co-error function, erfc(x), as the complement to the error function:

$$\mathrm{erfc}(x) \equiv 1 - \mathrm{erf}(x) = \frac{2}{\sqrt{\pi}} \int_x^\infty e^{-u^2}\, du. \tag{1.60}$$

Tables of the error function and co-error function are available, but the pervasive use of computers has made them all but obsolete. For calculational purposes, the integrand in **equation 1.58** or **equation 1.60** can be expanded using a series,

$$\mathrm{erf}(x) = \frac{2}{\sqrt{\pi}} \sum_{n=0}^{\infty} \frac{(-1)^n x^{2n+1}}{n!\,(2n+1)}, \tag{1.61}$$

and then the integration can be performed term by term. **Figure 1.16** plots erf(x) as a function of x.

appendix 1.4

The Center-of-Mass Frame

We show in this appendix that the total kinetic energy of two particles of velocities $\mathbf{v}_1$ and $\mathbf{v}_2$ is given by $\frac{1}{2}\mu v_r^2 + \frac{1}{2}M v_{\mathrm{com}}^2$, where $\mathbf{v}_r = \mathbf{v}_2 - \mathbf{v}_1$, and where $\mathbf{v}_{\mathrm{com}}$, the vector describing the velocity of the center of mass, is defined by the equation $(m_1 + m_2)\mathbf{v}_{\mathrm{com}} = m_1\mathbf{v}_1 + m_2\mathbf{v}_2$, and $M \equiv m_1 + m_2$. **Figure 1.17** shows the vector relationships.

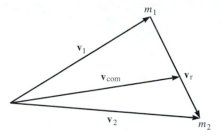

Figure 1.17

Vector diagram for center-of-mass conversion.

The virtue of this transformation is that the total momentum of the system $\mathbf{p} = m_1\mathbf{v}_1 + m_2\mathbf{v}_2$ is also equal to the momentum of the center of mass, defined as $M\mathbf{v}_{com}$. Because we assume that no external forces are acting on the system, $\mathbf{F} = M\mathbf{a}_{com} = (d\mathbf{p}_{com}/dt) = 0$, so that the momentum of the center of mass does not change during the interaction between the two particles.

Note that since $(m_1/M) + (m_2/M) = 1$ we can write

$$\mathbf{v}_2 - \mathbf{v}_{com} = \left(\frac{m_1}{M} + \frac{m_2}{M}\right)\mathbf{v}_2 - \mathbf{v}_{com}$$

$$= \frac{m_1}{M}\mathbf{v}_2 + \frac{m_2}{M}\mathbf{v}_2 - \mathbf{v}_{com}. \tag{1.62}$$

However,

$$m_1\mathbf{v}_1 + m_2\mathbf{v}_2 = M\mathbf{v}_{com}, \tag{1.63}$$

so that

$$-\frac{m_1\mathbf{v}_1}{M} = \frac{m_2\mathbf{v}_2}{M} - \mathbf{v}_{com}.$$

Consequently,

$$\mathbf{v}_2 - \mathbf{v}_{com} = \frac{m_1}{M}\mathbf{v}_2 - \frac{m_1}{M}\mathbf{v}_1$$

$$= \frac{m_1}{M}\mathbf{v}_r. \tag{1.64}$$

In a similar way, we find that

$$\mathbf{v}_{com} - \mathbf{v}_1 = \frac{m_2}{M}\mathbf{v}_r. \tag{1.65}$$

We now note an important point, that the velocities of the particles with respect to the center of mass are just given by the two pieces of the vector $\mathbf{v}_r$: $\mathbf{u}_1 = -(m_2/M)\mathbf{v}_r$, and $\mathbf{u}_2 = (m_1/M)\mathbf{v}_r$, as shown in **Figure 1.18.** Note also that in the moving frame of the center of mass, there is no net momentum for the particles; that is, $m_1\mathbf{u}_1 + m_2\mathbf{u}_2 = 0$. This important property enables us to calculate the velocity of one particle in the center-of-mass frame given just the mass and the velocity of the other particle.

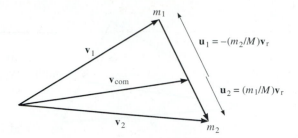

█ **Figure 1.18**

Vector diagram for center-of-mass conversion, showing the relative velocities in the center-of-mass frame for the two particles.

We can rearrange **equations 1.64** and **1.65** to get

$$\mathbf{v}_{com} - \frac{m_2}{M}\mathbf{v}_r = \mathbf{v}_1,$$

$$\mathbf{v}_{com} + \frac{m_1}{M}\mathbf{v}_r = \mathbf{v}_2. \tag{1.66}$$

The total energy is then

$$
\begin{aligned}
\frac{1}{2}m_1\mathbf{v}_1^2 + \frac{1}{2}m_2\mathbf{v}_2^2 &= \frac{1}{2}m_1\left(\mathbf{v}_{com} - \frac{m_2}{M}\mathbf{v}_r\right)^2 + \frac{1}{2}m_2\left(\mathbf{v}_{com} + \frac{m_1}{M}\mathbf{v}_r\right)^2 \\[2mm]
&= \frac{1}{2}m_1\mathbf{v}_{com}^2 - \frac{2m_1m_2}{2M}\mathbf{v}_r\cdot\mathbf{v}_{com} + \frac{m_1m_2^2}{2M^2}\mathbf{v}_r^2 \\[2mm]
&\quad + \frac{1}{2}m_2\mathbf{v}_{com}^2 + \frac{2m_2m_1}{2M}\mathbf{v}_r\cdot\mathbf{v}_{com} + \frac{m_2m_1^2}{2M^2}\mathbf{v}_r^2 \\[2mm]
&= \frac{1}{2}M\mathbf{v}_{com}^2 + \frac{m_1m_2M}{2M^2}\mathbf{v}_r^2 \\[2mm]
&= \frac{1}{2}Mv_{com}^2 + \frac{1}{2}\mu v_r^2.
\end{aligned}
\tag{1.67}
$$

It will often be useful to consider collisions in the center-of-mass frame. For example, we will make extensive use of this view in talking about molecular scattering in Section 8.4. Problem 1.12 shows how this result can be used to calculate the average relative velocity.

suggested readings

G. G. Hammes, *Principles of Chemical Kinetics* (Academic Press, New York, 1978).

W. Kauzmann, *Kinetic Theory of Gases,* Vol. 1, *Thermal Properties of Matter,* (Benjamin, Reading, MA, 1966)

E. H. Kennard, *Kinetic Theory of Gases* (McGraw-Hill, New York, 1938).

R. D. Present, *Kinetic Theory of Gases* (McGraw-Hill, New York, 1958).

R. E. Weston, Jr., and H. A. Schwarz, *Chemical Kinetics* (Prentice-Hall, Englewood Cliffs, NJ, 1972).

problems

1.1 Molecules all of mass m and speed v exert a pressure p on the walls of a vessel. If half the molecules are replaced by ones of another type all with mass $\frac{1}{2}m$ and speed $2v$, will the pressure (a) increase, (b) decrease, (c) remain constant?

1.2 Suppose the probability of obtaining a score between 0 and 100 on an exam increases monotonically between 0 and 1.00. Is the average score on the exam (a) greater than 50, (b) equal to 50, (c) less than 50?

1.3 Suppose some property q of a gas is proportional to $(0.326\ \text{s}^3\ \text{m}^{-3})v_x^3 + (\pi\ \text{s}^9\ \text{m}^{-9})v_x^9$. What is the average value of q?

1.4 Without referring to any formula, decide whether at constant density the mean free path (a) increases, (b) decreases, or (c) stays constant with increasing temperature and explain your answer.

1.5 Consider a deck of cards. With aces valued at one and jacks, queens, and kings valued at 11, 12, and 13, respectively, calculate the average value of a card drawn at random from a full deck.

1.6 The distribution of the grades S (where $0 \le S \le 100$) for a class containing a large number of students is given by the continuous function $P(S) = K(50 - |S - 50|)$, where $|x|$ is the absolute value of x and K is a normalization constant. Determine the normalization constant and find out what fraction of the students received grades greater than or equal to 90.

1.7 A pair of dancers is waltzing on a one-dimensional dance floor of length L. Since they tend to avoid the walls, the probability of finding them at a position x between walls at $x = 0$ and $x = L$ is proportional to $\sin^2(\pi x/L)$. What is the normalized distribution function for the position of the waltzers? Using this distribution function, calculate the most probable position for the waltzers. Calculate the average position of the waltzers. (*Hint:* The integral of $y\sin^2 y\ dy$ is $[y^2/4] - [(y\sin 2y)/4] - [(\cos 2y)/8]$; this is also the probability for finding a particle in a box at a particular position.)

1.8 By setting the derivative of the formula for the Maxwell-Boltzmann speed distribution equal to zero, show that the speed at which the distribution has its maximum is given by **equation 1.33.**

1.9 Show using **equations 1.16** and **1.37** that the average molecular energy is $3kT/2$.

1.10 Prove **equation 1.41** from **equation 1.40.** Integration can be accomplished by making the following change of variable. Let $\epsilon = kTx^2$, so that $d\epsilon = kT\ d(x^2)$ and $\epsilon^{1/2} = (kT)^{1/2}x$. Substitute these into **equation 1.40** and integrate by parts, recalling that since $d(uv) = u\ dv + v\ du$, then $\int d(uv) = \int u\ dv + \int v\ du$, so that $\int u\ dv = (uv)|_\text{limits} - \int v\ du$, where the notation $|_\text{limits}$ indicates that the product (uv) should be evaluated at the limits used for the integrals.

1.11 The Maxwell-Boltzmann distribution may not be quite valid! Calculate the fraction of N_2 molecules having speeds in excess of the speed of light.

1.12 The object of this problem is to show more rigorously that $\langle v_r \rangle = (8kT/\pi\mu)^{1/2}$, where μ, the reduced mass, is defined as $\mu \equiv m_1 m_2/(m_1 + m_2)$.

We have already learned in Appendix 1.4 that the total kinetic energy of two particles is given by $\frac{1}{2}\mu v_r^2 + \frac{1}{2}M v_{com}^2$, where $\mathbf{v}_r = \mathbf{v}_2 - \mathbf{v}_1$ and $\mathbf{v}_{com}$, the center-of-mass velocity vector, is defined by the equation $(m_1 + m_2)\mathbf{v}_{com} = m_1\mathbf{v}_1 + m_2\mathbf{v}_2$, and $M \equiv m_1 + m_2$.

a. Consider the probability of finding two molecules, one with velocity $\mathbf{v}_1$ and one with velocity $\mathbf{v}_2$. Using **equation 1.30,** we see that this probability is given by

$$F(v_{1x})F(v_{1y})F(v_{1z})F(v_{2x})F(v_{2y})F(v_{2z})\,dv_{1x}\,dv_{1y}\,dv_{1z}\,dv_{2x}\,dv_{2y}\,dv_{2z}$$

$$= \left(\frac{m_1}{2\pi kT}\right)^{3/2} \left(\frac{m_2}{2\pi kT}\right)^{3/2} \exp\left(-\frac{m_1\mathbf{v}_1^2}{2kT}\right)\exp\left(-\frac{m_2\mathbf{v}_2^2}{2kT}\right)$$

$$\times\, dv_{1x}\,dv_{1y}\,dv_{1z}\,dv_{2x}\,dv_{2y}\,dv_{2z}.$$

Use the result from Appendix 1.4 to show that this probability can also be written as

$$F(v_{rx})F(v_{ry})F(v_{rz})F(v_{comx})F(v_{comy})F(v_{comz})$$

$$\times\, dv_{rx}\,dv_{ry}\,dv_{rz}\,dv_{comx}\,dv_{comy}\,dv_{comz}$$

$$= \left(\frac{m_1}{2\pi kT}\right)^{3/2} \left(\frac{m_2}{2\pi kT}\right)^{3/2} \exp\left(-\frac{M\mathbf{v}_{com}^2}{2kT}\right)\exp\left(-\frac{\mu\mathbf{v}_r^2}{2kT}\right)$$

$$\times\, dv_{rx}\,dv_{ry}\,dv_{rz}\,dv_{comx}\,dv_{comy}\,dv_{comz}.$$

b. Now transform the Cartesian coordinates to spherical ones and show by integration over all coordinates that the average relative velocity $<v_r>$ is given by $(8kT/\pi\mu)^{1/2}$.

1.13 What is the ratio of the probability of finding a molecule moving with the average speed to the probability of finding a molecule moving with three times the average speed? How does this ratio depend on the temperature?

1.14 You are caught without an umbrella in the rain and wish to get to your dorm, 1 km away, in the driest possible condition. Should you walk or run? To answer this question, calculate the ratio of the rain drop collisions with your body under the two conditions. Assume that the cross section is independent of direction (i.e., that you are spherical), that you run at 8 m/s, you walk at 3 m/s, and that the rainfall is constant with a velocity of, say, 15 m/s.

1.15 Calculate the root-mean-squared deviation of the speed from its mean value: $[<(v - <v>)^2>]^{1/2}$.

1.16 Find $<v^4>$ for a gas of molecular weight M at temperature T.

1.17 A very expensive gas is sold by the molecule, and the price is proportional to the velocity of the individual molecule: price in \$ = $v/<v>$. If I buy a bulb of these gaseous molecules, what is the average price per molecule, and does the price depend on the temperature of the bulb?

1.18   In a group of molecules all traveling in the positive z direction, what is the probability that a molecule will be found with a z-component speed between 400 and 401 m/s if $m/(2kT) = 5.62 \times 10^{-6}$ s^2/m^2? (*Hint:* You need to find and normalize a one-dimensional distribution function first!)

1.19   We will see in Chapter 3, **equation 3.4,** that the rate constant for a reaction as a function of temperature is given by the average of $\sigma(\epsilon_r)v_r$ over the thermal energy distribution $G(\epsilon_r)$, where $\epsilon_r = \frac{1}{2}mv_r^2$ and $\sigma(\epsilon_r)$ is the energy-dependent cross section for the reaction. The thermal relative kinetic energy distribution $G(\epsilon_r)$ has the same functional form as the kinetic energy distribution $G(\epsilon)$ given in **equation 1.37,** except that all energies $\epsilon = \frac{1}{2}mv^2$ are replaced by relative kinetic energies $\epsilon_r = \frac{1}{2}\mu v_r^2$.

   a.   Suppose that for a particular reaction $\sigma(\epsilon_r) = c\epsilon_r^2$, where c is a constant. Calculate $k(T)$.
   b.   Suppose that for another reaction $\sigma(\epsilon_r) = c/\epsilon_r$; calculate $k(T)$.

Chapter Two

2

The Rates of Chemical Reactions

2.1   INTRODUCTION

The objective of this chapter is to obtain an empirical description of the rates of chemical reactions on a macroscopic level and to relate the laws describing those rates to mechanisms for reaction on the microscopic level. Experimentally, it is found that the rate of a reaction depends on a variety of factors: on the temperature, pressure, and volume of the reaction vessel; on the concentrations of the reactants and products; and on whether or not a catalyst is present. By observing how the rate changes with such parameters, an intelligent chemist can learn what might be happening at the molecular level. The goal, then, is to describe in as much detail as possible the reaction mechanism. This goal is achieved in several steps. First, in this chapter, we will learn how an overall mechanism can be described in terms of a series of *elementary steps*. In later chapters, we will continue our pursuit of a detailed description (1) by examining how to predict and interpret values for the rate constants in these elementary steps and (2) by examining how the elementary steps might depend on the type and distribution of energy among the available degrees of freedom. In addition to these lofty intellectual pursuits, of course, there are very good practical reasons for understanding how reactions take place, reasons ranging from the desire for control of synthetic pathways to the need for understanding of the chemistry of the Earth's atmosphere.

2.2   EMPIRICAL OBSERVATIONS: MEASUREMENT OF REACTION RATES

One of the most fundamental empirical observations that a chemist can make is how the concentrations of reactants and products vary with time. The first substantial quantitative study of the rate of a reaction was performed by L. Wilhelmy, who in 1850 studied the inversion of sucrose in acid solution with a polarimeter. There are many methods for making such observations: one might monitor the concentrations spectroscopically, through absorption, fluorescence, or light scattering; one might measure concentrations electrochemically, for example, by potentiometric determination of the pH; or one might monitor the total volume or pressure if these are related in a simple way to the concentrations. Whatever the method, the result is usually something like that illustrated in **Figure 2.1.**

In general, as is true in this figure, the reactant concentrations will decrease as time goes on, while the product concentrations will increase. There may also be "intermediates" in the reaction, species whose concentrations first grow and then decay with time. How can we describe these changes in quantitative mathematical terms?

2.3   RATES OF REACTIONS: DIFFERENTIAL AND INTEGRATED RATE LAWS

We define the *rate law* for a reaction in terms of the time rate of change in concentration of one of the reactants or products. In general, the rate of change of the chosen species will be a function of the concentrations of the reactant and product species as well as of external parameters such as the temperature. For example, in **Figure 2.1** the rate of change for a species at any time is proportional to the slope

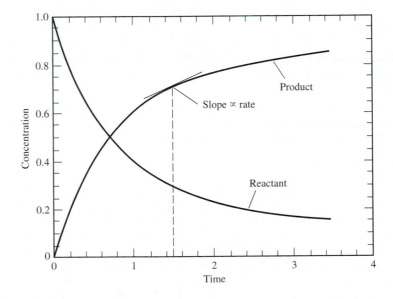

Figure 2.1

Concentration of reactant and product as a function of time.

of its concentration curve. The slope varies with time and generally approaches zero as the reaction approaches equilibrium. The stoichiometry of the reaction determines the proportionality constant. Consider the general reaction

$$aA + bB = cC + dD. \tag{2.1}$$

We will define the rate of change of [C] as rate = (1/c) d[C]/dt. This rate varies with time and is equal to some function of the concentrations: (1/c) d[C] dt = f([A], [B],[C],[D]). Of course, the time rates of change for the concentrations of the other species in the reaction are related to that of the first species by the stoichiometry of the reaction. For the example presented above, we find that

$$\frac{1}{c}\frac{d[C]}{dt} = \frac{1}{d}\frac{d[D]}{dt} = -\frac{1}{a}\frac{d[A]}{dt} = -\frac{1}{b}\frac{d[B]}{dt}. \tag{2.2}$$

By convention, since we would like the rate to be positive if the reaction proceeds from left to right, we choose positive derivatives for the products and negative ones for the reactants.

The equation (1/c) d[C]/dt = f([A],[B],[C],[D]) is called the *rate law* for the reaction. While f([A],[B],[C],[D]) might in general be a complicated function of the concentrations, it often occurs that f can be expressed as a simple product of a *rate constant, k,* and the concentrations each raised to some power:[a]

$$\frac{1}{c}\frac{d[C]}{dt} = k[A]^m[B]^n[C]^o[D]^p. \tag{2.3}$$

When the rate law can be written in this simple way, we define the *overall order* of the reaction as the sum of the powers, i.e., *overall order* $q = m + n + o + p$, and we define the *order of the reaction* with respect to a particular species as the power to which its concentration is raised in the rate law, e.g., *order with respect to [A] = m.* Note that since the left-hand side of the above equation has units of concentration per time, the rate constant will have units of time^{-1} concentration$^{-(q-1)}$. As we will see below, the form of the rate law and the order with respect to each species give us a clue to the mechanism of the reaction. In addition, of course, the rate law enables us to predict how the concentrations of the various species change with time.

An important distinction should be made from the outset: the overall order of a reaction cannot be obtained simply by looking at the overall reaction. For example, one might think (mistakenly) that the reaction

$$H_2 + Br_2 \rightarrow 2\,HBr \tag{2.4}$$

should be second order simply because the reaction consumes one molecule of H_2 and one molecule of Br_2. In fact, the rate law for this reaction is quite different:

$$\frac{1}{2}\frac{d[HBr]}{dt} = k[H_2][Br_2]^{1/2}. \tag{2.5}$$

[a]Note that both the rate constant and the Boltzmann constant have the same symbol, k. Normally, the context of the equation will make the meaning of k clear.

Thus the order of a reaction is not necessarily related to the stoichiometry of the reaction; it can be determined only by experiment.

Given a method for monitoring the concentrations of the reactants and products, how might one experimentally determine the order of the reaction? One technique is called the *method of initial slopes.* If we were to keep $[Br_2]$ fixed while monitoring how the initial rate of [HBr] production depended on the H_2 starting concentration, $[H_2]_0$, we would find, for example, that if we doubled $[H_2]_0$, the rate of HBr production would increase by a factor of 2. By contrast, were we to fix the starting concentration of H_2 and monitor how the initial rate of HBr appearance rate depended on the Br_2 starting concentration, $[Br_2]_0$, we would find that if we doubled $[Br_2]_0$, the HBr production rate would increase not by a factor of 2, but only by a factor of $\sqrt{2}$. Experiments such as these would thus show the reaction to be first order with respect to H_2 and half order with respect to Br_2.

While the rate law in its differential form describes in the simplest terms how the rate of the reaction depends on the concentrations, it will often be useful to determine how the concentrations themselves vary in time. Of course, if we know $d[C]/dt$, in principle we can find [C] as a function of time by integration. In practice, the equations are sometimes complicated, but it is useful to consider the differential and integrated rate laws for some of the simpler and more common reaction orders.

2.3.1   First-Order Reactions

Let us start by considering first-order reactions, A $\rightarrow$ products, for which the differential form of the rate law is

$$\frac{-d[A]}{dt} = k[A]. \tag{2.6}$$

Rearrangement of this equation yields

$$\frac{d[A]}{[A]} = -k\,dt.$$

Let [A(0)] be the initial concentration of A and let [A(t)] be the concentration at time t. Then integration yields

$$\int_{[A(0)]}^{[A(t)]} \frac{d[A]}{[A]} = -\int_0^t k\,dt,$$

$$\ln \frac{[A(t)]}{[A(0)]} = -kt, \tag{2.7}$$

or, exponentiating both sides of the equation,

$$[A(t)] = [A(0)]\exp(-kt). \tag{2.8}$$

Equation 2.8 is the *integrated rate law* corresponding to the *differential rate law* given in **equation 2.6**. While the differential rate law describes the rate of the reaction, the integrated rate law describes the concentrations.

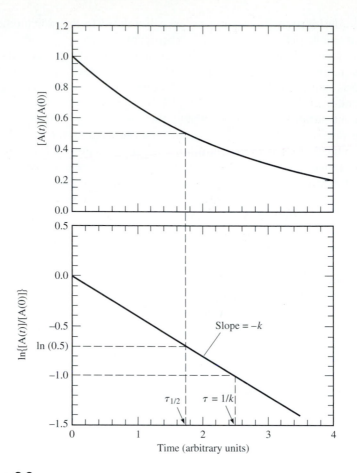

Figure 2.2

Decay of $[A(t)]$ for a first-order reaction.

Figure 2.2 plots the $[A(t)]/[A(0)]$ in the upper panel and the natural log of $[A(t)]/[A(0)]$ in the lower panel as a function of time for a first-order reaction. Note that the slope of the line in the lower panel is $-k$ and that the concentration falls to $1/e$ of its initial value after a time $\tau = 1/k$, often called the *lifetime* of the reactant. A related quantity is the time it takes for the concentration to fall to half of its value, obtained from

$$\frac{[A(t = \tau_{1/2})]}{[A(0)]} = \frac{1}{2}$$

$$= \exp(-k\tau_{1/2}), \tag{2.9}$$

$$\tau_{1/2} = \frac{\ln(2)}{k}.$$

The quantity $\tau_{1/2}$ is known as the *half-life* of the reactant.

An example of a first-order process is the radiative decay of an electronically excited species. **Figure 2.3** shows the time dependence of the fluorescence intensity

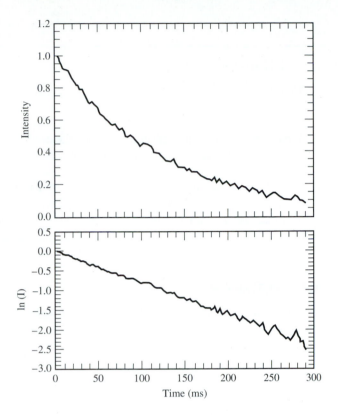

Figure 2.3

I^* fluorescence intensity as a function of time on linear (upper) and logarithmic (lower) scales.

for iodine atoms excited to their $^2P_{1/2}$ electronic state, denoted here as I^*. The chemical equation is

$$I^* \xrightarrow{k_{rad}} I + h\nu, \qquad\qquad (2.10)$$

where $\nu/c = 7603$ cm^{-1}. The deactivation of I^* is important because this level is the emitting level of the iodine laser. Since the reaction is first order, $-d[I^*]/dt = k_{rad}[I^*]$. From the stoichiometry of the photon production, $-d[I^*]/dt$ is also equal to $d(h\nu)/dt$, so that $d(h\nu)/dt = k_{rad}[I^*]$. Finally, the fluorescence intensity, I, is defined as the number of photons detected per unit time, $I = d(h\nu)/dt$, so that the intensity is directly proportional to the instantaneous concentration of $I*$: $I = k_{rad}[I^*]$. The top panel plots I as a function of t, while the lower panel plots $\ln(I)$ against the same time axis. It is clear that the fluorescence decay obeys first-order kinetics. The lifetime derived from these data is 126 ms, and the half-life is 87 ms as calculated in **Example 2.1.** The measurement of the radiative decay for I^* is actually quite difficult, and the data in **Figure 2.3** represent only a lower limit on the lifetime. The experimental problem is to keep the I^* from being deactivated by a method other than radiation, for example, by a collision with some other species. This process will be discussed in more detail later, after we have considered second-order reactions.

example 2.1

The Lifetime and Half-Life of I* Emission

Objective   Find the lifetime and half-life of I^* from the data given in **Figure 2.3.**

Method   First determine the rate constant k_{rad}. Then the lifetime is simply $\tau = 1/k_{rad}$, while the half-life, given in **equation 2.9**, is $\tau_{1/2} = \ln(2)/k_{rad}$.

Solution   The slope of the line in the bottom half of **Figure 2.3** can be determined to be $-7.94\ \text{s}^{-1}$. Thus, the lifetime is $1/(7.94\ \text{s}^{-1}) = 126$ ms, and the half-life is $\ln(2)/(7.94\ \text{s}^{-1}) = 87$ ms.

2.3.2   Second-Order Reactions

Second-order reactions are of two types, those that are second order in a single reactant and those that are first order in each of two reactants. Consider first the former case, for which the simplest overall reaction is

$$2\ A \rightarrow \text{products,} \tag{2.11}$$

with the differential rate law[b]

$$\frac{-d[A]}{dt} = k[A]^2. \tag{2.12}$$

Of course, a simple method for obtaining the integrated rate law would be to rearrange the differential law as

$$\frac{-d[A]}{[A]^2} = k\,dt \tag{2.13}$$

and to integrate from $t = 0$ when $[A] = [A(0)]$ to the final time when $[A] = [A(t)]$. We would obtain

$$\frac{1}{[A(t)]} - \frac{1}{[A(0)]} = kt. \tag{2.14}$$

However, to prepare the way for more complicated integrations, it is useful to perform the integration another way by introducing a change of variable. Let x be defined as the amount of A that has reacted at any given time. Then $[A(t)] = [A(0)] - x$, and

$$\frac{-d[A]}{dt} = \frac{dx}{dt} = k([A(0)] - x)^2. \tag{2.15}$$

---

[b]The alert student might notice that we have omitted the $\frac{1}{2}$ on the left-hand side of the equation, amounting to a redefinition of the rate constant for the reaction. The reason for temporarily abandoning our convention is so that second-order reactions of the type $2\ A \rightarrow$ products and those of the type $A + B \rightarrow$ products will have the same form.

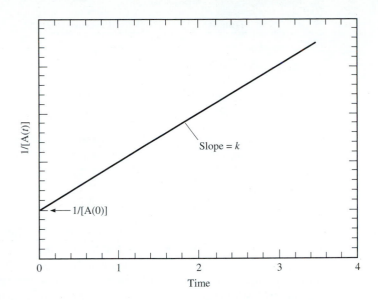

Figure 2.4

Variation of concentration with time for a second-order reaction of the type $2\,A \rightarrow$ products.

Rearrangement gives

$$\frac{dx}{([A(0)] - x)^2} = k\,dt, \qquad \textbf{(2.16)}$$

and integration yields

$$\int_0^x ([A(0)] - x)^{-2}\,dx = k\int_0^t dt,$$

$$([A(0)] - x)^{-1}\big|_0^x = kt,$$

$$\frac{1}{[A(0)] - x} - \frac{1}{[A(0)]} = kt, \qquad \textbf{(2.17)}$$

$$\frac{1}{[A(t)]} - \frac{1}{[A(0)]} = kt.$$

Note that the same answer is obtained using either method.

Equations 2.14 and **2.17** suggest that a plot of $1/[A(t)]$ as a function of time should yield a straight line whose intercept is $1/[A(0)]$ and whose slope is the rate constant k, as shown in **Figure 2.4.**

example 2.2

Diels-Alder Condensation of Butadiene, a Second-Order Reaction

Objective   Butadiene, C_4H_6, dimerizes in a Diels-Alder condensation to yield a substituted cyclohexene, C_8H_{12}. Given the data on the 400-K gas phase reaction below, show that the dimerization occurs as a second-order process and find the rate constant.

Time (s)	Total Pressure (torr)
0	626
750	579
1,500	545
2,460	510
3,425	485
4,280	465
5,140	450
6,000	440
7,500	425
9,000	410
10,500	405

Method

According to **equation 2.14,** when the reciprocal of the reactant pressure, $P(C_4H_6)$, is plotted as a function of time, a second-order process is characterized by a linear function whose slope is the rate constant. The complication here is that we are given the *total* pressure rather than the *reactant* pressure as a function of time. The reactant pressure is related to the total pressure through the stoichiometry of the reaction $2\ C_4H_6 \rightarrow C_8H_{12}$. Let $2x$ be the pressure of C_4H_6 that has reacted; then $P(C_8H_{12}) = x$, and $P(C_4H_6) = P_0 - 2x$, where P_0 is the initial pressure. The total pressure is thus $P_{tot} = P(C_4H_6) + P(C_8H_{12}) = P_0 - x$, or $x = P_0 - P_{tot}$. Consequently, $P(C_4H_6) = P_0 - 2(P_0 - P_{tot}) = 2P_{tot} - P_0$.

Solution

A plot of $1/(2P_{tot} - P_0)$ versus time is shown in **Figure 2.5.** A least-squares fit gives the slope of the line as $k = 3.8 \times 10^{-7}\ s^{-1}$ torr^{-1}. Recalling that 1 torr $= (1/760)$ atm and assuming ideal gas

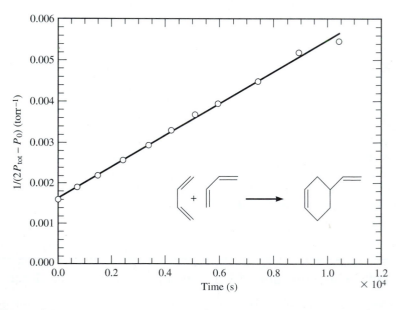

Figure 2.5

Plot of reciprocal C_4H_6 pressure as a function of time for the Diels-Alder condensation of butadiene.

behavior, we can express k in more conventional units: $k = (3.8 \times 10^{-7}\,s^{-1}\,torr^{-1}) \times (760\,torr/1\,atm) \times (82.06\,cm^3\,atm\,mol^{-1}\,K^{-1}) \times (400\,K) = 9.48\,cm^3\,mol^{-1}\,s^{-1}$. Thus, $-d[A]/dt = (9.48\,cm^3\,mol^{-1}\,s^{-1})[A]^2$.

We now turn to reactions that are second order overall but first order in each of two reactants. The simplest reaction of this form is

$$A + B \rightarrow products, \tag{2.18}$$

with the differential rate law

$$\frac{-d[A]}{dt} = k[A][B]. \tag{2.19}$$

Consider a starting mixture of A and B in their stoichiometric ratio, where $[A(0)] = [B(0)]$. Then, again letting x be the amount of A (or B) that has reacted at time t, we see that $[A(t)] = [A(0)] - x$ and that $[B(t)] = [B(0)] - x = [A(0)] - x$, where the last equality takes into account that we started with a stoichiometric mixture. Substituting into the differential rate law we obtain

$$\frac{dx}{dt} = k([A(0)] - x)^2, \tag{2.20}$$

just as in the case for the reaction $2A \rightarrow$ products. The solution is given by **equation 2.17,** and a similar equation could be derived for $1/[B(t)]$.

Suppose, however, that we had started with a nonstoichiometric ratio, $[B(0)] \neq [A(0)]$. Substitution into the differential rate law would then yield

$$\frac{dx}{dt} = k([A(0)] - x)([B(0)] - x), \tag{2.21}$$

or

$$\frac{dx}{([A(0)] - x)([B(0)] - x)} = k\,dt. \tag{2.22}$$

This equation can be integrated by using the method of partial fractions. We rewrite **equation 2.22** as (see Problem 2.15)

$$\frac{dx}{[B(0)] - [A(0)]}\left[\frac{1}{[A(0)] - x} - \frac{1}{[B(0)] - x}\right] = k\,dt. \tag{2.23}$$

Integrating, we find

$$\frac{1}{[B(0)] - [A(0)]}\int_{x=0}^{x}\left[\frac{1}{[A(0)] - x} - \frac{1}{[B(0)] - x}\right]dx = \int_{t=0}^{t} k\,dt, \tag{2.24}$$

or

$$\frac{1}{[B(0)] - [A(0)]}\left\{[-\ln([A(0)] - x)] - [-\ln([B(0)] - x)]\right\}\Big|_0^x = kt, \tag{2.25}$$

$$\ln \frac{\{[B(0)] - x\}[A(0)]}{[B(0)]\{[A(0)] - x\}} = \{[B(0)] - [A(0)]\} kt, \qquad (2.26)$$

$$\ln \frac{[B][A(0)]}{[A][B(0)]} = \{[B(0)] - [A(0)]\} kt. \qquad (2.27)$$

Thus, a plot of the left-hand side of **equation 2.27** versus t should thus give a straight line of slope $\{[B(0)] - [A(0)]\}k$.

2.3.3   Pseudo-First-Order Reactions

It often occurs for second-order reactions that the experimental conditions can be adjusted to make the reaction appear to be first order in one of the reactants and zero order in the other. Consider again the reaction

$$A + B \rightarrow products, \qquad (2.28)$$

with the differential rate law

$$\frac{-dA}{dt} = k[A][B]. \qquad (2.29)$$

We have already seen that the general solution for nonstoichiometric starting conditions is given by **equation 2.27.** Suppose, however, that the initial concentration of B is very much larger than that of A, so large that no matter how much A has reacted the concentration of B will be little affected. From the differential form of the rate law, we see that

$$\frac{d[A]}{[A]} = -k[B]dt, \qquad (2.30)$$

and, if $[B] = [B(0)]$ is essentially constant throughout the reaction, integration of both sides yields

$$\ln\left[\frac{[A(t)]}{[A(0)]}\right] = -k[B(0)]t, \qquad (2.31)$$

or

$$[A(t)] = [A(0)]\exp\{-k[B(0)]t\}. \qquad (2.32)$$

Note that this last equation is very similar to **equation 2.8,** except that the rate constant k has been replaced by the product of k and $[B(0)]$. Incidentally, it is easy to verify that **equation 2.32** can be obtained from the general solution for nonstoichiometric second-order reactions **equation 2.27** in the limit when $[B(0)] \gg [A(0)]$ (Problem 2.16).

Pseudo-first-order reactions are ubiquitous in chemical kinetics. An example illustrates their analysis. **Figure 2.6** shows the decay of the concentration of excited

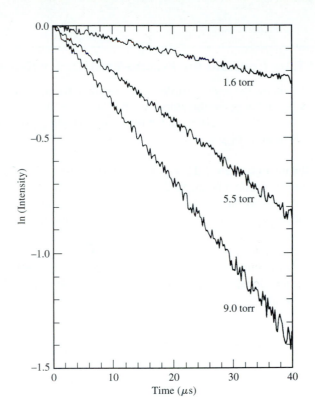

Figure 2.6

Variation of I^* concentration with time for various starting concentrations of $NO(v = 0)$.

$I(^2P_{1/2})$ atoms, here again called I^*, following their relaxation by $NO(v = 0)$ to the ground $I(^2P_{3/2})$ state, here called simply I:

$$I^* + NO(v = 0) \rightarrow I + NO(v > 0). \qquad (2.33)$$

Note that in this process the electronic energy of I^* is transferred to vibrational excitation of the NO; this type of process is often referred to as $E \rightarrow V$ transfer.[c] In this experiment, the I^* concentration was created at time zero by pulsed-laser photodissociation of I_2, $I_2 + h\nu \rightarrow I^* + I$, and the I^* concentration was monitored by its fluorescence intensity. Because the initial concentration of I^* is several orders of magnitude smaller than the concentration of $NO(v = 0)$, the latter hardly varies throughout the reaction, so the system can be treated as pseudo-first-order. Consequently, the data in **Figure 2.6** are plotted as ln of I^* fluorescence intensity versus time; a straight line is obtained for each initial concentration of $NO(v = 0)$. It is clear that the slope becomes steeper with increasing $NO(v = 0)$ concentration, as predicted by **equation 2.32**. The value of k can be determined from the variation as roughly 3.9×10^3 s^{-1} torr^{-1}, or 1.2×10^{-13} cm^3 molec^{-1} s^{-1}, as shown in **Example 2.3.**

[c]The data in the figure are taken from A. J. Grimley and P. L. Houston, *J. Chem. Phys.* **68**, 3366–3376 (1978). A review of this type of process appears in P. L. Houston, "Electronic to Vibrational Energy Transfer from Excited Halogen Atoms," in *Photoselective Chemistry,* Part 2, J. Jortner, Ed., (J. Wiley & Sons, New York, 1981), 381–418 (1981).

example 2.3

Evaluation of Rate Constant for Pseudo-First-Order Reaction

Objective   Evaluate the second-order rate constant from the data shown in **Figure 2.6** given the slopes of the lines as -0.627×10^{-2} for 1.6 torr, -0.213×10^{-1} for 5.5 torr, and -0.349×10^{-1} for 9.0 torr, all in units of μs^{-1}.

Method   From **equation 2.31** we see that the slope of $\ln([A(t)]/[A(0)])$ versus t should be the negative of the rate constant times the starting pressure of the constant component. Thus, k should be given by the negative of the slope divided by the pressure of the constant component.

Solution   For the three points given, $k = (0.00627/1.6) = 3.92 \times 10^{-3}$, $k = (0.0213/5.5) = 3.87 \times 10^{-3}$, and $k = (0.0349/9.0) = 3.88 \times 10^{-3}$. The average is 3.89×10^{-3} in units of μs^{-1} torr^{-1}, or $(3.89 \times 10^{-3}\ \mu s^{-1}\ torr^{-1}) \times (10^6\ \mu s)/(1\ s) = 3.89 \times 10^3\ s^{-1}\ torr^{-1}$.

Comment   A better method for solution would be to plot the negative of the slopes as a function of the pressure of the constant component and to determine the best line through the points. Such a plot is shown in **Figure 2.7**. The slope of the line is equal to $k[B(0)]$ over $[B(0)]$, i.e., the slope is equal to the rate constant. If the kinetic scheme is correct, the intercept of the line should be zero. A positive intercept would indicate deactivation of I* by some other process or species, for example radiative decay or deactivation by the remaining I_2 precursor.

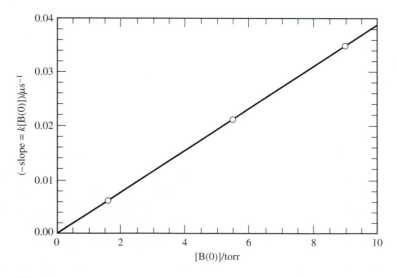

Figure 2.7

Figure showing alternative analysis of pseudo-first-order reaction.

Second-order rate constants are reported in a variety of units. As we have just seen, the units most directly related to the experiment are (time^{-1} pressure^{-1}), for example, s^{-1} torr^{-1}. However, reporting the rate constant in these units has the disadvantage that at different temperatures the rate constant is different both because of the inherent change in the constant with temperature and because the pressure changes as the temperature changes. The use of time^{-1} density^{-1} for rate constant units avoids this complication; the density is usually expressed either as molecules/cm^3 or in moles/L. At any given temperature, of course, the two sets of units can be related. For example, at 300 K, the ideal gas law can be used to determine that 1 torr is equivalent to 3.22×10^{16} molecules cm^{-3}. Thus, the rate constant for $I^* + NO(v = 0) \rightarrow I + NO(v > 0)$ listed above as 3.9×10^3 s^{-1} torr^{-1} is equivalent to $(3.9 \times 10^3$ s^{-1} torr$^{-1})/(3.22 \times 10^{16}$ molecules cm^{-3} torr$^{-1}) = 1.2 \times 10^{-13}$ cm^3 molec^{-1} s^{-1}, which in turn is equivalent to $(1.2 \times 10^{-13}$ cm^3 molec^{-1} s$^{-1})(6.02 \times 10^{23}$ molec/mole$) = 7.2 \times 10^{10}$ cm^3 mole^{-1} s^{-1}. If we multiply by (1 L/1000 cm^3), we see that it is also equal to 7.2×10^7 L mole^{-1} s^{-1}.

2.3.4   Higher-Order Reactions

For higher-order reactions, integration of the differential rate law equation becomes more complicated. For example, in an overall reaction

$$2\,A + B \rightarrow products, \qquad (2.34)$$

where the differential rate law is

$$-\frac{1}{2}\frac{d[A]}{dt} = -\frac{d[B]}{dt} = k[A]^2[B], \qquad (2.35)$$

the integrated rate expression for a nonstoichiometric starting mixture is

$$\frac{1}{[A(0)] - 2[B(0)]}\left[\frac{1}{[A(0)]} - \frac{1}{[A]}\right] + \frac{1}{[[A(0)] - 2[B(0)]]^2}\ln\frac{[A][B(0)]}{[A(0)][B]}$$

$$= \frac{1}{2}kt. \qquad (2.36)$$

However, such higher-order reactions usually take place under conditions where the concentration of one of the species is so large that it can be regarded as constant. An example might be the recombination of O atoms: $O + O + O_2 \rightarrow 2\,O_2$. Under conditions where $[O_2] >> [O]$, the third-order process becomes pseudo-second-order, and the integrated rate expression is simply related to expressions already derived. For example, in the reaction $2\,A + B \rightarrow$ products, for large [B(0)] the integrated rate expression is simply **equation 2.14** with k replaced by $k[B(0)]$. Alternatively, the reaction might become pseudo-first-order (see Section 2.3.3), as would be the case for $O + O_2 + O_2 \rightarrow O_2 + O_3$ with O_2 in excess. The differential rate law is $-d[O]/dt = k[O_2]^2[O]$. If the concentration of O_2 is very nearly constant throughout the reaction, the integrated rate law is an expression similar to **equation 2.8:** $[O] = [O]_0\exp(-k[O_2]_0^2 t/2)$. It can be verified that **equation 2.36** reduces to equations similar to **equation 2.8** or **equation 2.14** for the limiting cases when either [A(0)] or [B(0)] is very large, respectively (Problem 2.17).

2.3.5   Temperature Dependence of Rate Constants

The temperature dependence of the reaction rate, like the order of the reaction, is another empirical measurement that provides a basis for understanding reactions on a molecular level. Most rates for simple reactions increase sharply with increasing temperature; a rule of thumb is that the rate will double for every 10 K increase in temperature. Arrhenius[d] first proposed that the rate constant obeyed the law

$$k = A \exp\left(-\frac{E_a}{RT}\right), \tag{2.37}$$

where A is a temperature-independent constant, often called the preexponential factor, and E_a is called the activation energy. The physical basis for this law will be discussed in more detail in the next chapter, but we can note in passing that for a simple reaction in which two molecules collide and react, A is proportional to the number of collisions per unit time, and the exponential factor describes the fraction of collisions that have enough energy to lead to reaction.

The temperature dependence of a wide variety of reactions can be fit by this simple Arrhenius law. An alternate way of writing the law is to take the natural logarithm of both sides:

$$\ln k = \ln A - \frac{E_a}{RT}. \tag{2.38}$$

Thus, a plot of $\ln k$ as a function of $1/T$ should yield a straight line whose slope is $-E_a/R$ and whose intercept is $\ln A$. Such a plot is shown in **Figure 2.8** for the reaction of hydrogen atoms with O_2, one of the key reactions in combustion.

Plots such as these have been used to determine the "Arrhenius parameters," A and E_a for a wide number of reactions. A selection of Arrhenius parameters is provided for first-order reactions in **Table 2.1** and for second-order reactions in **Table 2.2**.[e]

The Arrhenius form of the rate constant allows us to calculate the rate constant k_2 at a new temperature T_2, provided that we know the activation energy and the rate constant k_1 at a specific temperature, T_1. Since the Arrhenius A parameter is independent of temperature, subtraction of the Arrhenius form **equation 2.38** for k_1, $\ln k_1 = \ln A - E_a/RT_1$, from that for k_2, $\ln k_2 = \ln A - E_a/RT_2$, yields

$$\ln \frac{k_2}{k_1} = -\frac{E_a}{R}\left[\frac{1}{T_2} - \frac{1}{T_1}\right]. \tag{2.39}$$

Example 2.4 illustrates the use of **equation 2.39.**

---

[d]Svante August Arrhenius was born in Uppsala, Sweden, in 1859 and died in Stockholm in 1927. He is best known for his theory that electrolytes are dissociated in solution. He nearly turned away from chemistry twice in his career, once as a undergraduate and once when his Ph.D. thesis was awarded only a "fourth class," but his work on electrolytic solutions was eventually rewarded with a Nobel Prize in Chemistry in 1903. His paper on activation energies was published in *Z. physik. Chem.* **4,** 226 (1889).

[e]Data taken from *NIST Chemical Kinetics Database Version 4.0,* W. Gary Mallard et al., Chemical Kinetics Data Center, National Institute of Standards and Technology, Gaithersburg, MD 20899. The database covers thermal gas-phase kinetics and includes over 15,800 reaction records for over 5700 reactant pairs and a total of more than 7400 distinct reactions. There are more than 4000 literature references through 1990.

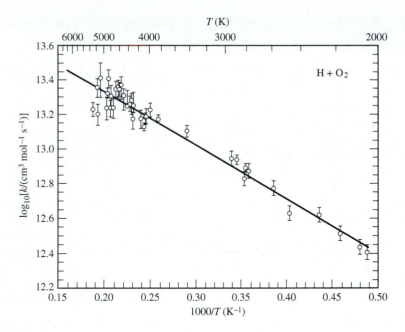

Figure 2.8

Arrhenius plot of the reaction $H + O_2 \rightarrow OH + O$.

The data are taken from H. Du and J. P. Hessler, *J. Chem. Phys* **96,** 1077 (1992).

TABLE 2.1	Arrhenius Parameters for Some First-Order Gas Phase Reactions		
Reaction	**A (s^{-1})**	**E_a/R (K)**	**Range (K)**
$CH_3CHO \rightarrow CH_3 + HCO$	3.8×10^{15}	39,500	500–2,000
$C_2H_6 \rightarrow CH_3 + CH_3$	2.5×10^{16}	44,210	200–2,500
$O_3 \rightarrow O_2 + O$	7.6×10^{12}	12,329	383–413
$HNO_3 \rightarrow OH + NO_2$	1.1×10^{15}	24,000	295–1,200
$CH_3OH \rightarrow CH_3 + OH$	6.5×10^{16}	46,520	300–2,500
$N_2O \rightarrow N_2 + O$	6.1×10^{14}	28,210	1,000–3,600
$N_2O_5 \rightarrow NO_3 + NO_2$	1.6×10^{15}	11,220	200–384
$C_2H_5 \rightarrow C_2H_4 + H$	2.8×10^{13}	20,480	250–2,500
$CH_3NC \rightarrow CH_3CN$	3.8×10^{13}	19,370	393–600
$H_2S \rightarrow SH + H$	6.0×10^{14}	40,820	1,965–19,700
cyclopropane $\rightarrow$ propene	2.9×10^{14}	31,410	897–1,450
cyclobutane $\rightarrow$ 2 C_2H_4	2.3×10^{15}	30,490	891–1,280

TABLE 2.2	Arrhenius Parameters for Some Second-Order Gas Phase Reactions		
Reaction	A (cm^3 mol^{-1} s^{-1})	E_a/R (K)	Range (K)
$CH_3 + H_2 \rightarrow CH_4 + H$	6.7×10^{12}	6,250	300–2,500
$CH_3 + O_2 \rightarrow CH_3O + O$	2.4×10^{13}	14,520	298–3,000
$F + H_2 \rightarrow HF + H$	8.4×10^{13}	500	200–300
$H + H_2O \rightarrow OH + H_2$	9.7×10^{13}	10,340	250–3,000
$H + CH_4 \rightarrow CH_3 + H$	1.9×10^{14}	6,110	298–2,500
$H + O_2 \rightarrow OH + H$	1.4×10^{14}	8,000	250–3,370
$H + NO \rightarrow OH + N$	1.3×10^{14}	24,230	1,750–4,500
$H + N_2O \rightarrow N_2 + OH$	7.6×10^{13}	7,600	700–2,500
$H + CO_2 \rightarrow CO + OH$	1.5×10^{14}	13,300	300–2,500
$H + O_3 \rightarrow O_2 + OH$	8.4×10^{13}	480	220–360
$O + H_2 \rightarrow OH + H$	5.2×10^{13}	5,000	293–2,800
$O + CH_4 \rightarrow OH + CH_3$	1.0×10^{14}	5,090	298–2,575
$O + NO_2 \rightarrow NO + O_2$	3.9×10^{12}	−120	230–350
$O + C_2H_2 \rightarrow$ products	3.4×10^{13}	1,760	195–2,600
$O + O_3 \rightarrow O_2 + O_2$	7.5×10^{12}	2,140	197–2,000
$O + C_2H_4 \rightarrow$ products	9.5×10^{12}	946	200–2,300
$OH + H_2 \rightarrow H + H_2O$	4.6×10^{12}	2,100	200–450
$OH + CH_4 \rightarrow H_2O + CH_3$	2.2×10^{12}	1,820	240–300
$OH + O \rightarrow O_2 + H$	1.4×10^{13}	−110	220–500

example 2.4

Relationship between the Rate Constants at Two Temperatures

Objective Given that the rate constant for the $H + O_2 \rightarrow OH + O$ reaction is 4.7×10^{10} cm^3 mol^{-1} s^{-1} at 1000 K and that the activation energy is 66.5 kJ/mol, determine the rate constant at 2000 K.

Method Use **equation 2.39** with $k_1 = 4.7 \times 10^{10}$ and $E_a = 66.5$.

Solution $\ln(k_2/4.7 \times 10^{10}) = -(66.5$ kJ mol^{-1}/8.314 J mol^{-1} K$^{-1}) \times \{(1/2000$ K$) - (1/1000$ K$)\}$; $\ln k_2 = \ln(4.7 \times 10^{10}) - 8000 \times \{0.0005 - 0.001\} = 24.6 + 4.00 = 28.6$; $k_2 = \exp(28.6) = 2.64 \times 10^{12}$ cm^3 mol^{-1} s^{-1}.

Comment An alternative method would be to determine A from the equation $k_1 = A \exp(-E_a/RT)$ with $E_a = 66.5$ kJ/mol; then one could use the Arrhenius parameters to determine the rate constant at the new temperature. Note from **Figure 2.8** that 2000 K is at the edge of the graph, where $\log_{10}k_2 \cong 12.42$, or $k_2 = 2.63 \times 10^{12}$ cm^3 mol^{-1} s^{-1}, in good agreement with the rate determined by the first method.

At this point, a word of caution is necessary. Not all reactions obey the simple Arrhenius form. For example, a recent review[f] of the measured rate constants for the $OH + CO \rightarrow H + CO_2$ reaction in the temperature range from 300 to 2000 K recommends $k(T) = (3.25 \times 10^{10} \text{ cm}^3 \text{ mol}^{-1} \text{ sec}^{-1})(T^{1.5})\exp(+250/T)$. Note that the preexponential factor is not independent of temperature, and that the "activation energy" is negative. We will see in the next chapter why the Arrhenius preexponential factor might depend on temperature. For reactions with very low activation energies, such as this one where there is no barrier to formation of an HOCO complex, the rate of the reaction might actually go down with increasing temperature, because at high temperatures, the two species might not stay in one another's vicinity long enough for the attractive force between them to draw them together. Thus, many reactions may exhibit a more complicated dependence of k on T than given by the simple Arrhenius form.

2.4   REACTION MECHANISMS

As we have seen in the example of the $Br_2 + H_2$ reaction, **equation 2.5,** many rate laws do not have the form that might be supposed from the overall stoichiometry. The value of the rate law, in fact, is that it gives us a clue to what might be happening on the molecular level. Here we start to leave empiricism and try to account for our experimental observations on a more chemical and microscopic level. What could be the behavior of the molecules that would lead to the observed macroscopic measurements? We postulate that the macroscopic rate law is the consequence of a *mechanism* consisting of *elementary steps,* each one of which describes a process that takes place on the microscopic level. Three types of microscopic processes account for essentially all reaction mechanisms: these are *unimolecular, bimolecular,* and *termolecular* reactions. As we will see, for such elementary steps the order of the reaction is equal to the molecularity; i.e., a unimolecular reaction follows first-order kinetics, a bimolecular reaction follows second-order kinetics, and a termolecular reaction follows third-order kinetics.

Unimolecular reactions involve the reaction of a single (energized) molecule:

$$A \rightarrow \text{products.} \qquad (2.40)$$

Since the number of product molecules produced per unit time will be proportional to the number of A species, $d[\text{products}]/dt = -d[A]/dt = k[A]$, so the order of this unimolecular reaction is unity; i.e., unimolecular reactions are first order. We have already seen an example of such a unimolecular process, namely the first-order radiative decay of excited iodine atoms discussed in Section 2.3.1 and **Example 2.1.**

Bimolecular reactions involve the collision between two species:

$$A + B \rightarrow \text{products.} \qquad (2.41)$$

Since the number of products per unit time will be proportional to the number of collisions between A and B, the rate of this bimolecular reaction should be proportional to the product of the concentrations of A and B (see Section 1.7): $d[\text{products}]/dt = k[A][B]$. Consequently, bimolecular reactions follow second-order kinetics.

[f]D. L. Baulch, C. J. Cobos, R. A. Cox, C. Esser, P. Frank, Th. Just, J. A. Kerr, M. J. Pilling, J. Troe, R. W. Walker, and J. Warnatz, *J. Phys. Chem. Ref. Data* **21,** 411 (1992).

Termolecular reactions are encountered less frequently than unimolecular or bimolecular reactions, but they are important at high pressures or in condensed phases. A termolecular reaction is one that requires the collision of three species:

$$A + B + C \rightarrow \text{products.} \qquad (2.42)$$

By a simple extension of our arguments in Chapter 1, the number of termolecular collisions will be proportional to the product of the concentrations of A, B, and C, so that termolecular reactions follow third-order kinetics.

One might well wonder how often three molecules could converge from separate directions to a single point in space.[g] In fact, most termolecular reactions can reasonably be viewed as two consecutive bimolecular collisions. The termolecular process might start with a "sticky" bimolecular collision, one in which the collision pair, say A and B, remain in close proximity for a finite period. If the pair is struck during this period by the third species, then products are formed; if not, then the reactants A and B separate. Whether one regards the overall process as two bimolecular collisions or as a termolecular one depends, in part, on the length of time that A and B spend together. If the time is no longer than, say, 1–100 vibrational periods, the current practice is to call the process a termolecular one. Whether one regards what happens as two bimolecular collisions or as one termolecular collision, the overall process will be proportional to the product of the pressures of the three species, i.e., third order.

In summary, we see that unimolecular processes follow first-order kinetics, bimolecular processes follow second-order kinetics, and termolecular processes follow third-order kinetics. In general, *the order of an **elementary** process is equal to its molecularity,* but this statement is true only for elementary processes and not necessarily for an overall process consisting of several elementary steps. We now proceed to combine elementary steps into mechanisms so as to determine the overall reaction and its order.

2.4.1 Opposing Reactions, Equilibrium

Perhaps the most rudimentary "mechanism" is a reaction that can proceed either in the forward or reverse direction. For simplicity, we consider opposing first-order elementary reactions, $A \rightleftharpoons B$, with rate constants k_1 in the forward direction and k_{-1} in the reverse direction. The differential rate law for the reaction is

$$-\frac{d[A]}{dt} = \frac{d[B]}{dt} = k_1[A] - k_{-1}[B]. \qquad (2.43)$$

If the initial concentrations of A and B are [A(0)] and [B(0)], respectively, then $[A(t)] = [A(0)] - x$ and $[B(t)] = [B(0)] + x$, where x is the concentration of A that has reacted to B at any time.

Before demonstrating by integration that A and B approach their equilibrium values exponentially, we digress for a moment to state two important principles. The first is the *principle of microscopic reversibility,* which, as noted by Tolman,[h] is a consequence of the time-reversal symmetry of classical or quantum mechanics. Consider a system that has reached some final state from an initial one by some path. If all the molecular momenta (internal and translational) are reversed, the system will return by

---

[g]Extension of this argument immediately suggests why higher-than-termolecular processes are never observed.

[h]R. C. Tolman, *Phys. Rev.* **23,** 699 (1924); *The Principles of Statistical Mechanics,* Clarendon Press, Oxford, p. 163.

the same path to its initial state. A macroscopic manifestation of this principle is called the *principle of detailed balance:* in a system at equilibrium, any process and its reverse proceed at the same rate. Although the principle of detailed balance is most powerful when applied to multiequation equilibria, it is still useful in the simple system under study, where it implies that at equilibrium $k_1A_e = k_{-1}B_e$, where A_e and B_e are the equilibrium concentrations. A proof of this relationship between the forward and reverse rates follows from the fact that, at equilibrium, the concentrations of the species do not change. Hence, $d[A]/dt = d[B]/dt = 0$ when $[A] = A_e$ and $[B] = B_e$, so that the left-hand side of **equation 2.43** is zero, and $k_1A_e = k_{-1}B_e$. Note also that, since the equilibrium constant for the reaction is $K_e = B_e/A_e = k_1/k_{-1}$, the principle of detailed balance also tells us that the equilibrium constant for opposing first-order reactions will be equal to the ratio of the forward and reverse rate constants.

Returning to the integration of **equation 2.43,** we rewrite the equation in terms of the single variable x:

$$\frac{dx}{dt} = k_1([A(0)] - x) - k_{-1}([B(0)] + x). \tag{2.44}$$

However, since $A_e = [A(0)] - x_e$ and $B_e = [B(0)] + x_e$, we can rewrite **equation 2.44** by substituting for $[A(0)]$ and $[B(0)]$:

$$\frac{dx}{dt} = k_1(A_e + x_e - x) - k_{-1}(B_e - x_e + x), \tag{2.45}$$

or

$$\frac{dx}{dt} = \{(k_1A_e - k_{-1}B_e) + k_1(x_e - x) + k_{-1}(x_e - x)\}$$
$$= (k_1 + k_{-1})(x_e - x). \tag{2.46}$$

For this last equality, we have used the fact that $k_1A_e = k_{-1}B_e$. Rearrangement gives

$$\frac{dx}{x_e - x} = (k_1 + k_{-1})\,dt. \tag{2.47}$$

Integrating both sides, we find

$$-\int_0^x d\ln(x_e - x) = (k_1 + k_{-1})\int_0^t dt, \tag{2.48}$$

or

$$-\ln\frac{x_e - x}{x_e} = (k_1 + k_{-1})t. \tag{2.49}$$

Exponentiation of both sides gives

$$x = x_e\{1 - \exp[-(k_1 + k_{-1})t]\}, \tag{2.50}$$

or, after subtracting both sides from $[A(0)]$,

$$[A(t)] = A_e + x_e\exp[-(k_1 + k_{-1})t]. \tag{2.51}$$

Thus, $[A(t)]$ starts at $A_e + x_e = [A(0)]$ and decreases to A_e exponentially with a rate constant equal to the sum of the forward and reverse rates. Similarly, $[B(t)]$ starts at $B_e - x_e = [B(0)]$ and increases to B_e exponentially with the same rate constant.

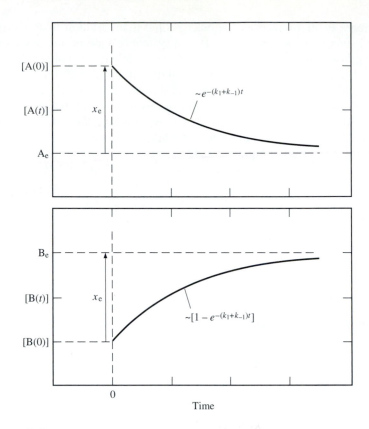

Figure 2.9

Concentrations as a function of time for opposing reactions.

These relationships are shown in **Figure 2.9.** Of course, in the limit where the forward rate constant is much larger than the reverse, $k_1 \gg k_{-1}$, the completed reaction produces nearly all B. Consequently, $A_e = 0$ and $x_e = [A(0)]$, so that **equation 2.51** reduces to $[A(t)] = [A(0)]\exp(-k_1 t)$, which is simply **equation 2.8**.

For opposing second-order reactions, the mathematics is somewhat more complicated, but the general result is the same in the limit when x_e is small: the system will approach equilibrium exponentially with a rate constant equal to the sum of the rate constants for the forward and reverse reactions. We will return to this result when considering temperature-jump techniques for measuring fast reaction rates in solution in Chapter 5.

2.4.2   Parallel Reactions

A common occurrence in real systems is that a species can react by more than one pathway. Consideration of such parallel reactions provides insight into the phenomenon of competition and clears up some common misconceptions. For simplicity we will model the parallel reactions as competing first-order processes:

$$A \xrightarrow{k_1} B,$$

$$A \xrightarrow{k_2} C.$$

(2.52)

The differential equations describing the parallel reactions are

$$\frac{d[A]}{dt} = -k_1[A] - k_2[A] = -[k_1 + k_2][A],$$

$$\frac{d[B]}{dt} = k_1[A], \tag{2.53}$$

$$\frac{d[C]}{dt} = k_2[A].$$

The solution to the first of these equations, for a starting concentration of [A(0)], is obtained by straightforward integration:

$$[A(t)] = [A(0)]\exp\{-[k_1 + k_2]t\}. \tag{2.54}$$

Substitution of this equation for A into the second differential equation leads to

$$\frac{d[B]}{dt} = k_1[A(0)]\exp\{-[k_1 + k_2]t\},$$

$$\int_{[B]=0}^{[B(t)]} d[B] = k_1[A(0)]\int_0^t \exp\{-[k_1 + k_2]t\}\,dt, \tag{2.55}$$

$$[B(t)] = \frac{k_1}{k_1 + k_2}[A(0)][1 - \exp\{-[k_1 + k_2]t\}].$$

A similar equation can be derived for [C(t)]:

$$[C(t)] = \frac{k_2}{k_1 + k_2}[A(0)][1 - \exp\{-[k_1 + k_2]t\}]. \tag{2.56}$$

There are two important points to note in comparing **equations 2.55** and **2.56.** First, both B and C rise exponentially with a rate constant equal to $[k_1 + k_2]$, as shown schematically in **Figure 2.10.** The reason for this behavior is that the rates of production of both B and C depend on the concentration of A, which decreases exponentially with a rate constant equal to the sum $[k_1 + k_2]$. Second, the ratio of products, sometimes called the *branching ratio,* is $[B(t)]/[C(t)] = k_1/k_2$ at all times. Thus, while the magnitude of [B] relative to [C] is constant and not generally unity, the time constant $[k_1 + k_2]$ with which each concentration approaches its final value is the same for both [B] and [C]. The situation is analogous to a bucket of water, A, leaking through two holes of different sizes. Buckets B and C collect the leaks. Of course, the relative amount of water collected in buckets B and C will depend on the ratio of the areas of the two holes, but the level of water in either bucket will rise toward its final value with the same rate, a rate equal to the rate at which water disappears from bucket A.

Parallel reactions are often encountered; we have already seen an example. The data of **Figure 2.3** and **Example 2.1** clearly show that one mechanism for deactivation of I^* is radiative decay. On the other hand, the data of **Figure 2.6** and **Example 2.3** show that I^* can also be deactivated by collision with NO. In this second example, we actually monitor the I^* concentration by its fluorescence intensity, so clearly both radiative decay and collisional decay are occurring in parallel. The

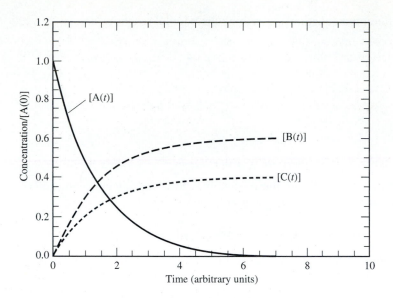

Figure 2.10

Concentrations as a function of time for the parallel reactions in **equation 2.52.**

solution for the fluorescence intensity I, analogous to **equation 2.54,** is $\ln(I/I_0) = -(k_{rad} + k[NO])t$. However, for the NO pressures used, $k[NO] >> k_{rad}$, so that little error was made in **Example 2.3.** For example, the branching ratio for decay by radiation versus decay by collision in 1.6 torr of NO is $k_{rad}/(k[NO]) = (7.94 \text{ s}^{-1})/[(3.9 \times 10^{-3} \, \mu\text{s}^{-1} \text{ torr}^{-1})(1.6 \text{ torr})] = 1.27 \times 10^{-3}$; i.e., nearly all of the deactivation is caused by collisions. However, the radiation is still important—it enables us to monitor the deactivation!

2.4.3   Consecutive Reactions and the Steady-State Approximation

Another common mechanism, which also leads to the important *steady-state approximation*, involves consecutive reactions. An example is the sequence of first-order processes

$$A \xrightarrow{k_1} B, \tag{2.57}$$

$$B \xrightarrow{k_2} C, \tag{2.58}$$

for which the differential equations are

$$\frac{d[A]}{dt} = -k_1[A], \tag{2.59}$$

$$\frac{d[B]}{dt} = k_1[A] - k_2[B], \tag{2.60}$$

$$\frac{d[C]}{dt} = k_2[B]. \tag{2.61}$$

The goal of our study of this mechanism is to develop a method, called the steady-state approximation, that we can use to simplify the analysis of consecutive reactions. To see the limitations of this approximation, we must first look at the exact solution for the simple system above. We assume that the initial concentration of A is [A(0)] and that [B(0)] = [C(0)] = 0. Integration of the first of these three differential equations then gives the time dependent concentration of A:

$$[A] = [A(0)]\exp(-k_1 t). \qquad (2.62)$$

Substitution of this solution into **equation 2.60** yields the differential equation

$$\frac{d[B]}{dt} = k_1[A(0)]e^{-k_1 t} - k_2[B]. \qquad (2.63)$$

The solution to this equation, as can be verified by direct differentiation (Problem 2.18), is[i]

$$B(t) = \frac{k_1}{k_2 - k_1}[A(0)](e^{-k_1 t} - e^{-k_2 t}). \qquad (2.64)$$

Finally, since by mass balance [A(0)] = [A] + [B] + [C],

$$[C] = [A(0)] - [B] - [A]$$

$$= [A(0)]\left\{1 - \frac{k_1}{k_2 - k_1}[e^{-k_1 t} - e^{-k_2 t}] - e^{-k_1 t}\right\}. \qquad (2.65)$$

While the mathematical complexity of this solution may at first seem daunting, this exact solution will enable us to see how to make some very useful and simplifying approximations. We first consider the case when k_1 is much larger than k_2. From the scheme in **equations 2.57** and **2.58**, we can see that in this limit the reaction A → B occurs first and goes nearly to completion before the reaction B → C takes place. Thus, we expect that nearly all the A is converted to the intermediate B before any appreciable conversion of B to C occurs. Indeed, in the limit of $k_1 \gg k_2$, **equation 2.64** reduces to

$$[B(t)] = [A(0)](e^{-k_2 t} - e^{-k_1 t}). \qquad (2.66)$$

Because $k_1 \gg k_2$, the second term in the parentheses rapidly approaches zero while the first term is still near unity. Consequently, the concentration of the intermediate B rapidly reaches a value nearly equal to [A(0)], and then during most of the reaction B decays slowly according to $[B(t)] = [A(0)]\exp(-k_2 t)$.

The exact solution for $k_1 = 10k_2$ is shown in **Figure 2.11**. The situation is analogous to three buckets located above one another. Imagine that the top bucket has a large hole and leaks into the second, which itself leaks through a much smaller hole into a third. Water placed in the first bucket would thus flow rapidly into the second bucket, from which it would then leak slowly into the third.

A more instructive limit for the consecutive reactions is when $k_2 \gg k_1$. In this case, the water leaks out of the middle bucket faster than it comes in, so one might

[i]A solution of this differential equation by the method of Laplace transformation is described in Appendix 6.1.

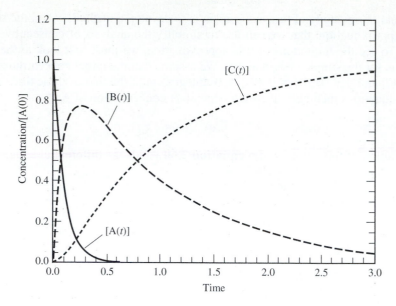

Figure 2.11

Concentrations in a consecutive reaction when $k_1 = 10k_2$.

guess that the water level in this bucket never rises very high. Indeed, in the limit when $k_2 \gg k_1$, the solution for [B] given by **equation 2.64** reduces to

$$[B(t)] = \frac{k_1}{k_2}[A(0)](e^{-k_1 t} - e^{-k_2 t}). \qquad \textbf{(2.67)}$$

Since $k_2 \gg k_1$, the second term in the parentheses rapidly approaches zero, while the first term is still close to unity. Consequently, the concentration of B rapidly approaches $(k_1/k_2)[A(0)]$ and then decays more slowly according to $[B] = (k_1/k_2)[A(0)]\exp(-k_1 t)$. Because k_1/k_2 is very small, the maximum concentration of B is much less than [A(0)]. **Figure 2.12** shows the exact concentrations for $k_2 = 10k_1$.

We now come to the major point of this section. It would be extremely tedious if we had to integrate the differential equations whenever we encountered a set of consecutive reactions. Fortunately, in most situations there is an easier method. Consider again the case illustrated in **Figure 2.12,** for which $k_2 \gg k_1$. After an initial transient rise, called the *induction period,* the concentration of B is very close to $[B] \approx (k_1/k_2)[A(0)]\exp(-k_1 t) = (k_1/k_2)A$. Rearrangement of this last expression yields $k_2[B] \approx k_1[A]$, or, after inserting this approximation into **equation 2.60,** we find that $d[B]/dt \approx 0$. Recall that B is the "intermediate" in the consecutive reaction, and that, because k_1/k_2 is small, its concentration is always much less than [A(0)]. The steady-state approximation can then be summarized as follows:

> After an initial induction period, the concentration of any intermediate species in a consecutive reaction can be calculated by setting its time derivative equal to zero, provided that the concentration of the intermediate is always small compared to the starting concentrations.

The first qualifier, "after an initial induction period," reminds us that [B] has to build up before the approximation can hold, as shown in **Figure 2.12.** The second qualifier, "provided that the concentration of the intermediate is always small," is equivalent to stating that $k_2 \gg k_1$.

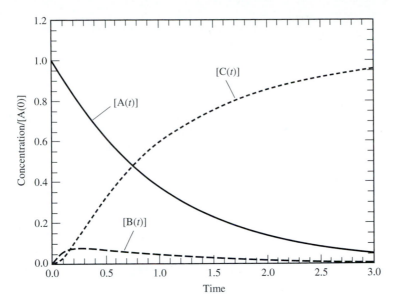

Figure 2.12

Concentrations in consecutive reaction when $k_2 = 10k_1$.

Having obtained results for the consecutive reactions via the difficult method of integration, it is instructive to see how easily the solution could have been obtained using the steady-state approximation. If we had set $d[B]/dt = 0$ in **equation 2.60,** we would have found immediately that $[B] = (k_1/k_2)[A]$. Since the concentration of A as a function of time is obtained easily as the solution to **equation 2.59,** we would have readily found that $[B] = (k_1/k_2)[A(0)]\exp(-k_1t)$. Finally, **equation 2.65** gives [C] as $[A(0)] - [A] - [B]$. Alternatively, [C] can be found by inserting the solution for [B] into **equation 2.61** and then integrating. In the sections that follow, we will make extensive use of the steady-state approximation.

example 2.5

Destruction of Stratospheric Ozone as Determined by Using the Steady-State Approximation

Background   The balance of ozone in the stratosphere is of critical concern because this molecule absorbs ultraviolet light that would be harmful to life at Earth's surface. The principal production mechanism for ozone is recombination of O atoms with O_2. The principal destruction mechanism is that given below. There is increasing concern over alternative destruction mechanisms involving molecules introduced into the stratosphere by human activity. These will be discussed in detail in Section 7.4.

Objective   Determine the destruction rate of ozone in the following mechanism, which is very similar to the Lindemann mechanism:

$$O_3 + M \underset{k_{-1}}{\overset{k_1}{\rightleftharpoons}} O_2 + O + M$$

$$O_3 + O \xrightarrow{k_2} 2\,O_2$$

Method   First, determine the rate law for the destruction of ozone, i.e., an expression for $-d[O_3]/dt$. Then use the steady-state approximation to solve for the concentration of the intermediate, O. Finally, substitute the O atom concentration into the ozone rate law and simplify.

Solution   The rate law for the destruction of ozone is

$$\frac{-d[O_3]}{dt} = k_1[O_3][M] - k_{-1}[O_2][O][M] + k_2[O_3][O].$$

The steady-state equation for [O] is

$$\frac{d[O]}{dt} = 0 = k_1[O_3][M] - k_{-1}[O_2][O][M] - k_2[O_3][O].$$

Some algebra can be avoided by subtracting these two to obtain

$$\frac{-d[O_3]}{dt} = 2k_2[O_3][O].$$

Solution of the steady-state equation gives

$$[O] = \frac{k_1[O_3][M]}{k_2[O_3] + k_{-1}[O_2][M]}.$$

Substitution of this equation into the simplified ozone destruction equation gives the final answer:

$$\frac{-d[O_3]}{dt} = \frac{2k_2k_1[M][O_3]^2}{k_2[O_3] + k_1[O_2][M]}.$$

Comment   Note that at high values of the pressure, [M] will be large enough so that the second term in the denominator will be large compared to the first. The result will then simplify to $-d[O_3]/dt = (2k_2k_1/k_{-1}) \cdot [O_3]^2/[O_2]$.

2.4.4   Unimolecular Decomposition: The Lindemann Mechanism

As an example of the use of the steady-state approximation, we consider in detail the mechanism of unimolecular decomposition. The overall reaction is A → products, and under high-pressure conditions the rate law is $-d[A]/dt = k_{ap}[A]$, where k_{ap} is the apparent rate constant. A question that begs an answer is how the A molecules obtain enough energy to decompose. The matter was debated vigorously in the early 1900s.[j] It was F. A. Lindemann who first suggested in 1922[k] that the reactants obtained the necessary energy from collisions. In its simplest form, the mechanism he proposed is shown below:

$$A + M \underset{k_{-1}}{\overset{k_1}{\rightleftharpoons}} A^* + M,$$

$$A^* \overset{k_2}{\rightarrow} P.$$

(2.68)

In these equations P stands for the products and M represents any molecule that can energize A by collision; M might be A itself, or it might be a nonreactive molecule in which the reactant is mixed.

The overall rate of the reaction is $-d[A]/dt$, or equivalently by $d[P]/dt$:

$$-\frac{d[A]}{dt} = \frac{d[P]}{dt} = k_2[A^*].$$

(2.69)

Since A^* is an intermediate in the mechanism, it will be useful to apply the steady-state approximation:

$$\frac{d[A^*]}{dt} = k_1[A][M] - k_{-1}[A^*][M] - k_2[A^*] = 0.$$

(2.70)

Here, the time dependence of A^* is equal to a production term, $k_1[A][M]$, and two destruction terms, $k_{-1}[A^*][M]$ and $k_2[A^*]$. We can then solve this last equation for the steady-state concentration of A^* to obtain

$$[A^*] = \frac{k_1[A][M]}{k_{-1}[M] + k_2}.$$

(2.71)

Substitution of this last equation into **equation 2.69** yields the solution

$$\frac{d[P]}{dt} = \frac{k_2 k_1[A][M]}{k_{-1}[M] + k_2}.$$

(2.72)

Recall that use of the steady-state approximation assumes that the concentration of the intermediate is small compared to the concentration of the starting material. This assumption is almost always valid for the system under consideration. Rearrangement of **equation 2.71** shows that $[A^*]/[A]$ is much smaller than unity when $k_1[M]/(k_{-1}[M] + k_2) \ll 1$. However, even if k_2 were zero, this last expression would still be satisfied since k_1/k_{-1} is simply the equilibrium constant for the first reaction, and this equilibrium constant must be smaller than unity because A^* has much more energy than A. In addition, for A^* of sufficiently high energy, k_2 is usually very rapid, so that the inequality $k_1[M]/(k_{-1}M + k_2) \ll 1$ is ensured.

Having convinced ourselves that the steady-state approximation is valid for the Lindemann mechanism, **equation 2.68,** it is instructive to examine the solution, **equation 2.72,** under two limiting conditions. Let us first consider the "high-pressure

[j]For an interesting discussion of the history of this problem, see J. I. Steinfeld, J. S. Francisco, and William L. Hase, *Chemical Kinetics and Dynamics,* 2nd ed. (Prentice-Hall, Englewood Cliffs, NJ, 1999), Section 11.3.

[k]F. A. Lindemann, *Trans. Faraday Soc.* **17,** 598 (1922).

limit," for which $k_{-1}[M] >> k_2$. In this limit, the denominator of **equation 2.72** can be approximated by its first term, and division of numerator and denominator by [M] gives $d[P]/dt = (k_2k_1/k_{-1})[A]$. Thus, in this limit the [M] cancels and the reaction is first order. Physically, in the high-pressure limit A^* is rapidly being created and destroyed, and only a small fraction goes on to form products.

In the "low-pressure" limit, when $k_{-1}[M] << k_2$, the second term in the denominator of **equation 2.72** dominates, and $d[P]/dt = k_1[A][M]$. In this limit the reaction is second order. Physically, in this limit most of the A^* that is formed lasts long enough to react to form P, and very little gets deactivated.

This behavior is shown in **Figure 2.13,** which plots a theoretical curve on a log-log plot for the apparent first-order rate constant, defined by $k_{ap} \equiv (1/[A])\, d[P]/dt$, as a function of [M] for the isomerization of *cis*-but-3-ene to *trans*-but-2-ene. The log-log plot is necessary to show both extremes in pressure. In the high-pressure limit, we have seen that $d[P]/dt = (k_2k_1/k_{-1})[A]$ so that $\log(k_{ap})$ should be a constant. At high pressures, the apparent rate constant is, indeed, found to be constant, but below about 10^6 torr the apparent rate constant is linearly proportional to pressure. This is because, in the low-pressure limit $d[P]/dt = k_1[A][M]$ so that $\log(k_{ap})$ should be equal to $\log(k_1[M])$; i.e., it should vary linearly with $\log[M]$, as observed in the plot. An important practical application of the Lindemann mechanism is given in **Example 2.5.**

The high-pressure result for the Lindemann mechanism also illustrates an important point about the temperature dependence of the overall rate constant in complex reactions. In the high-pressure limit $d[P]/dt = k_{ap}[A] = (k_2k_1/k_{-1})[A]$. Thus, the temperature dependence of the overall rate constant k_{ap} depends on how k_1, k_{-1}, and k_2 depend on temperature. Suppose that each of the rate constants for the elementary processes can be expressed in Arrhenius form: $k_i = A_i \exp(-E_i/kT)$. Simple multiplication and division of exponentials shows that

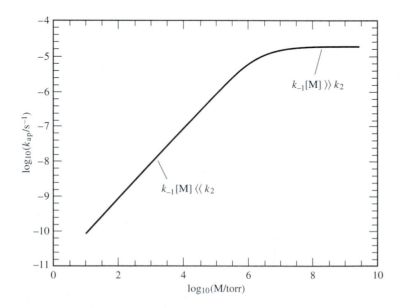

Figure 2.13

Theoretical dependence of the Lindemann apparent rate constant with pressure for the isomerization of *cis*-but-2-ene.

$$k_{ap} = \frac{k_2 k_1}{k_{-1}}$$

$$= \frac{A_2 A_1}{A_{-1}} \exp[-(E_2 + E_1 - E_{-1})/kT] \qquad \textbf{(2.73)}$$

$$= A_{ap} \exp(-E_{ap}/kT),$$

where $A_{ap} = A_2 A_1/A_{-1}$ and $E_{ap} = E_2 + E_1 - E_{-1}$. The general form of this result holds for very complicated reactions, even, as described in Problem 2.26, for reactions as complicated as those that control the rate of firefly flashing!

Like most good experimental or theoretical advances, the Lindemann mechanism raised more questions than it answered. For example, how do the rate constants k_1 and k_2 depend on how much energy the A^* molecule has? And what kind of energy is important in energizing A? We will defer these questions until Chapter 7; they form the basis for much of the current research in physical chemistry.

2.5   HOMOGENEOUS CATALYSIS

A catalyst, by definition, is a substance that is neither created nor consumed in the reaction but that increases the rate of the reaction. In most catalytic mechanisms, the catalyst transforms the reactants through a series of intermediates to products, but the catalyst is regenerated in the process of making the product. Since the intermediates are usually of much lower concentration than the starting material, catalytic mechanisms can usually be analyzed using the steady-state approximation. As an example, we will study in detail below the use of the steady-state approximation to analyze enzyme reactions. We will concentrate here on forms of homogeneous catalysis, while leaving the important area of heterogeneous catalysis until Chapter 6.

2.5.1   Acid-Base Catalysis

One prevalent form of catalysis is acid or base catalyzed hydrolysis. For example, an ester might be hydrolyzed by the following mechanism in the presence of an acid:

The same reaction can be base catalyzed:

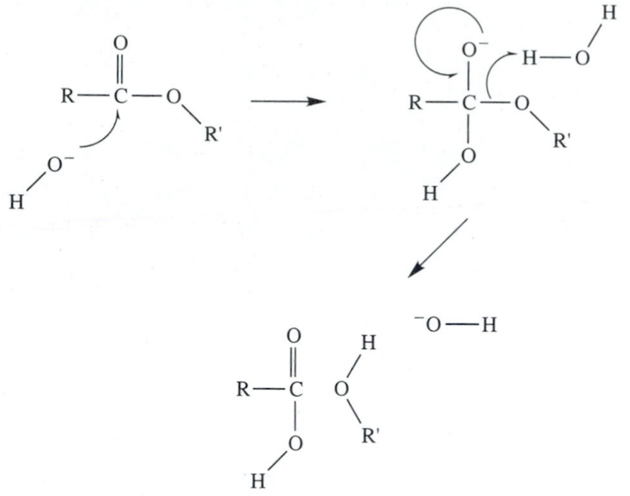

Since neither H_3O^+ nor OH^- is consumed in these processes, the overall rate law for this mechanism would be first order in each of the two reactants, the ester and water. However, because these reactions are always carried out in aqueous solution, the concentration of water can be taken as constant and incorporated into the rate constant; i.e., the system follows pseudo-first-order kinetics. The species H_3O^+ or OH^- is a catalyst in the reaction; it is neither created nor consumed but provided another, more favorable, mechanism for the reaction.

2.5.2 Enzyme Catalysis

Enzymes are macromolecules (MW $= 10^4 - 10^6$) that are remarkable in their efficiency and specificity in catalyzing reactions of biological significance. For example, the enzyme invertase (β-fructofuranidase), a component of yeast, catalyzes the conversion of sucrose to fructose and glucose by the hydrolysis reaction:

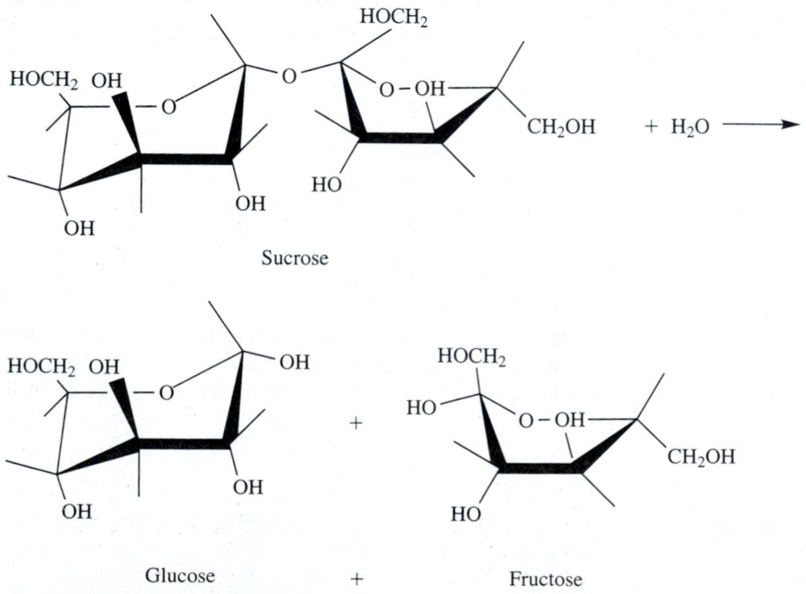

Sucrose

Glucose   +   Fructose

While the reaction can be catalyzed simply by H^+ ions, it proceeds much faster if the yeast enzyme is present. The name for the enzyme comes from the fact that the conversion produces an inversion in the direction of rotation of plane polarized light by the sugar solution, noted in 1832 by Persoz. Michaelis and Menten used the yeast enzyme responsible for this conversion in their classic 1914 studies that resulted in the following proposed mechanism for enzymatic action, known, not surprisingly, as the Michaelis-Menten mechanism.[1]

Consider the following reaction sequence:

$$E + S \underset{k_{-1}}{\overset{k_1}{\rightleftharpoons}} X,$$

$$X \overset{k_2}{\rightarrow} P + E. \tag{2.74}$$

In this mechanism, the enzyme E can reversibly bind to a substrate S to yield the intermediate X. Once bound, the enzyme can also convert the substrate to products P, which it releases while returning to its original state. The enzyme is thus available to convert more substrate. How does the rate of the reaction depend on the amount of substrate and the amount of enzyme?

We analyze the sequence using the steady-state approximation for the intermediate complex X:

$$\frac{d[X]}{dt} = k_1[E][S] - k_{-1}[X] - k_2[X] = 0. \tag{2.75}$$

Let the original concentration of enzyme be E_0. Then by mass balance $E_0 = [E] + [X]$. In a similar manner, if the initial concentration of substrate is S_0, then by mass balance $S_0 = [S] + [X] + [P]$. Substituting these into **equation 2.75,** we obtain

$$\frac{d[X]}{dt} = k_1(E_0 - [X])(S_0 - [X] - [P]) - k_{-1}[X] - k_2[X] = 0. \tag{2.76}$$

We now suppose, as is generally the case, that the initial concentration of enzyme is much smaller than the initial concentration of substrate, $E_0 \ll S_0$. Since [X] can never be larger than E_0, it follows that $[X] \ll S_0$, so that we may safely ignore it in the term $S_0 - [X] - [P]$. We now consider the initial rate of the reaction, $v_0 \equiv d[P]/dt$ as $t \rightarrow 0$, where $[P] \approx 0$. Solving **equation 2.76** for [X] gives

$$[X] = \frac{k_1 E_0 S_0}{k_1 S_0 + k_{-1} + k_2}. \tag{2.77}$$

[1]Jean-Francois Persoz (1805–1868) was a professor at the Sorbonne; Leonor Michaelis (1875–1949) was a German-born physician and biochemist who did research at the Rockefeller Foundation later in life. In addition to developing with Menten his famous equation, he is responsible for finding that keratin, the chief ingredient of hair, is soluble in glycolic acid, a discovery that made possible the development of the permanent. Maude Menten received her B.A. in 1904 and her M.D. in 1911 from the University of Toronto but had to leave Canada to pursue a career as a research scientist, because in those days women were not allowed to do research in Canadian universities. After study at the Rockefeller Institute and the Western Reserve University, she went to Berlin to study with Michaelis, where the two developed the Michaelis-Menten equation. Ultimately, she received a Ph.D. in biochemistry from the University of Chicago and became a professor at the University of Pittsburgh School of Medicine.

Finally, the initial rate of the reaction is

$$v_0 = \left.\frac{d[P]}{dt}\right|_{t \to 0} = k_2[X]$$

$$= \frac{k_1 k_2 E_0}{k_1 + \dfrac{k_{-1} + k_2}{S_0}}$$

$$= \frac{V_m}{1 + \dfrac{K_m}{S_0}},$$

(2.78)

where $V_m = k_2 E_0$ and $K_m = (k_{-1} + k_2)/k_1$. Note that the rate is proportional to the initial concentration of enzyme and to the rate constant k_2, sometimes called the *turnover number* of the enzyme. At very high initial substrate concentrations, the initial rate approaches $d[P]/dt = V_m = k_2 E_0$, or $d([P]/E_0)/dt = k_2$. The turnover number, k_2, is thus the number of molecules of product per molecule of enzyme that can be created per unit time. Typical values are 10^2–10^3 per second, but values as large as 10^6 per second have been observed.

A plot of **equation 2.78** is shown in **Figure 2.14,** in which reduced units have been used. The rate at first increases linearly with S_0 and then levels off to an asymptote equal to V_m. While **Figure 2.14** is useful in showing the behavior of the initial rate, it is not very useful for determining the rate constants. An alternate method of analysis is to take the reciprocal of both sides of **equation 2.78** to obtain the Lineweaver-Burk form

$$\frac{1}{v_0} = \frac{1}{V_m} + \frac{K_m}{V_m S_0}.$$

(2.79)

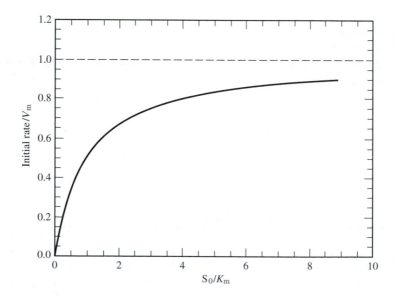

Figure 2.14

A plot of the initial rate, in units of V_m, as a function of the initial substrate concentration, in units of K_m, for the Michaelis-Menten mechanism.

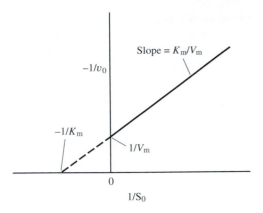

Figure 2.15

Lineweaver-Burk plot for and enzyme reaction obeying the Michaelis-Menten mechanism.

This form shows that a plot of the reciprocal of the initial rate as a function of $(S_0)^{-1}$ should yield a straight line whose intercept is V_m^{-1} and whose slope is K_m/V_m, as shown in **Figure 2.15**.

An example of an enzyme whose kinetics obey the Michaelis-Menten mechanism is lactase, the enzyme responsible for catalyzing the hydrolysis of lactose to D-glucose and D-galactose. Most adults of northern European background have sufficient enzyme to digest the milk that they consume, but many of other backgrounds do not possess this enzyme, the lack of which leads to breakdown of lactose by microbial action in the large intestine rather than by the enzyme in the small intestine. The resulting fermentation in the large intestine leads to diarrhea. One strategy to make milk more digestible for those who might otherwise suffer is to add *Lactobacillus acidophilus* to the milk. **Example 2.6** examines the enzyme kinetics of lactase in the hydrolysis of a synthetic substrate similar to lactose.[m]

example 2.6

Determination of the Michaelis-Menten Constant for the Lactase Catalyzed Hydrolysis of a Synthetic Substrate, o-nitrophenyl-β-D-galactopyranoside

Objective   Calculate the Michaelis-Menten constant, K_m, given the following data pairs for $1/v_0$, in arbitrary units, and $1/S_0$, in units of $10^3\ M^{-1}$: (2.8,0.7), (3.2,0.9), (4.2,1.3), (6.2,2.2), and (9.0,3.3).

Method   Plot $1/v_0$ as a function of $1/S_0$ and determine the slope $(= K_m/V_m)$ and the intercept $(= 1/V_m)$. The value of K_m is given by the ratio of the former to the latter.

Solution   A least squares fit of the data to a line gives a slope of 0.24×10^{-2} and an intercept of 1.08. The ratio is $2.2 \times 10^{-3}\ M$.

[m]S. F. Russo and L. Moothart, *J. Chem. Ed.* **63**, 242 (1986).

An important mediator in the catalytic action of enzymes is the phenomenon of enzyme inhibition. An inhibitor is a compound that decreases the enzyme-catalyzed reaction by reacting with the enzyme itself or the enzyme substrate complex.

Competitive inhibition occurs when the inhibitor competes with the substrate for binding at the active site of the enzyme. The mechanism can be represented by

$$E + S \underset{k_{-1}}{\overset{k_1}{\rightleftharpoons}} X,$$

$$X \overset{k_2}{\rightarrow} P + E \tag{2.80}$$

$$E + I \rightleftharpoons EI,$$

where it is assumed that the last reaction is always at equilibrium and that the complex EI cannot catalyze the reaction. Application of the steady-state approximation yields (Problem 2.20)

$$v_0 = \frac{V_m}{1 + \dfrac{K_m}{S_0}\left[1 + \dfrac{[I]}{K_I}\right]}, \tag{2.81}$$

where $K_I = [E][I]/[EI]$ is the equilibrium constant for the reverse of the last reaction in **equation 2.80.** The Lineweaver-Burk equation then becomes

$$\frac{1}{v_0} = \frac{1}{V_m} + \left[1 + \frac{[I]}{K_I}\right]\frac{K_m}{V_m S_0}, \tag{2.82}$$

so that a plot of the inverse of the initial rate as a function of the inverse of the initial substrate concentration gives the same intercept as in the absence of inhibitor, but a different slope, as shown in **Figure 2.16A.** An example of a competitive inhibitor is malonic acid, $CH_2(COOH)_2$, which resembles succinic acid, $(COOH)CH_2CH_2(COOH)$ closely enough to bind to the enzyme succinic dehydrogenase and inhibit it from converting succinic acid to fumaric acid, $(COOH)CH=CH(COOH)$.

Noncompetitive inhibition occurs when the inhibitor does not bind to the active site of the enzyme but still inhibits product formation:

$$E + S \underset{k_{-1}}{\overset{k_1}{\rightleftharpoons}} X,$$

$$X \overset{k_2}{\rightarrow} P + E \tag{2.83}$$

$$E + I \rightleftharpoons EI$$
$$X + I \rightleftharpoons XI,$$

where the last two reactions are assumed to be in equilibrium with the same equilibrium constant for their reverse: $K_I = [E][I]/[EI] = [X][I]/[XI]$. The initial rate of reaction is (Problem 2.21)

$$v_0 = \frac{V_m S_0}{[S_0 + K_m]\left[1 + \dfrac{[I]}{K_I}\right]}, \tag{2.84}$$

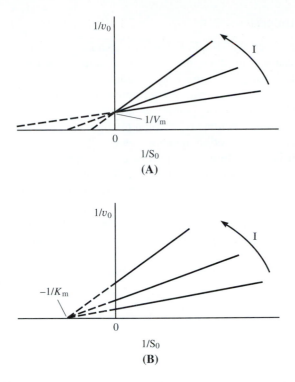

Figure 2.16

Effect of (A) competitive inhibition and (B) noncompetitive inhibition on the initial reaction rate.

and the Lineweaver-Burk equation is

$$\frac{1}{v_0} = \left[\frac{1}{V_{\mathrm{m}}} + \frac{K_{\mathrm{m}}}{V_{\mathrm{m}}S_0} \right]\left[1 + \frac{[\mathrm{I}]}{K_I} \right]. \tag{2.85}$$

Note that the presence of the inhibitor affects both the slope and intercept of the Lineweaver-Burk plot for noncompetitive inhibition, as shown in **Figure 2.16B.**

Methemoglobin reductase is an example of an enzyme that uses NADH to maintain hemoglobin in the active oxygen-carrying form. The salicylate ion is a noncompetitive inhibitor with respect to the substrate NADH.[n]

NADH
(nicotinamide adenine dinucleotide)

Salicylate Ion

[n]A. G. Splittgerber, K. Mitchell, G. Dahle, M. Puffer, and K. Blomquist, *J. Chem. Ed.* **52,** 680 (1975).

2.5.3 Autocatalysis

Not all reactions obey the simple behavior illustrated in **Figure 2.1,** in which the reactants decrease and products increase monotonically in time. Increasing attention has recently been paid to reactions in which species concentrations oscillate. Indeed, discontinuities in the rate of oscillation for such reactions as the concentrations change have helped to open a new area of science sometimes called chaos theory.[o] All oscillating reactions have some element of "autocatalysis," catalysis in which a product of one reaction appears as a catalyst in the same or another coupled reaction. It is interesting, and appropriate, to examine briefly such reactions here since they illustrate an interesting form of catalysis, provide another example for use of the steady-state approximation, and are fascinating in their own right.

Although the "Lotka" mechanism does not, so far as anyone knows, correspond to any observed chemical system, its simple mechanism illustrates the basic principles in more complex oscillatory systems. This mechanism has three steps:

$$A + X \xrightarrow{k_1} 2X,$$

$$X + Y \xrightarrow{k_2} 2Y, \tag{2.86}$$

$$Y \xrightarrow{k_3} Z.$$

Suppose that reactant is continually supplied to the system so that the concentration of A does not change from its initial value [A(0)]. Then the concentrations of the intermediates X and Y will reach a steady state, while the concentration of product Z will increase with time. The steady-state equations for X and Y are

$$\frac{d[X]}{dt} = 0 = k_1[X][A(0)] - k_2[X][Y],$$

$$\frac{d[Y]}{dt} = 0 = k_2[X][Y] - k_3[Y]. \tag{2.87}$$

The steady-state values for [X] and [Y] are thus given by

$$k_2 Y_{ss} = k_1[A(0)],$$
$$k_2 X_{ss} = k_3. \tag{2.88}$$

We now consider perturbing the system in some way so that the concentrations of the intermediates change. Let the new concentrations X and Y be different from their steady-state values by time-dependent differences x and y:

$$[X] = x + X_{ss},$$
$$[Y] = y + Y_{ss}. \tag{2.89}$$

We wish to determine how x and y vary with time. Substitution of **equation 2.89** into **equation 2.87** leads to

---

[o]See, for example, the interesting book *Chaos: Making a New Science,* by James Gleick (Viking, New York, 1987).

$$\frac{dx}{dt} = k_1[A(0)][x + X_{ss}] - k_2[x + X_{ss}][y + Y_{ss}],$$

$$\frac{dy}{dt} = k_2[x + X_{ss}][y + Y_{ss}] - k_3[y + Y_{ss}],$$

(2.90)

or

$$\frac{dx}{dt} = x[k_1[A(0)] - k_2Y_{ss}] + [k_1X_{ss}[A(0)] - k_2X_{ss}Y_{ss}]$$

$$-k_2X_{ss}y - k_2xy,$$

$$\frac{dy}{dt} = y[k_2X_{ss} - k_3] + [k_2X_{ss}Y_{ss} - k_3Y_{ss}]$$

$$+ k_2Y_{ss}x + k_2xy.$$

(2.91)

Note that **equations 2.87** and **2.88** can be used to show that the terms of **equation 2.91** in square brackets are zero. If x and y are small displacements, then terms like xy can be neglected with respect to terms like xY_{ss} or yX_{ss}, so that

$$\frac{dx}{dt} = -k_2X_{ss}y,$$

$$\frac{dy}{dt} = k_2Y_{ss}x.$$

(2.92)

If we take the derivative of both sides of the first equation in **equation 2.92** and then use the second equation to substitute for the derivative of y, we obtain the second-order differential equation

$$\frac{d^2x}{dt^2} + k_2^2X_{ss}Y_{ss}x = 0.$$

(2.93)

A specific solution to this equation that allows $x = x_0$ and $y = 0$ at $t = 0$ is

$$x = x_0\cos \omega t,$$

(2.94)

where

$$\omega^2 = k_2^2X_{ss}Y_{ss} = k_1k_3[A(0)].$$

(2.95)

A similar solution can be obtained for y. We thus see that, rather than decaying exponentially, the concentration displacements will oscillate indefinitely. The oscillation frequency will depend on the concentration to which A is maintained.

Many actual chemical reactions have been observed to oscillate. An often-cited but quite complex example is called the Belousov-Zhabotinsky reaction; a variant called the Briggs-Rauscher reaction exhibits an oscillating color change and is often used as a class demonstration.[p]

[p]B. P. Belousov, *Ref. Radiats. Med.* **145**, 1958 (1959); A. M. Zhabotinsky, *Dokl. Akad. Nauk SSSR* **157**, 392 (1964); T. S. Briggs and W. C. Rauscher, *J. Chem. Educ.* **50**, 496 (1973); see also "Oscillating Chemical Reactions," by E. S. Scott, R. Schreiner, L. R. Sharpe, B. Z. Shakhashiri, and G. E. Dirreen, in *Chemical Demonstrations, A Handbook for Teachers of Chemistry*, B. Z. Shakhashiri, Vol. 2, Chapter 7 (University of Wisconsin Press, Madison, 1985).

2.6   FREE RADICAL REACTIONS: CHAINS AND BRANCHED CHAINS

Many gas-phase reactions take place by so-called "chain" mechanisms involving free radical carriers, molecules, or atoms with one or more unpaired electrons. These mechanisms always consist of at least three steps: an *initiation* step creates the radicals that carry the chain; one or more *chain* steps convert reactants to products using the radical(s) in the role of a catalyst; and a *termination* step stops the chain by consuming the chain carrying radicals. Since the radical species are usually in small concentration and are intermediates, the steady-state approximation can be used to determine the overall rate law, which sometimes takes a complex form. An example is the $H_2 + Br_2 \rightarrow 2\,HBr$ reaction, which we have already noted has the overall rate law $\frac{1}{2} d[HBr]/dt = k[H_2][Br_2]^{1/2}$. We now consider this reaction in detail.

2.6.1 $H_2 + Br_2$

A chain mechanism is consistent with the overall rate law for this reaction. The initiation step is the collisional production of Br atoms:

$$Br_2 + M \xrightarrow{k_1} 2\,Br + M. \tag{2.96}$$

There are two chain steps:

$$Br + H_2 \xrightarrow{k_2} HBr + H,$$
$$H + Br_2 \xrightarrow{k_3} HBr + Br. \tag{2.97}$$

Note that the net result of the chain is to convert one molecule of Br_2 and one of H_2 into two of HBr while regenerating the radicals so that the chain can continue. The termination step in this case is simply the reverse of the initiation step:

$$2\,Br + M \xrightarrow{k_{-1}} Br_2 + M. \tag{2.98}$$

The key point to realize is that the chain steps can occur many times for every initiation or termination and that these steps are principally responsible for conversion of reactants into products. We would thus expect that

$$\frac{1}{2}\frac{d[HBr]}{dt} = \frac{1}{2}(k_2[Br][H_2] + k_3[H][Br_2]). \tag{2.99}$$

We use the steady-state approximation to solve for the concentration of Br and H radicals:

$$\frac{d[Br]}{dt} = 0$$

$$= 2k_1[Br_2][M] - 2k_{-1}[Br]^2[M] - k_2[Br][H_2] + k_3[H][Br_2], \tag{2.100}$$

$$\frac{d[H]}{dt} = 0 = \qquad\qquad\qquad + k_2[Br][H_2] - k_3[H][Br_2]$$

Addition of these two equations and then solution for [Br] gives

$$[\text{Br}] = \left(\frac{k_1}{k_{-1}}\right)^{1/2} [\text{Br}_2]^{1/2}. \tag{2.101}$$

We also note that the second of these equations implies that the two bracketed terms on the right-hand side of **equation 2.99** are equal, so that

$$\frac{1}{2}\frac{d[\text{HBr}]}{dt} = \frac{1}{2}\left(2k_2[\text{Br}][\text{H}_2]\right)$$

$$= k_2\left(\frac{k_1}{k_{-1}}\right)^{1/2}[\text{H}_2][\text{Br}_2]^{1/2}. \tag{2.102}$$

Note that this expression is consistent with the overall rate law provided that $k = k_2(k_1/k_{-1})^{1/2}$. In actuality, this reaction is somewhat more complicated, as discussed in Problem 2.23.

2.6.2   Rice-Herzfeld Mechanism

Many organic reactions also occur by free radical processes. For example, in 1934 F. O. Rice and K. F. Herzfeld showed that the decomposition of ethane to ethylene and hydrogen, while following first-order kinetics, actually has a rather more complicated mechanism. There are two initiation steps:

$$\text{C}_2\text{H}_6 \xrightarrow{k_1} 2\,\text{CH}_3, \tag{2.103}$$

and

$$\text{CH}_3 + \text{C}_2\text{H}_6 \xrightarrow{k_2} \text{CH}_4 + \text{C}_2\text{H}_5. \tag{2.104}$$

There are then two propagation steps:

$$\text{C}_2\text{H}_5 \xrightarrow{k_3} \text{C}_2\text{H}_4 + \text{H},$$

$$\text{H} + \text{C}_2\text{H}_6 \xrightarrow{k_4} \text{C}_2\text{H}_5 + \text{H}_2. \tag{2.105}$$

Termination typically takes place by

$$\text{H} + \text{C}_2\text{H}_5 \xrightarrow{k_5} \text{C}_2\text{H}_6. \tag{2.106}$$

The steady-state approximation can be made on the radicals CH_3, C_2H_5, and H, and, after some straightforward but rather tedious algebra (Problem 2.24), we obtain

$$-\frac{d[\text{C}_2\text{H}_6]}{dt} = \left[\frac{3}{2}k_1 + \left(\frac{k_1^2}{4} + \frac{k_1 k_3 k_4}{k_5}\right)^{1/2}\right][\text{C}_2\text{H}_6]. \tag{2.107}$$

Note that this agrees with the observation that the overall reaction is first order in C_2H_6.

A point of confusion often arises when considering radical reactions of this type. While the $\text{H}_2 + \text{Br}_2$ reaction mechanism gave no products other than the expected HBr, the ethane decomposition mechanism produces CH_4 in addition to the expected H_2 and C_2H_4. However, since the main chain is carried by the C_2H_5 radical in steps 3 and 4, the amount of CH_4 produced per molecule of H_2 or C_2H_4 is negligibly small. **Example 2.7** provides another illustration of a Rice-Herzfeld mechanism, in this case for the decomposition of acetaldehyde.

2.6.3   Branched Chain Reactions: Explosions

Explosions are simply reactions whose rates become more and more rapid as time proceeds. There are normally two causes for a rapid increase in rate, giving rise to two types of explosions. A *thermal explosion* takes place when the rate of heat conduction out of a vessel surrounding an exothermic series of reactions is too small to prevent a significant temperature rise in the vessel. According to the Arrhenius expression, most rates increase with increasing temperature, so that as the temperature rises, more reactants are consumed per unit time, creating more heat from the exothermic process, and causing the temperature to rise further. For a spherical vessel, the rate of heat loss due to thermal conductivity is proportional to the surface area of the vessel, while the rate of heat production is proportional to the amount of reactants, i.e., to the volume of the vessel, assuming the reactant concentrations are the same in both size vessels. Since the volume increases with the cube of radius while the surface area increases as the square, heat production will overcome heat loss in a large enough vessel. Explosions can thus occur unexpectedly if an exothermic reaction that ran smoothly in a small vessel is scaled up to obtain more product without proper attention to the thermal consequences.

example 2.7

The Rice-Herzfeld Mechanism for the Decomposition of Acetaldehyde

Objective   Show that the Rice-Herzfeld mechanism for the decomposition of acetaldehyde, listed below, is consistent with the observation that the overall rate of decomposition is (3/2)-order in acetaldehyde, and determine how the overall rate constant is related to those for the individual steps. The overall reaction is $CH_3CHO \rightarrow CH_4 + CO$, while the Rice-Herzfeld mechanism is

$$\text{Initiation:} \quad CH_3CHO \xrightarrow{k_1} CH_3 + CHO,$$

$$\text{Chain:} \quad CH_3 + CH_3CHO \xrightarrow{k_2} CH_4 + CO + CH_3, \quad \textbf{(2.108)}$$

$$\text{Termination:} \quad CH_3 + CH_3 \xrightarrow{k_3} C_2H_6.$$

Method   Determine the production rate of CH_4 and then use the steady-state approximation for the CH_3 intermediate.

Solution   The production rate of CH_4 is

$$\frac{d[CH_4]}{dt} = k_2[CH_3][CH_3CHO]. \quad \textbf{(2.109)}$$

The concentration of $[CH_3]$ can be found from the steady state approximation:

$$\frac{d[CH_3]}{dt} = 0 = k_1[CH_3CHO] - 2k_3[CH_3]^2. \quad \textbf{(2.110)}$$

Note that there is no net contribution to [CH$_3$] from the second step, since for every mole consumed one mole is produced. Rearrangement of the steady-state equation yields

$$[CH_3] = \left(\frac{k_1}{2k_3}\right)^{1/2}[CH_3CHO]^{1/2}, \qquad (2.111)$$

and substitution of this into **equation 2.109** yields

$$\frac{d[CH_4]}{dt} = k_2\left(\frac{k_1}{2k_3}\right)^{1/2}[CH_3CHO]^{3/2}. \qquad (2.112)$$

Consequently, the overall rate constant is $k_2(k_1/2k_3)^{1/2}$, and the mechanism is consistent with an overall order of 3/2.

The second type of explosion is caused by a *branched chain reaction*. The mechanism for the oxygen-hydrogen reaction provides a good example. The overall reaction is, of course, 2 H$_2$ + O$_2$ → 2 H$_2$O. The mechanism is complicated, but the following steps are the most important:

Initiation:

$$H_2 + O_2 \rightarrow HO_2 + H,$$
$$O_2 \rightarrow 2\,O, \qquad (2.113)$$
$$H_2 \rightarrow 2\,H.$$

Chain Branching:

$$H + O_2 \rightarrow OH + O,$$
$$O + H_2 \rightarrow OH + H. \qquad (2.114)$$

Chain Propagation:

$$OH + H_2 \rightarrow H_2O + H. \qquad (2.115)$$

Termination:

$$H + wall \rightarrow \tfrac{1}{2}H_2,$$
$$O + wall \rightarrow \tfrac{1}{2}O_2, \qquad (2.116)$$
$$H + O_2 + M \rightarrow HO_2 + M.$$

Like other chain reactions, this mechanism has initiation, propagation, and termination steps. Initiation of the reaction might be caused, for example, by the brief exposure of the reactant mixture to rapid heating from a spark or flame, in short, any reaction that creates free oxygen or hydrogen atoms. For example, the initiation of the hydrogen-oxygen explosion on the space shuttle *Challenger* was the heating of the external oxygen tank by flames from the propulsion rocket that escaped through the o-ring seals. Termination is caused either by collision of radicals with the wall, where they stick and recombine, or by recombination assisted by a third body, here called M. (While technically a radical, HO$_2$ is unreactive relative to O, H, and OH; it is treated here as a stable compound.) The propagation step is a normal one, where one radical is consumed and another produced. But there are also chain branching steps, steps in which the net number of radicals increases. For example, in the first reaction of **equation 2.114** an H atom is consumed but an O atom and an OH radical are produced. If reactions such as those in **equation 2.114** are prevalent, the rate of the overall reaction can increase rapidly.

The overall reaction rate for a stoichiometric mixture of $2 H_2 + O_2$ is a complicated function of temperature, pressure, vessel size, and material. The rates for each step depend on temperature. The rate of termination due to diffusion of O and H to the walls depends on pressure and vessel size. The sticking probability for these radicals when they reach the wall depends on the vessel material. Finally, the rate of $H + O_2$ recombination depends on pressure. The bottom panel of **Figure 2.17** shows the dividing line in the T-p plane between the explosion and steady reaction regimes for a typical system, while the top panel schematically indicates the rate of the reaction as a function of total pressure for a temperature of 800 K. It is constructive to consider the dominant processes at 800 K as the pressure is increased.

For low pressures, the mean free path of the radicals O and H is large enough so that they reach the walls of the reaction vessel with high probability. Under these conditions, the termination steps dominate and the reaction proceeds in a controlled fashion. As the pressure increases, however, the chain branching steps start to dominate, and a branched chain explosion occurs. It is interesting to note, however, that a reaction mixture at higher total pressure would produce a steady reaction, a so-called hydrogen-oxygen flame. In this region, the pressure is sufficiently high that

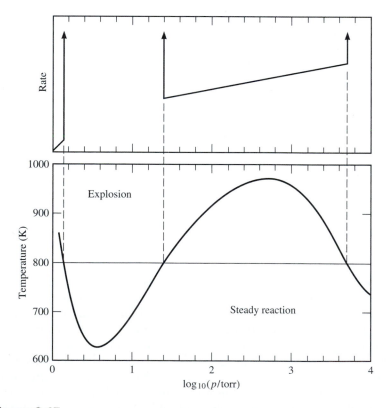

Figure 2.17

Explosion limits for a stoichiometric mixture of H_2 and O_2.

the termination step, $H + O_2 + M \rightarrow HO_2 + M$, occurs with good probability. In this region, then, the high rate of chain termination prevents an explosion. At higher pressures still, the heat due to the exothermic reaction cannot be carried away as fast as it is produced, so that a thermal explosion occurs.

2.7 DETERMINING MECHANISMS FROM RATE LAWS

Most of the examples we have used so far indicate how to determine whether a mechanism consisting of several elementary steps is consistent with an overall rate law. A much more difficult problem is how to figure out a reasonable mechanism of elementary steps given the rate law. In general, since several mechanisms are usually consistent with an overall rate law, it is not possible to obtain a single answer to this question. Nor is it really possible to obtain *any* answer without what amounts to an educated guess. In this section, we examine ways to make our guesses educated.

There are two basic concepts that we can use as a guide. The first is the realization that, in many reactions, there is *one* step in the mechanism whose rate is much smaller than any other step. In such a case, the overall rate of the reaction is usually controlled by this *rate-limiting step*. The rate law then contains in its numerator the rate of this rate-limiting step.

The second concept is one that we have already encountered. When the rates of steps in the mechanism are comparable, it is still usually possible to assume that the concentration of reaction intermediates is low and varies slowly. This is just the *steady-state* hypothesis of Section 2.4.3.

Even with a few principles in hand, it is ultimately experience that provides the best guide to guessing mechanisms from overall rate laws. Fortunately, we already have some experience. Let us look back at the systems we have studied to see if we can determine how we might apply the two basic concepts to obtain clues for determining a plausible mechanism from the overall rate law.

We start our excursion with the rate-law result of the Lindemann mechanism for the overall reaction $A \rightarrow$ products:

$$\frac{d[P]}{dt} = \frac{k_2 k_1 [A][M]}{k_{-1}[M] + k_2}. \tag{2.117}$$

One obvious clue to this reaction is that the rate law depends on something other than the reactants or products, namely, the overall pressure of the gas mixture and not just the partial pressure of the reactant. Another clue comes in the complex form of the denominator. Recall that a mechanism that leads to this rate law is

$$A + M \underset{k_{-1}}{\overset{k_1}{\rightleftharpoons}} A^* + M,$$
$$A^* \overset{k_2}{\rightarrow} P. \tag{2.118}$$

The complex form of the denominator comes from the fact that the intermediate in this reaction can disappear in two possible ways. Thus, we have already developed two clues: (1) complex denominators likely indicate that an intermediate in steady-state can disappear in more than one way and (2) the presence of something other than a reactant or product in the numerator (or, as it also turns out, in the denominator) likely indicates that an equilibrium step appears in the mechanism.

We gain a few more clues by examining the $H_2 + Br_2$ reaction studied in Section 2.6.1 and Problem 2.23. The overall rate law (for the more complicated mechanism of Problem 2.23) is of the form

$$\frac{1}{2}\frac{d[HBr]}{dt} = \frac{k_{\text{numerator}}[Br_2]^{1/2}[H_2]}{1 + \dfrac{k_{\text{denominator}}[HBr]}{[Br_2]}} \tag{2.119}$$

There are several clues to the mechanism in this rate law, which differs from that of Section 2.6.1 only by the inclusion of the denominator. First, the fractional order with respect to the reactant $[Br_2]$ indicates here and more generally that some sort of dissociation is taking place. Note that in the absence of products, the denominator is unity. If we recognize that Br_2 dissociation is taking place, it is then clear from the numerator that two likely reactions in the mechanism are

$$Br_2 + M \;\rightleftharpoons\; 2\,Br + M, \qquad \textit{fast equilibrium}$$

$$Br + H_2 \;\rightarrow\; HBr + H. \qquad \textit{slow reaction}$$

The numerator also suggests that the rate limiting step is the second of these reactions. What about the denominator? It suggests that production of [HBr] slows the reaction, so there must be a reaction going backward from this product toward the reactants. The obvious choice would be to include the reverse of the second reaction in our mechanism, which, listing this reaction separately, now becomes

$$Br_2 + M \;\rightleftharpoons\; 2\,Br + M, \qquad \textit{fast equilibrium}$$

$$Br + H_2 \;\rightarrow\; HBr + H, \qquad \textit{slow reaction}$$

$$H + HBr \;\rightarrow\; Br + H_2. \qquad \textit{fast}$$

Note that we have indicated that this reverse reaction is fast. The reason comes from chemical intuition. The second reaction is endothermic, while its reverse, the third reaction, is exothermic. This being the case, the activation energy for the second reaction will be higher than for the third, so, all other things being equal (equal A factors), the third reaction will be faster than the second.

So far so good, but the mechanism we are left with, while plausible for both the factors in the numerator and denominator, leaves us with a product, H, that is not part of the of the overall reaction. It cannot be simply a very minor product, because one H atom is formed for each HBr formed, so something must happen to it. The reaction of H with Br_2 is also exothermic, so we might extend our mechanism by a fourth reaction to get

$$Br_2 + M \;\rightleftharpoons\; 2\,Br + M, \qquad \textit{fast equilibrium}$$

$$Br + H_2 \;\rightarrow\; HBr + H, \qquad \textit{slow reaction}$$

$$H + HBr \;\rightarrow\; Br + H_2, \qquad \textit{slow}$$

$$H + Br_2 \;\rightarrow\; HBr + Br. \qquad \textit{fast}$$

We now have what we recognize as a chain mechanism, where the second and fourth steps form a chain that gives the overall stoichiometry of the reaction. The mechanism seems plausible, so the next step is to see if in detail it matches the overall rate law. Problem 2.23 shows that it does.

The new take-home clues from this example are (3) fractional powers likely indicate some sort of dissociation process, (4) the presence of a product in the denominator likely indicates an equilibrium step leading back from products toward reactants, and (5) the numerator often corresponds to what might be a rate-limiting step.

What other clues might we learn from previously studied examples? Let us consider the Rice-Herzfeld decomposition of acetaldehyde that we studied in **Example 2.7.** The overall reaction is $CH_3CHO \rightarrow CH_4 + CO$ and the rate law is

$$\frac{d[CH_4]}{dt} = k[CH_3CHO]^{3/2}. \tag{2.120}$$

As in the $H_2 + Br_2$ case, the fractional power suggests a dissociation mechanism, but the fraction is no longer just $\frac{1}{2}$. In fact, the order of the reaction with respect to the reactant is *larger* than the stoichiometry of the reaction would suggest. Thus, it seems that one molecule of CH_3CHO must dissociate, while a second molecule must participate in the reaction later on. Likely dissociation steps are either

$$CH_3CHO \quad \rightarrow \quad CH_3 + CHO, \qquad \textit{dissociation 1}$$

or

$$\rightarrow \quad CH_3CO + H. \qquad \textit{dissociation 2}$$

Likely second steps then could be either

$$CH_3 + CH_3CHO \quad \rightarrow \quad CH_4 + CH_3CO, \qquad \textit{reaction 1}$$

or

$$H + CH_3CHO \quad \rightarrow \quad CH_4 + CHO. \qquad \textit{reaction 2}$$

Either reaction 1 or reaction 2 leaves an unstable radical that is likely to dissociate further, giving CO and either a CH_3 radical in the first case or an H atom in the second. In either case, then, we have a possible chain mechanism, where either reaction 1 or 2 generates an unstable radical that decomposes to yield CO and the starting radical. Thus, either the mechanism

$$CH_3CHO \quad \rightarrow \quad CH_3 + CHO, \qquad \textit{dissociation 1}$$

$$CH_3 + CH_3CHO \quad \rightarrow \quad CH_4 + CH_3CO, \qquad \textit{reaction 1}$$

$$CH_3CO \quad \rightarrow \quad CH_3 + CO, \qquad \textit{radical dissociation 1}$$

or

$$CH_3CHO \quad \rightarrow \quad CH_3CO + H, \qquad \textit{dissociation 2}$$

$$H + CH_3CHO \quad \rightarrow \quad CH_4 + CHO, \qquad \textit{reaction 2}$$

$$CHO \quad \rightarrow \quad CO + H, \qquad \textit{radical dissociation 2}$$

are plausible. But what about the chain termination step? In mechanism 1 the termination would be

$$CH_3 + CH_3 \quad \rightarrow \quad C_2H_6, \qquad \textit{termination 1}$$

while in mechanism 2 it would be

$$H + H\,(+M) \quad \rightarrow \quad H_2\,(+M), \qquad \textit{termination 2}$$

or possibly

$$H + CHO\,(+M) \quad \rightarrow \quad H_2CO\,(+M) \qquad \textit{termination 2'}$$

In fact, either mechanism 1 or mechanism 2 with either of the termination steps 2 or 2' would lead the overall rate law for this reaction, so on that basis alone, either is possible. The fact that trace amounts of C_2H_6 are found (far in excess of the amounts of H_2 or H_2CO) indicates that the first mechanism is more likely to be correct.

What are the take-home clues from this example? In addition to (3) fractional powers usually indicate dissociation processes, we find a new one: (6) if the reaction order exceeds the stoichiometric coefficient for any reactant, then it is likely that additional molecules of the reactant are consumed by reactions with intermediates following the rate limiting step. Here, the rate limiting step is the initial dissociation.

A final clue comes from an example we have not yet studied. Consider the reaction

$$BrO_3^- + 5\,Br^- + 6\,H^+ \rightarrow 3\,Br_2 + 3\,H_2O,$$

for which the observed rate-law equation is

$$\frac{d[Br_2]}{dt} = k[BrO_3^-][Br^-][H^+]^2. \tag{2.121}$$

When the order of the reaction is greater than 3, as it is here, it is extremely unlikely that the reaction takes place without several steps; the probability of four or more species coming together to react is virtually zero. When there are charged species, such as here, it is also unlikely that an elementary step involves the association of two like charges, since they would repel one another. A plausible scheme for the mechanism, then, involves several equilibria prior to the rate limiting step. Because of the charges on the species and the unlikelihood that association of H^+ with Br^- would lead to products, a reasonable assumption for the first equilibrium is

$$BrO_3^- + H^+ \rightleftharpoons HBrO_3,$$

for which $K_1 = [HBrO_3]/[H^+][BrO_3^-]$. Note now that if the rate limiting step were proportional to $[HBrO_3] = K_1[H+][BrO_3^-]$, then two of the species in the numerator of the rate law would be accounted for. However, there are still two more! Let's try another equilibrium, say,

$$HBrO_3 + H^+ \rightleftharpoons H_2BrO_3^+,$$

for which $K_2 = [H_2BrO_3^+]/[HBrO_3][H^+]$. Now, if the rate limiting step were proportional to $[H_2BrO_3^+] = K_2[HBrO_3][H^+] = K_2K_1[H^+][BrO_3^-][H^+]$, then three of the species in the numerator of the rate law would be accounted for. The remaining species is $[Br^-]$, so we postulate a rate limiting step of

$$Br^- + H_2BrO_3^+ \rightarrow products.$$

The take home clue from this example is (7) if the overall reaction order is higher than 3, then it is very likely that one or more equilibria precede the rate limiting step.

We summarize the clues learned from these examples here:

1. Complex denominators likely indicate that an intermediate in steady state can disappear in more than one way.

2. The presence of something other than a reactant or product in the numerator (or, as it turns out, also the denominator) likely indicates that an equilibrium step appears in the mechanism.
3. Fractional powers usually indicate some sort of dissociation process.
4. The presence of a product in the denominator likely indicates an equilibrium step leading back from products toward reactants.
5. The numerator often corresponds to what might be a rate limiting step.
6. If the reaction order exceeds the stoichiometric coefficient for any reactant, then it is likely that additional molecules of the reactant are consumed by reactions with intermediates following the rate limiting step.
7. If the overall reaction order is higher than 3, then it is very likely that one or more equilibria precede the rate limiting step.

As we have seen, determination of a mechanism that is consistent with the observed phenomenon (stoichiometry, rate law, minor products, as well as other observables that we have not considered: temperature dependence, observed intermediates, etc.) involves application of a few basic principles, use of several clues, and, above all, chemical intuition. Problem 2.35 provides some practice. In the end, a reaction mechanism is actually just like any scientific hypothesis. It must be consistent with all the known facts, but a new observation may require a change in the hypothesis. There may be many hypotheses that fit the known facts. The good ones will suggest further tests that can distinguish between competing explanations.

2.8   SUMMARY

The general goals of this chapter have been to obtain an empirical description of chemical reaction rates and to relate that macroscopic description to a microscopic reaction mechanism. We started by describing the rate law and defining the order of reaction. In general, the order of a reaction is not related to its stoichiometry.

For first-order reactions we found that the differential rate law could be described by the equation

$$\frac{-\mathrm{d}A}{\mathrm{d}t} = k[A]. \tag{2.6}$$

Integration gave a useful form for describing the concentrations as a function of time:

$$[A(t)] = [A(0)]\exp(-kt). \tag{2.8}$$

For second-order reactions involving a single species we found the differential form of the rate law to be

$$\frac{-\mathrm{d}[A]}{\mathrm{d}t} = k[A]^2.$$

This equation can be integrated to yield

$$\frac{1}{[A(t)]} - \frac{1}{[A(0)]} = kt. \tag{2.14}$$

The form of the integrated rate law equation was found to be somewhat more complex for a second-order reaction involving two species, but it could be obtained by using the stoichiometry of the reaction to reduce the number of variables. We also noted that second-order reactions could be treated as "pseudo-first-order" reactions if the concentration of one of the reactants is high enough so as not to change appreciably during the course of the reaction.

The rate constants for many simple chemical reactions were found to be accurately described by the Arrhenius law,

$$k = A \exp\left(-\frac{E_a}{RT}\right). \tag{2.37}$$

This law leads to an equation which enables us to relate rate constants at different temperatures:

$$\ln\frac{k_2}{k_1} = -\frac{E_a}{R}\left[\frac{1}{T_2} - \frac{1}{T_1}\right]. \tag{2.39}$$

Having learned to describe reaction rate laws and rate constants empirically, we next considered how an overall, macroscopic reaction might be consistent with a microscopic mechanism composed of a series of elementary steps. Each elementary step was taken to describe a simple collisional process. We argued for such elementary steps that the order of the elementary reaction should be equal to the molecularity.

We considered several types of simple reaction mechanisms. For opposing reactions at equilibrium (Section 2.4.1), we found that the equilibrium constant is equal to the ratio of the forward and reverse rate constants. For parallel reactions (Section 2.4.2), we saw that the branching ratio between product channels at any given time is equal to the ratio of rate constants for the two channels. For consecutive reactions (Section 2.4.3), we found that the concentration of the intermediate species depends on the ratio of the rate constants for the consecutive steps. Most importantly, we developed a useful approximation, the steady-state approximation:

> After an initial induction period, the concentration of any intermediate species in a consecutive reaction can be calculated by setting its time derivative equal to zero, provided that the concentration of the intermediate is always small compared to the starting concentrations.

Several examples of the steady-state approximation followed. Unimolecular decomposition was described by a collisional activation process called the Lindemann mechanism, and the steady-state approximation was also used to simplify the mathematical description of several important catalysis mechanisms.

Homogeneous catalysis was illustrated by the Michaelis-Menten mechanism,

$$\mathrm{E} + \mathrm{S} \underset{k_{-1}}{\overset{k_1}{\rightleftharpoons}} \mathrm{X},$$
$$\mathrm{X} \overset{k_2}{\rightarrow} \mathrm{P} + \mathrm{E}, \tag{2.74}$$

which led to a formula for the initial rate of reaction

$$v_0 \equiv \left.\frac{d[\mathrm{P}]}{dt}\right|_{t\to 0} = \frac{V_m}{1 + \dfrac{K_m}{S_0}}, \tag{2.78}$$

where $V_m = k_2 E_0$ and $K_m = (k_{-1} + k_2)/k_1$. In addition, both competitive and noncompetitive inhibition of enzyme reactions were described using the steady-state approximation.

Free radical reactions such as the chain reaction between H_2 and Br_2 and several Rice-Herzfeld mechanisms for decomposition were also shown to be amenable to steady-state treatment. The steady-state approximation is not generally good, however, in the case of branched chain reactions, particularly near the explosion limits.

In general, it is much more difficult to find a reaction mechanism that is consistent with an overall reaction stoichiometry and rate law than it is to determine the overall reaction stoichiometry and rate law from a mechanism. Nonetheless, the concepts of a rate limiting step and of the steady-state approximation are useful, as are several clues provided in the rate law.

suggested readings

R. S. Berry, S. A. Rice, and J. Ross, *Physical Chemistry* (Wiley, New York, 1980).

K. A. Connors, *Chemical Kinetics: The Study of Reaction Rates in Solution* (VCH Publishers, New York/Weinheim, 1990).

J. Gleick, *Chaos: Making a New Science* (Viking, New York, 1987).

K. J. Laidler, *Chemical Kinetics,* 3rd ed. (Harper and Row, New York, 1987).

J. I. Steinfeld, J. S. Francisco, and W. L. Hase, *Chemical Kinetics and Dynamics* (Prentice-Hall, Englewood Cliffs, NJ, 1989).

R. W. Weston, Jr., and H. A. Schwarz, *Chemical Kinetics* (Prentice-Hall, Englewood Cliffs, NJ, 1972).

problems

2.1 The overall rate of a reaction between two species doubles when the concentration of the first reactant is doubled while that of the second is held constant. When the concentration of the first reactant is held constant the rate increases with the square root of the concentration of the other reactant. The overall order of the reaction is (a) 1/2, (b) 1, (c) 3/2, (d) −1/2, (e) zero.

2.2 An overall reaction is third order. Possible units for the rate constant are (a) $cm^3\ molec^{-1}\ s^{-1}$, (b) $(L/mol)^2\ s^{-2}$, (c) $(L/mol)^2\ s$, (d) $cm^6\ molec^{-2}\ s^{-1}$, (e) $cm^9\ molec^{-3}\ s^{-1}$.

2.3 For a certain reaction the length of time it takes for the concentration to drop by a specified fraction is independent of the starting concentration. What is the order of the reaction?

2.4 Consider a reaction which is third order overall but first order in each of three concentrations. If the concentrations of two of the species are very large compared to that of the third, with what functional form will the concentration of the third species decrease in time?

2.5 The preexponential factor in the Arrhenius rate law can be evaluated by which of the following: (a) measurement of the limit in the rate as the temperature approaches infinity, (b) measurement of the rate at zero temperature, (c) measurement of the slope of a plot of $\ln k$ versus $1/T$, (d) measurement of the intercept of a plot of $\ln k$ versus $1/T$, (e) measurement of the rate constant at two different temperatures, (f) more than one of the above? (Specify which.)

2.6   True or false? (a) For an elementary step, the order of the reaction is equal
       to the molecularity of the reaction. (b) For an elementary step, the order of
       the reaction is equal to the sum of the stoichiometric coefficients for the
       reaction. (c) The order of an overall reaction is equal to the molecularity of
       the reaction. (d) The molecularity of an overall reaction is equal to the sum
       of the stoichiometric coefficients for the reaction. (e) The order of a reaction
       can be determined from the units of the rate constant for the reaction.

2.7   If a system at equilibrium is perturbed in the forward direction, the system
       will approach equilibrium with a rate determined by (a) the forward rate
       constant, (b) the reverse rate constant, (c) a combination of the two, (d)
       neither.

2.8   A reactant disappears to two possible products via parallel reaction chan-
       nels. To determine the branching ratio it is sufficient to measure which of the
       following: (a) the time constant for the appearance of one of the products,
       (b) the time constant for the appearance for each of the products, (c) ratio of
       products at a given time, (d) the appearance time constant for one of the
       products and its concentration at a particular time relative to the starting
       concentration of the reactant, (e) the disappearance rate constant for the
       reactant, (f) more than one of the above? (Specify which.)

2.9   In a consecutive reaction A → B → C the concentration of the intermediate
       will be small relative to the starting concentration of A if (a) the rate con-
       stant for the first step is larger than that for the second, (b) the rate constant
       for the second step is larger than that for the first, (c) very few products have
       formed, (d) nearly all the reactant has disappeared, (e) more than one of the
       above. (Specify which.)

2.10   The steady-state approximation can be used for the Lindemann mechanism
       because (a) A^* always decomposes very rapidly to products, (b) A* is deac-
       tivated very efficiently to A, (c) the concentration of the collision partner
       [M] is always very large, (d) very few products are formed, (e) more than
       one of the above. (Specify which.)

2.11   Under what conditions will the rate of an enzyme reaction be proportional
       to the initial concentration of substrate?

2.12   The initial rate of enzyme reactions levels off with increasing substrate con-
       centration because (a) too much substrate inhibits the reaction, (b) at high
       substrate concentration nearly all the enzyme is tied up with the substrate,
       (c) the substrate reacts with the products at such high concentrations, (d) the
       steady-state approximation fails.

2.13   The difference between competitive and noncompetitive inhibition is that
       (a) in the former the inhibitor binds to the active enzyme site, while in the
       latter it binds elsewhere; (b) in the latter the inhibitor binds to the active
       enzyme site, while in the former it binds elsewhere; (c) in the latter the
       inhibitor can bind both to the enzyme and to the enzyme-substrate interme-
       diate; (d) more than one of the above. (Specify which.)

2.14   What are the three types of reaction common to all chain mechanisms?

2.15   Show that **equation 2.23** can be rearranged to **equation 2.22.**

2.16   Show that when [B(0)] $>>$ [A(0)], **equation 2.27** reduces to **equation 2.32.**

2.17   Show that **equation 2.36** reduces to **equation 2.8** with k replaced by $k[A(0)]^2/2$ when $[A(0)]$ is very large and to **equation 2.14** with k replaced by $k[B(0)]$ when $[B(0)]$ is very large.

2.18   Show by direct differentiation that **equation 2.64** is the solution to **equation 2.63.**

2.19   What happens when $k_1 = k_2$ in **equation 2.64?** Mathematically, it is clear that we get an undefined answer, so we need to develop another way to evaluate the expression in the limit when k_1 approaches k_2. (*Hint:* Consider l'Hôpital's rule.)

2.20   Use the steady state approximation and the equilibrium constant $K_I = [E][I]/[EI]$ to show that the initial rate of an enzyme reaction with competitive inhibition is **equation 2.81.**

2.21   Use the steady-state approximation and the equilibrium constant $K_I = [E][I]/[EI] = [X][I]/[XI]$ to show that the initial rate of an enzyme reaction with noncompetitive inhibition is **equation 2.84.**

2.22 α-Chymotrypsin, produced in the pancreas, is one of a family of proteins that helps to digest proteins. Experiments examining its catalysis of the hydrolysis of p-nitrophenyl-acetate to p-nitrophenol plus acetate show that the mechanism for enzyme action follows the scheme:

$$E + S \underset{k_{-1}}{\overset{k_1}{\rightleftharpoons}} ES \overset{k_2}{\rightarrow} ES' + P_1 \overset{k_3}{\rightarrow} E + P_2,$$

where P_1 is p-nitrophenol and P_2 is an acetate ion. The solution of this kinetic scheme is lengthy but not complicated [see M. L. Bender, F. J. Kézdy, and F. C. Wedler, *J. Chem. Ed.* **44**, 84 (1967)]. The result is that

$$[P_1] = At + B(1 - e^{-bt}),$$

where A, B, and b are constants that depend on the rate constants and the initial values E_0 and S_0:

$$A = \frac{\dfrac{k_2 k_3}{k_2 + k_3} E_0 S_0}{S_0 + \dfrac{K_S k_3}{k_2 + k_3}},$$

$$B \approx E_0,$$

$$b \approx \frac{(k_2 + k_3) S_0}{S_0 + K_S},$$

and

$$K_S = \frac{[E][S]}{[ES]}.$$

Note that at long times the solution approaches $[P_1] = At + B$, a straight line. The table gives some typical data, where the concentration of p-nitrophenyl was monitored by its absorbance, Abs, at 400 nm.

t (s)	Abs	t (s)	Abs
20	0.032	400	0.176
40	0.059	500	0.180
60	0.081	600	0.182
80	0.098	700	0.184
100	0.113	800	0.186
120	0.124	900	0.188
140	0.133	1000	0.190
160	0.142	1100	0.192
180	0.148	1200	0.194
200	0.154	1300	0.196
250	0.163	1500	0.199
300	0.169	1600	0.201
350	0.173		

a. Given that the initial concentration of enzyme was 1.13×10^{-5} M, determine the values of A, B, and b, from the data.

b. Noting that P_1 is produced rapidly compared to P_2, we can surmise that $k_2 \gg k_3$. Calculate the value of k_3 assuming that k_2 is so much larger than k_3 that $S_0 > K_S k_3/(k_2 + k_3)$.

c. Given that $K_S = 1.6 \times 10^{-3}$ M and that $S_0 = 5.72 \times 10^{-5}$ M, calculate k_2.

d. Check the assumption made in (b).

2.23   The $H_2 + Br_2$ reaction mechanism is actually somewhat more complicated than indicated in Section 2.6.1 because the HBr product inhibits the reaction by reacting with H atoms in the reverse of step 2:

$$H + HBr \xrightarrow{k_{-2}} H_2 + Br.$$

Show that the revised rate law including this step is simply the former one divided by $\{1 + (k_{-2}[HBr])/(k_3[Br_2])\}$.

2.24   Use the steady-state approximation on the intermediates in the ethane decomposition reaction to derive **equation 2.107.**

2.25   A proposed mechanism for the reaction $N_2O_5 \rightarrow 2\,NO_2 + \frac{1}{2}O_2$ is

$$N_2O_5 \underset{k_{-1}}{\overset{k_1}{\rightleftharpoons}} NO_2 + NO_3,$$

$$NO_2 + NO_3 \xrightarrow{k_2} NO + O_2 + NO_2$$

$$NO + NO_3 \xrightarrow{k_3} 2\,NO_2.$$

Apply the steady-state approximation to show that the overall reaction rate is

$$-\frac{d[N_2O_5]}{dt} = k[N_2O_5].$$

In the process you should evaluate k in terms of k_1, k_{-1}, k_2, and k_3.

2.26   The following table gives the frequency of firefly flashing as a function of temperature. Find the activation energy for the process.[q]

Frequency (min^{-1})	T (K)
15.9	302
14.8	301
12.5	300
12.0	299
11.5	297
10.0	296
8.0	292

2.27   According to the information supplied on some milk cartons, homogenized milk will keep for 1/3 day at 80°F, for 1/2 day at 70°F, for 1 day at 60°F, for 2 days at 50°F, for 10 days at 40°F, and for 24 days at 32°F. Calculate the activation energy for the process that spoils milk.

2.28   The rate of the reaction of atomic chlorine with methyl iodide has been studied under pseudo-first-order conditions in the temperature range from 218–694 K by Ayhens, Nicovich, McKee, and Wine [*J. Phys. Chem.* **101,** 9382–9390 (1997)].

   a.   The table below gives the slope of a ln[Cl] versus time plot for $T = 532$ K, $[Cl]_0 \approx 6 \times 10^{10}$ atoms/cm^3, and varying densities of CH_3I.

$[CH_3I]$ 10^{14} molec/cm^3	τ_{decay} s^{-1}
0	37
0.740	418
1.64	887
4.70	2410

   At this temperature, what is the rate constant for the reaction?

   b.   Similar data was obtained at other temperatures, as shown in the table below.

Temperature (K)	Reaction Rate Constant (10^{-12} cm^3 molec^{-1} s^{-1})
364	1.73
396	2.65
399	2.42
423	2.67
424	2.75
522	4.62
532	5.04
690	9.20
694	9.30

   What is the activation energy for the reaction and what is the preexponential factor?

---

[q]This and many other amusing examples come from K. J. Laidler, *J. Chem. Ed.* **49,** 343 (1972).

2.29   A general way to discuss branched chain reactions is in terms of the following mechanism:

$$nA \xrightarrow{k_1} R,$$

$$R + A \xrightarrow{k_2} P + \alpha R,$$

$$R \xrightarrow{k_3} nA.$$

In these equations, A is a substance decomposing to give product, P, and R is a reactive chain carrier. Step two is a branching step if $\alpha > 1$, i.e., if more radicals are produced than consumed.

    a.   Assuming the concentration of radicals to be in steady state, what is the rate law for production of P?

    b.   From your answer to part (a) determine the condition on α that predicts explosion, i.e., that predicts $dP/dt \to \infty$.

2.30   Consider the autocatalytic reaction

$$A \xrightarrow{k_1} B,$$

$$B + A \xrightarrow{k_2} C + A.$$

    a.   Give the differential equations which describe the rate of change of [A], [B], and [C].

    b.   Assuming that the steady-state approximation is valid and that the initial concentration of A is [A(0)] and those of B and C are zero, develop expressions for [B], [A], and [C].

    c.   Under what limiting condition will the steady-state approximation be valid?

    d.   *Without* assuming the steady-state approximation to be valid, develop expressions for [A] and [B] as functions of time.

    e.   From your general solution in part (d), verify the steady-state solution in part (b) for [B] as $t \to \infty$.

2.31   The overall reaction $(CH_3)_2CO + X_2 \to CH_2XCOCH_3 + HX$ is thought to proceed by the following mechanism, in which HA represents any proton donating acid and X_2 is a halogen molecule:

$$(CH_3)_2CO + HA \underset{k_{-1}}{\overset{k_1}{\rightleftharpoons}} (CH_3)_2COH^+ + A^-,$$

$$(CH_3)_2COH^+ + A^- \xrightarrow{k_2} CH_2 = C(OH)CH_3 + HA,$$

$$CH_2 = C(OH)CH_3 + X_2 \xrightarrow{k_3} CH_2XC(OH)CH_3^+ + X^-,$$

$$CH_2XC(OH)CH_3^+ + X^- \xrightarrow{k_4} CH_2XCOCH_3 + HX.$$

    a.   Use the steady-state approximation to find the rate expression.

b. From the rate expression, predict the relative rate of bromination versus iodination.

c. What is the rate limiting step if $k_2 \gg k_{-1}$? (d) What is the rate limiting step if $k_{-1} \gg k_2$?

2.32 Excited bromine atoms Br^*, where the star indicates $Br(4\ ^2P_{1/2})$, are deactivated electronically by HCl, which accepts the energy and becomes vibrationally excited, a process called electronic-to-vibrational energy transfer. The transfer takes place according to the following equations, where Br is $Br(4\ ^2P_{3/2})$, the ground state bromine atom, and where HCl without further notation is understood to be $HCl(v = 0)$. It should be noted that pulsed laser photolysis of Br_2 was used to create an initial concentration of Br^*, given as $[Br^*]_0$, but that only a negligibly small fraction of Br_2 was dissociated.

$$Br^* + HCl \xrightarrow{k_1} Br + HCl(v = 1),$$

$$Br^* + HCl \xrightarrow{k_2} Br + HCl,$$

$$Br^* + Br_2 \xrightarrow{k_{Br_2}} Br + Br_2,$$

$$HCl(v = 1) + HCl \xrightarrow{k_v} 2\ HCl.$$

The energy transfer was monitored by observing either the decay of Br^* fluorescence or the rise and decay of $HCl(v = 1)$ fluorescence. The following data were obtained from the HCl signal, where the amplitude gives the maximum concentration of $HCl(v = 1)$ relative to $[Br^*]_0$ and the signal was of the form $Signal(t) = Amplitude[\exp(-t/\tau_{decay}) - \exp(-t/\tau_{rise})]$.

Run	P(Br$_2$) (torr)	P(HCl) (torr)	Amplitude	τ_{rise} (μs)	τ_{decay} (μs)
A	0.5	0.61	0.056	97	1370
B	0.5	0.97	0.055	62	1000
C	0.5	1.32	0.054	50	750
D	0.5	2.10	0.054	32	525
E	0.5	3.25	0.054	21	360

a. Given that the Br^* fluorescence decays in about 1 ms for a mixture of Br_2 with about 1 torr of HCl, find k_1, k_2, k_{Br_2}, and k_v.

b. Does the data indicate the occurrence of any process other than those listed in the mechanism? If so, what is the process and what is its rate constant?

2.33 Reactants A and B react to give products C and D, but not necessarily in this stoichiometry. The reaction mechanism is

$$A \underset{k_{-1}}{\overset{k_1}{\rightleftharpoons}} I,$$

$$I + B \xrightarrow{k_2} C,$$

$$I + B \xrightarrow{k_3} D.$$

a. Derive an expression for the overall disappearance rate of A, $-d[A]/dt$.

b. At the end of the reaction, what will be the ratio of [C] to [D] if each of these concentrations was zero at the start of the reaction?

c. For each mole of A that is consumed, 1 mole of B is consumed. How many moles of C are produced for each mole of A and B consumed? How many moles of D are produced for each mole of A and B consumed?

d. Does the stoichiometry depend on temperature? Discuss.

2.34 The thermal decomposition of dimethyether has been studied by measuring the increase in pressure with time:

$$(CH_3)_2O \rightarrow CH_4 + H_2 + CO$$

Time (s)	390	780	1195	2000	3155
Pressure Increase (torr)	96	179	250	363	467

Show that the reaction is first order and find the rate constant.

2.35 Suggest plausible mechanisms for the following two reactions given the overall reaction rate law.

a. $I^- + OCl^- \rightarrow Cl^- + OI^-$; $-d[I]^-/dt = k[I^-][OCl^-]/[OH^-]$

b. $C(CH_3)_3Cl + OH^- \rightarrow C(CH_3)_3OH + Cl^-$; $-d[C(CH_3)_3Cl]/dt = k[C(CH_3)_3Cl][OH^-]/\{k'[Cl^-] + k''[OH^-]\}$

2.36 With the availability of modern computers and programs, it is now a relatively simple procedure to solve a system of differential equations by numerical integration. The general scheme is to write down the differential equation for each concentration of interest. Each equation will have the form $d[C]/dt = f(\text{rate constants, concentrations})$, where $f(. . .)$ is a function. By approximating dt as a small time interval, we obtain that $\Delta[C] = f(\text{rate constants, concentrations}) \times \Delta t$, where Δt is the small, fixed time interval. If we know the starting concentrations, we can use the approximate equations for each of the species to calculate the new concentrations after the first time interval. Then, starting with these new concentrations, we can calculate the concentrations after the second time interval, and so on.

a. Using a spread sheet program (Excel, Lotus 1-2-3, etc.), integrate the equations for two parallel, first-order reactions and show that the conclusions of **Figure 2.10** are justified. Vary the rate constants and confirm that the ratio of products changes as expected.

b. Do the same exercise for consecutive reactions and confirm the results of **Figure 2.11** and **Figure 2.12.**

c. Program the spread sheet to integrate the equations for the Lotka mechanism. Use starting concentrations for [X] and [Y] that are close, but not exactly equal to the steady-state concentrations and show that the concentrations oscillate with time. How does the amplitude and frequency of the oscillation depend on the rate constants and the steady concentration [A(0)]?

3

Theories of Chemical Reactions

Chapter Outline

3.1   INTRODUCTION

The goal of this chapter is to provide a theoretical basis for understanding and predicting reaction rates. We begin by considering an encounter between two gas-phase reactants, and we assume that on the microscopic level, the outcome and rate of the reaction are determined by the forces acting between atoms making up the reactants. We would like to know what those forces are, or equivalently, what the potential is, since the force is simply proportional to the slope of the potential, $F_x = -\mathrm{d}V/\mathrm{d}x$. Because we will rarely be able to know the potential in complete detail, it will also be of interest to determine which of its features are most responsible for determining the reaction rate, or conversely, which features are determined most accurately by measuring the reaction rate.

Two simple theories of reaction rates will be considered in detail: collision theory and activated complex theory. Both of these theories make approximations about the potential energy function controlling the reaction, and both theories relate the rate constant to simple features of the potential energy function. In its simplest form, collision theory concentrates primarily on the barrier to reaction, while activated complex theory includes not only the barrier but also some information about the "width" of the channel leading from reactants to products. Our goal here will be to explore these general properties of the potential energy function and to learn what approximations are made in these simple theories of reaction rates. We leave until Chapter 8, Molecular Reaction Dynamics, the question of how more detailed investigations can provide closer approximations to the potential energy function. A thermodynamic formulation of activated complex theory will provide us with an association between the entropy change between reactants and activated complexes and the orientational requirements for reaction.

3.2 POTENTIAL ENERGY SURFACES

Consider the reaction between two species having a total of N atoms. For any fixed geometry of these nuclei, quantum mechanics can in principle be used to calculate the energy of the system. What will be of interest in our attempt to calculate reaction rates is the way in which this energy varies with geometry. We first determine the number of coordinates on which the energy depends.

In general, the position of each nucleus can be described by three coordinates, so the position of all atoms relative to an arbitrary origin of the coordinate system can be described by $3N$ coordinates. However, if we assume that no forces act on the system other than the forces between the atoms, then the potential energy of the system depends on the relative coordinates between the atoms but neither on the position of the system as a whole nor on its overall orientation. Since three coordinates describe the position of the system as a whole and three (two for a linear system) describe its orientation, the potential energy of the system in the absence of external forces is a function of $3N - 6$ variables ($3N - 5$ for linear systems). Thus, if the coordinates are called q_i, we have $V = V(q_1, q_2, \ldots, q_k)$, where $k = 3N - 6$ for nonlinear or $3N - 5$ for linear systems.

Even for a very simple reaction involving just three atoms, $A + BC \rightarrow AB + C$, we see that in the general case the potential energy will be a function of $3 \times 3 - 6 = 3$ coordinates. We would like to picture the energy as a function of these coordinates. If one dimension is used to describe the energy, then a complete description of the way in which V varies with the three coordinates would require a four-dimensional figure, one not easily visualized by inhabitants of three-dimensional space. A way to circumvent this deficiency of human perception is to hold all but two of the coordinates at fixed values and to plot V as a function of those two coordinates.

For example, in the well-studied hydrogen atom exchange reaction, $D + H_2 \rightarrow HD + H$,[a] if we constrain all three atoms to be collinear, then the resulting potential energy diagram looks like that in **Figure 3.1.** A somewhat different way of viewing it is from the top, as in **Figure 3.2** where each contour line represents a change in energy of 0.2 eV.

Of course, for nonlinear angles between the three atoms, the potential surface would look different—for example, like that in **Figure 3.3** for a D–H–H angle of 40°. Note that the barrier to reaction is higher than that in **Figure 3.2.**

Returning to **Figure 3.2,** we consider the path of an exchange reaction. In the "entrance" valley on the lower left side of the figure, the H–H distance is small, given by the equilibrium distance for an H_2 molecule, and the D–H distance is large; these conditions describe the reactants. In this valley, motion along the H–H

---

[a]The potential is described in A. J. C. Varandas, F. B. Brown, D. G. Truhlar, C. A. Mead, and N. C. Blais, *J. Chem. Phys.* **86,** 6258 (1987); D. G. Truhlar and C. J. Horowitz, *J. Chem. Phys.* **68,** 2466 (1978); *J. Chem. Phys.* **71,** 1514 (1979).

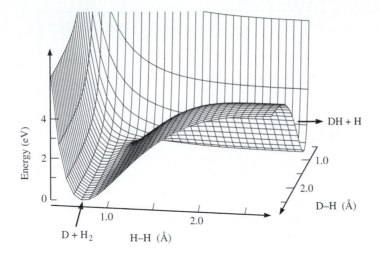

Figure 3.1

Potential energy surface for the collinear reaction D + H$_2$ → DH + H.

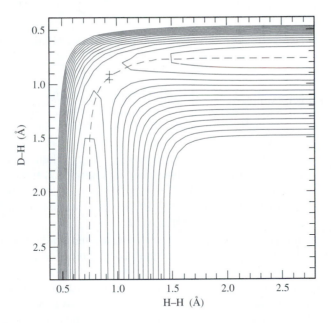

Figure 3.2

Contour diagram of the collinear D + H$_2$ → DH + H reaction.

coordinate corresponds to vibration of the H$_2$, whereas motion along the D–H coordinate represents the approach of the D and H$_2$ reactants. In the "exit" valley on the upper right side of the figure, the D–H distance is small and the H–H distance is large; these conditions describe the products. In this valley, motion along the D–H coordinate corresponds to vibration of the H–D product, whereas motion along the H–H coordinate corresponds to the separation of the H and D–H products. The saddle point on the contour diagram, marked by a ‡, is called the *activated complex* and is the highest point along the minimum energy path separating the reactants from

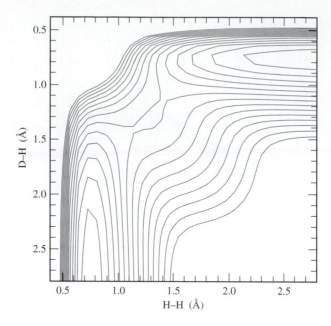

Figure 3.3

Contour diagram of the potential for $D + H_2 \rightarrow DH + H$ where the D–H–H angle is fixed at 40°.

the products. A typical reaction path would proceed from the valley on the lower left over the saddle point and into the valley on the upper right. We define the "reaction coordinate" as the minimum energy path along which the reaction can take place. Suppose for the moment, as is the case for $D + H_2$, that the collinear approach of the atoms is the one of least potential energy. Then the reaction coordinate would simply be given by the dashed line in **Figure 3.2.** If we straighten out this line and plot the energy at each point as a function of the position along the reaction coordinate, then for a generic $A + BC$ reaction we obtain the familiar picture for a chemical reaction given in **Figure 3.4.**

Even when the reaction involves several atoms, the reaction coordinate can always be viewed as the minimum energy path for the reaction along the valley of *some* two-dimensional contour map. However, because several coordinates may be changing at once in order to minimize the energy, the "distance" along the reaction path may not correspond to an easily visualized change in geometry. Thus, while the ordinate of **Figure 3.4** is a well-defined energy axis, the units along the abscissa are usually omitted.

We define the configuration of atoms at the geometry corresponding to the highest energy along the reaction path as the *activated complex,* usually denoted by a double dagger, e.g., ABC‡. The difference in energy between the zero-point energy of the reactants and that of the activated complex is called the *threshold energy,* ϵ^*, while the difference in energy between the zero-point energy of the reactants and products is $\Delta\epsilon$ for the reaction. Recall that the zero-point energy of an ensemble of atoms is $\epsilon_{zp} \equiv \Sigma \frac{1}{2}h\nu_i$, where the sum is over all vibrational frequencies ν_i.

In principle, of course, it should also be possible to calculate from quantum mechanics the detailed potential energy surface. And, in principle, as noted in the introduction, a detailed knowledge of the potential energy surface should be all that

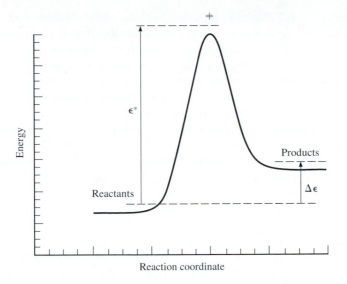

Figure 3.4

Energy as a function of reaction coordinate for an endothermic reaction.

we need to determine the rate constant. In practice, however, most systems are sufficiently complicated that neither of these steps is straightforward; drastic approximations are often needed, even using modern computational techniques. The philosophy of simple theories of reaction rates is to avoid these complicated calculations by concentrating on a few key features of the potential surface and by learning how these affect the rate. Simple collision theory, to be considered next, concentrates exclusively on the threshold energy ϵ^*. Activated complex theory, to be considered in Section 3.4, considers as well how the "width" or structure of the potential energy valley changes as reactants proceed through the activated complex.

3.3   COLLISION THEORY

3.3.1   Simple Collision Theory

The simplest theory of chemical reactions, already partially developed in Chapter 1, makes use of only one parameter of the potential energy surface, the height of the barrier to reaction, ϵ^*. The method of presentation is in three steps. First we draw on the results of Chapter 1 to show that the rate constant as a function of energy is $k(\epsilon_r) = \sigma(\epsilon_r)v_r$, where $\sigma(\epsilon_r)$ is an energy-dependent cross section and v_r is the relative velocity at energy ϵ_r. Next we argue from a simple model that the functional dependence of the cross section on energy should be $\sigma(\epsilon_r) = \pi b_{max}^2(1 - \epsilon^*/\epsilon_r)$, where ϵ^* is a minimum energy needed for reaction and ϵ_r is the collision energy. Finally, we obtain the dependence of the rate constant on temperature, $k(T)$, by averaging $k(\epsilon_r)$ over the energy distribution.

To develop this theory we return to **equation 1.44,** the equation for the collision rate between two dissimilar species, repeated here:

$$Z_{12} = \pi b_{max}^2 v_r n_1^* n_2^*, \qquad\qquad (3.1)$$

where $b_{max} = r_1 + r_2$ is the sum of the hard-sphere molecular radii and v_r is the relative velocity. This equation tells us that the collision rate, Z_{12}, is equal to a cross section, here πb_{max}^2 for hard-sphere collisions, times the relative velocity and the product of molecular densities. To account for the fact that not all collisions will lead to reaction, we need to modify what we mean by the cross section. We suppose that the cross section for a reactive collision will not be a constant πb_{max}^2 but rather will depend on the relative energy of the collision. (We recall from Chapter 1 that the energy corresponding to the motion of the center of mass is conserved, so that the only energy available for overcoming a barrier to reaction is the *relative* energy.) Only collisions with enough relative energy will react. Let us recognize the relative energy dependence of the cross section by writing it as $\sigma(\epsilon_r)$. For reactions at a particular relative energy ϵ_r corresponding to a relative velocity v_r, we can rewrite **equation 3.1** as reaction rate $= k(\epsilon_r)n_1^* n_2^*$, where $k(\epsilon_r) = \sigma(\epsilon_r)v_r$ is the rate constant for reaction at $\epsilon_r = \frac{1}{2}\mu v_r^2$.

We now model the reactants as spheres and assume that a reaction will occur only if the energy, when evaluated using the relative velocity along the line between the centers of the spheres, exceeds a particular value, ϵ^*. It seems a reasonable assumption even for spherical reactants that glancing collisions for a given velocity will be less effective in causing reaction than head-on collisions. Parameterization of the cross section in terms of the energy calculated from the velocity along the line of centers is one way to incorporate this assumption into our model.

Figure 3.5 helps in deriving the functional dependence of $\sigma(\epsilon_r)$ on ϵ_r. For simplicity, we suppose that one sphere is standing still and that the other is approaching it with relative velocity v_r at impact parameter b. The figure shows that the velocity along the line of centers is $v_{lc} = v_r \cos \alpha$, where α is the angle indicated in the figure. The value of $\cos \alpha$ is also related to b and the distance b_{max} along the line of centers by $\sin \alpha = b/b_{max}$. The energy associated with motion along the line of centers is then $\epsilon_{lc} = \frac{1}{2}\mu v_{lc}^2 = \frac{1}{2}\mu v_r^2 \cos^2 \alpha = (\frac{1}{2}\mu v_r^2)(1 - \sin^2 \alpha) = \epsilon_r(1 - b^2/b_{max}^2)$, where ϵ_r is the relative translational energy and μ is the reduced mass of the two reactants.

Returning to the assumption that $\epsilon_{lc} > \epsilon^*$ for reaction, we assert that the probability for reaction, $P(\epsilon_r,b)$, is unity when $\epsilon_{lc} = \epsilon_r(1 - b^2/b_{max}^2) > \epsilon^*$ and zero otherwise. The cross section for the reaction is calculated by averaging over the impact parameter b. **Figure 3.6** shows a narrow annulus of area = (circumference) × (width) = $(2\pi b) \times (db)$. If the probability of reaction in this annulus and for energy

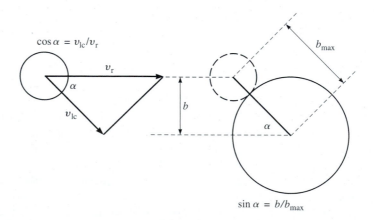

Figure 3.5

Simple model for obtaining energy dependence of the reaction cross section.

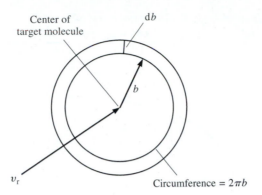

Center of target molecule

db

b

v_r

Circumference $= 2\pi b$

Figure 3.6

The average probability is the integral of $P(\epsilon_r, b)$ over b weighted by the area of the corresponding annulus.

ϵ_r is $P(\epsilon_r, b)$, then the average probability is simply the integral of $P(\epsilon_r, b)$ over the possible values of b, from $b = 0$ to $b = \infty$, each weighted by the area of the appropriate annulus.

$$\sigma(\epsilon_r) = \int_0^{\infty} P(\epsilon_r, b) 2\pi b \, db,$$

$$= \int_0^{b'} P(\epsilon_r, b) 2\pi b \, db, \tag{3.2}$$

where b' is the impact parameter above which $P(\epsilon_r, b) = 0$. Note that if the probability were unity for all impact parameters up to b_{max}, the integral would simply give the hard-sphere cross section πb_{max}^2. But the requirement $\epsilon_{lc} = \epsilon_r(1 - b^2/b_{max}^2) > \epsilon^*$ for $P(\epsilon_r, b) = 1$ implies $(b')^2 < b_{max}^2(1 - \epsilon^*/\epsilon_r)$, so that the range of integration is from 0 to b', where $(b')^2 = b_{max}^2(1 - \epsilon^*/\epsilon_r)$. The integral is easily evaluated to yield $\sigma(\epsilon_r) = \pi b_{max}^2(1 - \epsilon^*/\epsilon_r)$, provided of course that the relative energy ϵ_r exceeds ϵ^*; i.e., $\sigma(\epsilon_r) = 0$ when $\epsilon_r \leq \epsilon^*$.

The solid line in **Figure 3.7** (plotted against the left and lower axes) shows the energy dependence of $\sigma(\epsilon_r)$ in this simple model. When $\epsilon_r \gg \epsilon^*$, the reaction occurs for nearly every hard-sphere collision, and $\sigma(\epsilon_r) \approx \pi b_{max}^2$. The cross section drops rapidly to zero as $\epsilon_r \to \epsilon^*$. For reference, the points in this figure (plotted against the right and top axes) give experimental measurements of the $\sigma(\epsilon_r)$ for the $H + D_2$ reaction. The functional form of the solid line, while not exact, can be seen to be in qualitative agreement with the actual cross sections.

Given the above functional form for $\sigma(\epsilon_r)$, we now calculate $k(T)$ by averaging over the Boltzmann energy distribution in **equation 1.37** and repeated here:

$$G(\epsilon_r) \, d\epsilon_r = 2\pi \left(\frac{1}{\pi kT}\right)^{3/2} \sqrt{\epsilon_r} \exp\left(-\frac{\epsilon_r}{kT}\right) d\epsilon_r. \tag{3.3}$$

From **equation 1.16,** we know that the average of any quantity is the integral of that quantity times the distribution function of the variable on which it depends, $<Q> = \int QP(Q) \, dQ$. Here, the quantity we want to average is the reactive rate constant at a particular energy, $k(\epsilon_r) = \sigma(\epsilon_r) v_r$. Thus, $k(T) = \int k(\epsilon_r) G(\epsilon_r) \, d\epsilon_r = \int \sigma(\epsilon_r) v_r G(\epsilon_r) \, d\epsilon_r$:

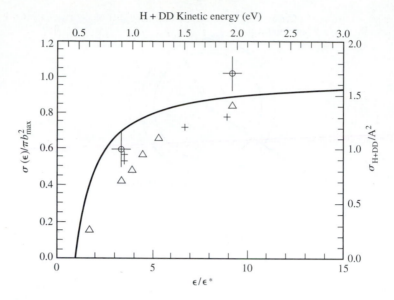

Figure 3.7

Dependence of reaction cross section on energy for the simple collision model. The solid line plotted against the bottom and left axes shows the prediction of the simple model. Data for H + D$_2$ are shown as points plotted against the top and right axes. The points with error bars are experimental results taken from K. Tsukiyama, B. Katz, and R. Bersohn, *J. Chem. Phys.* **84,** 1934 (1986), while the triangles are from the classical trajectory calculations of N. C. Blais and D. G. Truhlar, *J. Chem. Phys.* **83,** 2201 (1985), and the crosses are experimental points taken from R. A. Brownsword, M. Hillenkamp, T. Laurent, H.-R. Volpp, J. Wolfrum, R. K. Vatsa, and H.-S. Yoo, *J. Phys. Chem.* **101,** 6448 (1997).

$$k(T) = 2\pi \left(\frac{1}{\pi kT}\right)^{3/2} \int_0^\infty \sigma(\epsilon_r) v_r \sqrt{\epsilon_r} \, \exp\left(-\frac{\epsilon_r}{kT}\right) \, d\epsilon_r. \tag{3.4}$$

We now substitute for $\sigma(\epsilon_r)$ and v_r, noting that $\frac{1}{2}\mu v_r^2 = \epsilon_r$ so that $v_r = (2\epsilon_r/\mu)^{1/2}$:

$$k(T) = 2\pi \left(\frac{1}{\pi kT}\right)^{3/2} \int_{\epsilon^*}^\infty \pi b_{\max}^2 \left(1 - \frac{\epsilon^*}{\epsilon_r}\right) \sqrt{\frac{2\epsilon_r}{\mu}} \sqrt{\epsilon_r} \, \exp\left(-\frac{\epsilon_r}{kT}\right) d\epsilon_r$$

$$= 2\pi^2 b_{\max}^2 \left(\frac{1}{\pi kT}\right)^{3/2} \left(\frac{2}{\mu}\right)^{1/2} \int_{\epsilon^*}^\infty [\epsilon_r - \epsilon^*] \exp\left(-\frac{\epsilon_r}{kT}\right) d\epsilon_r \tag{3.5}$$

$$= \pi b_{\max}^2 \left(\frac{8kT}{\pi\mu}\right)^{1/2} \int_{\epsilon^*}^\infty \frac{[\epsilon_r - \epsilon^*]}{kT} \, \exp\left(-\frac{\epsilon_r}{kT}\right) \frac{d\epsilon_r}{kT}.$$

If we premultiply the integral by $\exp(-\epsilon^*/kT)$, multiply the integrand by $\exp(+\epsilon^*/kT)$, and transform variables by letting $x \equiv [\epsilon_r - \epsilon^*]/(kT)$, we obtain

$$k(T) = \pi b_{\max}^2 \left(\frac{8kT}{\pi\mu}\right)^{1/2} \exp\left(-\frac{\epsilon^*}{kT}\right) \int_{\epsilon^*}^\infty \frac{[\epsilon_r - \epsilon^*]}{kT} \exp\left(-\frac{\epsilon_r - \epsilon^*}{kT}\right) \frac{d\epsilon_r}{kT}$$

$$= \pi b_{\max}^2 \left(\frac{8kT}{\pi\mu}\right)^{1/2} \exp\left(-\frac{\epsilon^*}{kT}\right) \int_0^\infty x \exp(-x) \, dx. \tag{3.6}$$

The integral in **equation 3.6** can be shown to be unity either by integration by parts or by substitution of $x = y^2$ and use of **Table 1.1.** The result is thus

$$k(T) = \pi b_{max}^2 v_r \exp\left(-\frac{\epsilon^*}{kT}\right), \tag{3.7}$$

where $v_r = (8kT/\pi\mu)^{1/2}$ is the average relative velocity.

This surprisingly simple result can be interpreted (and remembered!) as follows. The quantity $\pi b_{max}^2 v_r$ is simply the rate constant for hard-sphere collisions and is equal to the average of the cross section times the relative velocity. The Boltzmann factor $\exp(-\epsilon^*/kT)$ can be interpreted as the fraction of collisions that provide energy greater than ϵ^*. The reaction rate constant is thus simply the rate constant for hard-sphere collisions times the fraction of those collisions with enough energy to react. This hand-waving "derivation" could have bypassed the calculus of **equations 3.1** through **3.7**, but would not have introduced the key ideas of the probability of reaction and the cross section. These ideas are necessary to extend simple collision theory to incorporate the modifications described below and in Chapter 8. A summary of the key steps in the derivation we pursued is given in **Figure 3.8.**

Before extending the theory, we note that the result in **equation 3.7** is very similar to the Arrhenius form, $k = A \exp(-E_a/kT)$, except that since v_r depends weakly on T the activation energy E_a is not exactly equal to ϵ^*. For our purposes, we will take the threshold energy ϵ^* to be equal to the activation energy E_a, although Problem 3.10 shows that they differ slightly. Finally, we note that although this development of collision theory has assumed that the energy necessary to overcome the barrier to reaction comes from the relative translational energy of the reactants, it is also possible to develop a theory in which the energy comes from the internal vibrational energy of the reactants. The final result is the same as that given in **equation 3.7.**

3.3.2   Modified Simple Collision Theory

Simple collision theory suffers, however, in that it usually overestimates the absolute magnitude of k. The reason is that for most reactions, the reactants must have favorable orientations for reaction, even when the collision supplies sufficient energy along the line of centers. We have already seen in comparing **Figure 3.2**

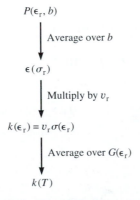

■ **Figure 3.8**

Summary of the key steps in the derivation of $k(T)$ from $P(\epsilon_r, b)$.

with **Figure 3.3** that the theoretically determined threshold energy for the $D + H_2$ reaction depends on the angle of D atom approach; the barrier to reaction is much higher for an approach angle of 40° than for the collinear approach. Direct experiments testing reactivity as a function of approach angle confirm the importance of orientation. For example, Richard Bernstein and coworkers determined that the reaction $Rb + CH_3I \rightarrow RbI + CH_3$ proceeds only if the attacking Rb atom approaches the I end of CH_3I within a cone of about 127°; attack within 53° of the CH_3 end gives a nonreactive collision (a more complete description is provided in Section 8.7.1).[b] While most reactions are neither as well understood theoretically as $D + H_2$ nor as well studied experimentally as the $Rb + CH_3I$ reaction, it is certain that orientation plays a quite general role in determining the magnitude of the rate constant. For this reason, simple collision theory is often modified by including a so-called *steric factor*, *p*, with $p < 1$, to account for the orientation requirement:

$$k(T) = p\pi b_{max}^2 v_r \exp\left(-\frac{\epsilon^*}{kT}\right). \tag{3.8}$$

Example 3.1 illustrates how collision theory might be used to estimate the value of *p*.

example 3.1

Estimation of the Steric Factor *p*

Objective   Consider the reaction $CH_3 + H_2 \rightarrow CH_4 + H$. Assuming that the cross-sectional areas ($\sigma = \pi r^2$) of CH_3 and H_2 are 4.0 and 2.7×10^{-19} m², respectively, use collision theory to calculate the steric factor *p* at 300 K given that the measured preexponential factor is 6.7×10^{12} cm³ mol⁻¹ s⁻¹.

Method   Use **equation 3.8** recognizing that $b_{max} = r_{CH3} + r_{H2}$ and that the preexponential factor is simply $A = p \pi b_{max}^2 v_r$.

Solution   i)   Calculate πb_{max}^2:

$$r_{CH_3} = \left(\frac{4.0 \times 10^{-19} \text{ m}^2}{\pi}\right)^{1/2}$$

$$= 3.6 \times 10^{10} \text{ m}, \tag{3.9}$$

$$r_{H_2} = \left(\frac{2.7 \times 10^{-19}}{\pi}\right)^{1/2} = 2.9 \times 10^{-10} \text{ m},$$

$$b_{max} = (3.6 + 2.9) \times 10^{-10} \text{ m},$$

---

[b]See D. H. Parker, K. K. Chakravorty, and R. B. Bernstein, *J. Phys. Chem.* **85,** 466 (1981) and S. E. Choi and R. B. Bernstein, *J. Chem. Phys.* **83,** 4463 (1985); also R. D. Levine and R. B. Bernstein, *Molecular Reaction Dynamics and Chemical Reactivity* (Oxford University Press, New York, 1987), p. 56.

so that

$$\pi b_{max}^2 = \pi(6.5 \times 10^{-10}\,\text{m})^2 = 1.3 \times 10^{-18}\,\text{m}^2.$$

ii)   Calculate v_r:

$$\mu = \frac{(15\,\text{amu})(2\,\text{amu})}{(15 + 2\,\text{amu})}$$

$$= 1.76\,\text{amu},$$

$$v_r = \left(\frac{8kT}{\pi\mu}\right)^{1/2} \qquad\qquad \textbf{(3.10)}$$

$$= \left[\frac{8(1.38 \times 10^{-23}\,\text{J K}^{-1})(300\,\text{K})}{\pi\,1.76/(6.02 \times 10^{23} \times 1000)}\right]^{1/2}$$

$$= 1900\,\text{m/s}.$$

iii)   Calculate p:

$$p = \frac{A}{\pi b_{max}^2 v_r}$$

$$= \frac{(6.7 \times 10^{12}\,\text{cm}^3\,\text{mol}^{-1}\,\text{s}^{-1})(10^{-6}\,\text{m}^3/\text{cm}^3)}{(1.3 \times 10^{-18}\,\text{m}^2/\text{molecule})(1900\,\text{m/s})(6.02 \times 10^{23}\,\text{molecule/mol})}$$

$$\textbf{(3.11)}$$

$$= \frac{6.7 \times 10^6}{1.49 \times 10^9}$$

$$= 4.5 \times 10^{-3}.$$

The steric factor p clearly does little more than embody our ignorance of the details of the reaction into a single factor. For many reactions, p is found to lie in the range from 0.001 to 0.1, reflecting the range of probabilities that the reactants will be in the correct configuration for reaction. A somewhat more sophisticated account of the orientation dependence of reactions has recently been introduced[c] and is described below.

The basic assumption of the modification is that the threshold energy depends on orientation. Suppose not only that the energy along the line of centers needs to be above some minimum energy ϵ^* for reaction, but also that the minimum energy

[c]This extension of collision theory was proposed independently by I. W. M. Smith [*Kinetics and Dynamics of Elementary Gas Reactions* (Butterworths, Boston, 1980), pp. 78 ff; *J. Chem. Ed.* **59**, 9 (1982)] and by R. D. Levine and R. B. Bernstein [*Chem. Phys. Lett.* **105**, 467 (1984)]. As pointed out by Levine [*J. Phys. Chem.* **94**, 8872 (1990)], both are similar to an earlier formulation by H. Pelzer and E. Wigner [*Z. Phys. Chem. B* **15**, 445 (1932)], as described in H. Johnston, *Gas Phase Reaction Rate Theory* (Ronald Press, New York, 1966). A more advanced version is given by A. Miklavc, M. Perdih, and I. W. M. Smith, *Chem. Phys. Lett.* **241**, 415 (1995).

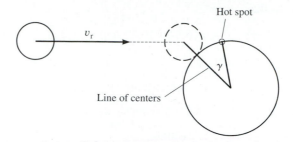

■ **Figure 3.9**

Simple model for a modified collision theory, one for which the effective activation energy
depends on angle of approach.

depends on the orientation of one reactant with respect to another. In simple terms,
suppose, as shown in **Figure 3.9** that the sphere representing one of the reactants has
a "hot spot" toward reactivity at angle $\gamma = 0$ but that the reactivity falls off as the
hot spot is rotated away from the point of impact. A simple model embodying this
idea is to assume that the energy along the line of centers needs to be bigger than $\epsilon^* + \epsilon'(1 - \cos \gamma)$. When γ is zero, the effective threshold energy is simply ϵ^*, but the
effective threshold energy rises to $\epsilon^* + \epsilon'$ for $\gamma = 90°$ and reaches $\epsilon^* + 2\epsilon'$ for $\gamma =$
$180°$. Following the argument above, we require $\epsilon_{lc} \geq \epsilon^* + \epsilon(1 - \cos \gamma)$, where ϵ_{lc}
is still given by $\epsilon_r(1 - b^2/b_{max}^2)$. Problem 3.7 outlines the solution, which, when compared to **equation 3.8**, shows that p can be identified as kT/ϵ'. Reasonable values of
ϵ' are in the range from 10 to 1000 kT, so this modified collision theory finds values
of p in the range 0.001–0.1, in fair agreement with experiment.

    Note that the modified (and improved) collision theory achieves better agreement
with experiment by incorporating one more feature of the potential energy function
into the model. In this case, the additional feature is the way in which the barrier to
reaction increases as the angle between the reactants is varied; in other words, the
additional feature is the "width" of the barrier. We will see below that activated complex theory incorporates a similar parameter, although in a different manner.

3.4   ACTIVATED COMPLEX THEORY (ACT)

While simple collision theory models the potential energy surface with only the
threshold energy ϵ^*, activated complex theory goes somewhat beyond this naive
view, but only by one step. As we will see, the additional feature of the surface
incorporated by ACT is related to the "width" or structure of the reaction channel
leading over the activation barrier. In this sense, it is similar to the modified collision theory just described, but the development of ACT provides new insights into
the meaning of the preexponential factor.

    We start by modeling the chemical reaction by the following scheme:

$$A + B \underset{k_{-1}}{\overset{k_1}{\rightleftharpoons}} AB^{\ddagger} \overset{k_2}{\rightarrow} \text{products,} \tag{3.12}$$

where A and B each might represent some collection of atoms. Under the steady-
state assumption, we can write the overall rate of the reaction as

$$\frac{d[\text{products}]}{dt} = k_2[AB^{\ddagger}]$$

$$= k_2 \frac{k_1}{k_{-1} + k_2}[\text{A}][\text{B}] \qquad \textbf{(3.13)}$$

$$\approx k_2 \frac{k_1}{k_{-1}}[\text{A}][\text{B}],$$

where the last equation was obtained by assuming that $k_2 \ll k_{-1}$.

What remains is to express the combination of rate constants $k_2 k_1 / k_{-1}$ in terms of properties of the reactants and the potential energy surface. We start with the ratio k_1 / k_{-1} and recognize that this ratio is simply the equilibrium constant for the "reaction" $\text{A} + \text{B} \rightleftharpoons \text{AB}^{\ddagger}$. From statistical mechanics, we know that an equilibrium constant can be written as the ratio of partition functions per unit volume, but we must be careful to take into account the offset in energy between the zero-point energies of the reactants and that of the activated complex. Thus, if we denote the partition function per unit volume for the activated complex by $q^{\ddagger}$ and the product of the partition functions per unit volume for the reactants by $q_A q_B$, then k_1 / k_{-1} is equal to $q^{\ddagger}/q_A q_B$ and the rate of the reaction becomes

$$\frac{d[\text{products}]}{dt} = k_2 \frac{q^{\ddagger}}{q_A q_B} \exp\left(-\frac{\epsilon^*}{kT}\right)[\text{A}][\text{B}], \qquad \textbf{(3.14)}$$

where each partition function per unit volume depends on temperature and gives the number of states available to the system at that temperature. The factor $\exp(-\epsilon^*/kT)$ accounts for the offset in energy, ϵ^*, between the zero points of the reactants and that of the activated complex. This offset is shown in **Figure 3.10.**

A reminder about the physical interpretation of partition functions may help to clarify the meaning of **equation 3.14.** The partition function is simply a number which describes how many states are available to the system at a given temperature; it is defined as $z \equiv \Sigma \exp(-\epsilon_n/kT)$, where the sum is over all states n of the system, each with energy ϵ_n. If the system were a single quantum particle in a

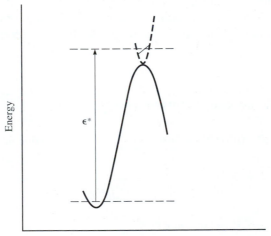

Reaction coordinate

■ Figure 3.10

The energy ϵ^* in ACT is the difference in energy between the zero points of the reactants and the activated complex.

one-dimensional box, for example, the possible states of the system are at the energies $\epsilon_n = n^2h^2/8mL^2$, where h is Planck's constant, m is the mass of the particle, and L is the length of the box. For a particle in a one-dimensional box, Problem 3.8 shows that this sum is $(2\pi mkT)^{1/2}L/h$. If, instead, the particle is translating freely in a *three*-dimensional box of volume V, the partition function is $z_t = (2\pi mkT)^{3/2}V/h^3$, where the subscript t denotes that this is the translational contribution to the partition function. For this example, the quantity required in **equation 3.14** is the partition function per unit volume: $q_t = z_t/V = (2\pi mkT)^{3/2}/h^3$.

Of course, if a particle has vibrational and rotational degrees of freedom, more states will be available at a given temperature. For independent degrees of freedom, the partition function for a single particle is simply the product of the partition functions for each type of motion: $z = z_t \times z_r \times z_v \times z_e$, where the additional subscripts denote rotation, vibration, and electronic motion, respectively. Partition functions for these degrees of motion are summarized in **Table 3.1.** At this stage, we will simply use the listed results. Note that only the translational partition function depends on volume.

It is clear from **equation 3.14** that the overall rate constant for the bimolecular reaction A + B → products is $k = k_2(q^{\ddagger}/q_Aq_B)\exp(-\epsilon^*/kT)$. The partition functions q_A and q_B can be calculated with the aid of **Table 3.1.** We then have left two tasks— to determine k_2 and to learn how to calculate $q^{\ddagger}$.

The rate constant k_2 describes the unimolecular decomposition rate of $AB^{\ddagger}$ going to products. We will take the approach that the rate of $AB^{\ddagger}$ decomposition is simply equal to the frequency of vibration of the complex along the path of the reaction coordinate, denoted $\nu^{\ddagger}$. Thus, the complex decomposes to products on every vibration, and the reaction rate simplifies to

$$k(T) = \nu^{\ddagger}\frac{q^{\ddagger}}{q_Aq_B}\exp\left(-\frac{\epsilon^*}{kT}\right). \tag{3.15}$$

There are admittedly two predicaments with this approach. First, it is difficult to imagine a vibrational frequency along a coordinate for which the potential energy function is not bound. We will see shortly, however, that the frequency $\nu^{\ddagger}$ actually cancels in the final result. Furthermore, Problem 3.9 shows that the same final result is obtained if the motion is treated as a translation rather than a vibration. Second, we have assumed in our model leading to **equation 3.13** that the

TABLE 3.1   Partition Functions for Molecular Degrees of Freedom

Type of Motion	Partition Function	Order of Magnitude
Translational, three-dimensional, per unit volume	$z_t/V = (2\pi mkT)^{3/2}/h^3$	10^{24} cm^{-3}
Rotational, linear molecule	$z_r = 8\pi^2IkT/h^2\sigma$	10^2
Rotational, nonlinear molecule	$z_r = [\pi^{1/2}(8\pi^2kT)^{3/2}/h^3\sigma] \times (I_AI_BI_C)^{1/2}$	10^3
Vibrational, each degree of freedom, measured from the lowest (zero point) vibrational level	$z_v = [1 - \exp(-h\nu/kT)]^{-1}$	1 (high ν) 10 (low ν)
Electronic	$z_e = \Sigma\, g_i\exp(-\epsilon_i^e/kT)$	1

Notes: I is the moment of inertia; ν is the vibrational frequency; ϵ_i^e are the electronic energy levels, g_i is the degeneracy, and σ is a symmetry number (e.g., $\sigma = 2$ for a homonuclear diatomic molecule).

decomposition of the complex is slow enough not to perturb the equilibrium A + B $\rightleftharpoons$ AB‡, yet we are now assuming that AB‡ decomposes on every vibration. Despite this apparent contradiction, a more careful examination, somewhat beyond the level of this text,[d] shows that for $\epsilon^* > 3kT$ the approximations are still valid, partly because the overall rate is small compared to $\nu^\ddagger$ under these conditions.

We now separate the partition function $q^\ddagger$ into two factors, a partition function for the vibration along the reaction coordinate, $z_v^\ddagger$, and a partition function for the remaining degrees of freedom, $q^{\ddagger\prime}$, where the prime notation reminds us that the latter partition function is missing one degree of freedom (the vibration). According to **Table 3.1**, $z_v^\ddagger$, can be calculated as

$$
\begin{aligned}
z_v^\ddagger &= \frac{1}{1 - \exp(-h\nu^\ddagger/kT)} \\[2mm]
&\approx \frac{1}{1 - [1 - (h\nu^\ddagger/kT)]} \\[2mm]
&= \frac{kT}{h\nu^\ddagger},
\end{aligned}
\tag{3.16}
$$

where the approximation uses only the first two terms in the Taylor expansion of the exponential since the frequency $\nu^\ddagger$ is assumed to be small. Substitution of **equation 3.16** into **equation 3.15** gives

$$
k(T) = \frac{kT}{h} \frac{q^{\ddagger\prime}}{q_A q_B} \exp\left(-\frac{\epsilon^*}{kT}\right).
\tag{3.17}
$$

Note that the ambiguous frequency $\nu^\ddagger$ cancels in this final result. **Example 3.2** illustrates how **equation 3.17** can be used to estimate rate constants.

example 3.2

Estimating Rate Constant Orders of Magnitude with ACT

Objective   Estimate the order of magnitude of the room-temperature rate constant for a reaction in which an atom and a heteronuclear diatomic molecule react through a nonlinear activated complex; leave the answer as a factor times $\exp(-\epsilon^*/kT)$.

Method   Use **equation 3.17** and the third column of **Table 3.1**.

Solution   The nonlinear activated complex of three atoms has three translational degrees of freedom, three rotational degrees of freedom, and $3N - 6 = 3$ vibrational degrees of freedom. One of these, the asymmetric stretch, is the reaction coordinate; the symmetric stretch is usually a high frequency while the bend is a low

[d]P. J. Robinson and K. A. Holbrook, *Unimolecular Reactions* (Wiley-Interscience, London, 1972), Section 4.12.

frequency. Thus, we may estimate $q^{\ddagger\prime} \approx (10^{24} \text{ cm}^{-3})(10^3)(10) = 10^{28} \text{ cm}^{-3}$. The partition function for the atom has only translational contributions: $q_A \approx 10^{24} \text{ cm}^{-3}$. The partition function for the diatom has three translational, two rotational, and one (probably high-frequency) vibrational degree of freedom: $q_B \approx (10^{24} \text{ cm}^{-3}) \cdot (10^2)(1) = 10^{26} \text{ cm}^{-3}$. Recalling that $kT/h \approx 6 \times 10^{12} \text{ s}^{-1}$ at room temperature, we find that $k \approx (6 \times 10^{12} \text{ s}^{-1})(10^{28} \text{ cm}^{-3})/[(10^{24} \text{ cm}^{-3})(10^{26} \text{ cm}^{-3})]\exp(-\epsilon^*/kT) = (6 \times 10^{-10} \text{ cm}^3 \text{ molecule}^{-1} \text{ s}^{-1})\exp(-\epsilon^*/kT)$.

Equation 3.17 is sometimes written as

$$k(T) = \frac{kT}{h} K^{\ddagger\prime}, \tag{3.18}$$

where

$$K^{\ddagger\prime} \equiv \frac{q^{\ddagger\prime}}{q_A q_B} \exp\left(-\frac{\epsilon^*}{kT}\right) \tag{3.19}$$

is a sort of equilibrium constant describing the ratio of activated complexes (less one degree of freedom along the reaction coordinate) to reactants.

Let us step back from **equation 3.17** for a moment to consider its physical meaning. The factor of kT/h, equal to $6.25 \times 10^{12} \text{ s}^{-1}$ at room temperature, is the frequency at which the reactants attempt to get to the activated complex. The factors $q^{\ddagger\prime}/(q_A q_B)\exp(-\epsilon^*/kT)$ determine the ratio of the number of states available to the activated complex (less one degree of freedom) divided by the number of states available to the reactants. The reaction rate is just the frequency times this ratio, since we have assumed that the activated complexes proceed to products on every vibration along the reaction coordinate.

Note that each of the three partition functions in the ratio $q^{\ddagger\prime}/(q_A q_B)$ has units of volume^{-1}, so that the units of the rate constant for this bimolecular reaction are volume per second (per molecule), as required. The Boltzmann factor $\exp(-\epsilon^*/kT)$ accounts for the difference in energy between the activated complex and the reactants; if the zero of energy had been the same for each of these species, this factor would have been unnecessary.

A view from the reactants' point of view of two possible A + B reactions is shown in **Figure 3.11,** where only one of the possible modes of vibration perpendicular to the reaction path is shown in each panel. Two colliding molecules would find it easier to react on the potential energy surface of **Figure 3.11A** rather than that of **Figure 3.11B.** The reason is that the vibrational levels are more closely spaced on the "wider" potential energy function, so that more states of the activated complex are accessible at any given temperature in this case. The reaction rate should depend both on the threshold energy and on the "width" of the channel leading over the barrier. Note, however, that the "width" in most reactions is multidimensional, so that it is really more accurate to refer to the "structure" of the activated complex state or, even better, to the partition function for the activated complex; for an activated complex with $N \geq 3$, there are $3N - 6 - 1$ degrees of freedom perpendicular to the reaction coordinate. For example, in a triatomic activated complex, there are $9 - 7 = 2$ perpendicular degrees of freedom. The

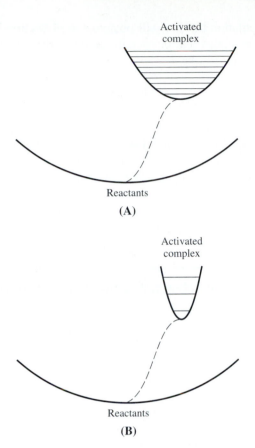

■   Figure 3.11

The reactants' view of two possible reactions.

asymmetric stretching motion of the triatomic system (A →← B–C →) is the reaction coordinate, while the symmetric stretch (← A–B–C →) and the bending vibration (↓ A–B↑ –C↓) are the two orthogonal vibrational modes.

   A connection between ACT and simple collision theory can be made calculating the ACT rate constant for the reaction of two atoms of average diameter b_{max}. As one might hope, the resulting expression, derived in **Example 3.3,** is exactly that calculated by collision theory, given in **equation 3.7.** We remarked in Section 3.3.2 that the steric factor, p, was typically less than unity. Activated complex theory gives an explicit formula for p; it is equal to the ratio between the value of $q^{\ddagger\prime}/(q_A q_B)$ for the reaction under consideration and the value of $q^{\ddagger\prime}/(q_A q_B)$ for a reaction between two atoms whose average diameter is equal to that of the reactants under consideration.

   It should be noted that ACT is not restricted to bimolecular reactions. Returning to **equation 3.17,** we could write the temperature-dependent rate of a unimolecular reaction as

$$k(T) = \frac{kT}{h}\frac{q^{\ddagger\prime}}{q_A}\,\exp\!\left(-\frac{\epsilon^*}{kT}\right). \qquad (3.20)$$

Of somewhat more interest is to ask what the rate for unimolecular reaction would be if all the reactant molecules had the same energy. The answer involves a somewhat more complex calculation that we will postpone until Section 7.5.4 when we discuss unimolecular reactions in more detail. However, the final result is reasonably simple:

$$k_a(E^*) = \frac{W(E^+)}{hN^*(E^*)},$$ (3.21)

where $k_a(E^*)$ is the rate constant for unimolecular decay of molecules with energy E^*, $W(E^+)$ is the number of vibrational states of the activated complex accessible at that same total energy, and $N^*(E^*)$ is the density of states (the number per unit energy interval) of the reactant at energy E^*. The energy E^+ is equal to $E^* - E_0$, where E_0 is the activation energy. Note that since $hN^*(E^*)$ has units of time and since $W(E^*)$ is dimensionless, $k_a(E^*)$ has units of s^{-1}, as required for a first-order reaction.

example 3.3

ACT Calculation of the Reaction Rate of Two Atoms

Objective   Show that the ACT rate constant for the reaction of two atoms whose average diameter is b_{max} gives the same result as collision theory, **equation 3.7.**

Method   Use **equation 3.17** with column 2 of **Table 3.1.** Note $q^{\ddagger\prime}$ is the partition function of the activated complex, but without the degree of freedom corresponding to the reaction coordinate. For the reaction of two atoms, the reaction coordinate corresponds to the vibration of the diatomic activated complex.

Solution   The exponential factors, $\exp(-\epsilon^*/kT)$, are the same for ACT and collision theory, so that our task is to show that the pre-exponential factor in ACT, $(kT/h)(q^{\ddagger\prime}/q_A q_B)$, is equivalent to the preexponential factor in **equation 3.7.** Using **Table 3.1** and recalling that the moment of inertia in this case is μb_{max}^2, we find that the ACT pre-exponential factor is

$$\frac{kT}{h} \frac{q^{\ddagger\prime}}{q_A q_B} = \frac{kT}{h} \frac{[2\pi(m_A + m_B)kT/h^2]^{3/2}[2IkT/\hbar^2]}{[2\pi m_A kT/h^2]^{3/2}[2\pi m_B kT/h^2]^{3/2}}$$

$$= \frac{kT}{h} \frac{[2IkT/\hbar^2]}{[2\pi(m_A m_B/(m_A + m_B))kT/h^2]^{3/2}}$$ (3.22)

$$= \frac{kT}{h} \frac{[2(\mu b_{max}^2)kT/\hbar^2]}{[2\pi\mu kT/h^2]^{3/2}} = \frac{(kT)^{1/2}(2\pi)^{1/2}2b_{max}^2}{\mu^{1/2}}$$

$$= \left(\frac{8kT}{\pi\mu}\right)^{1/2} \pi b_{max}^2 = v_r \pi b_{max}^2.$$

This is just the result of **equation 3.7.**

3.5   THERMODYNAMIC INTERPRETATION OF ACT

We now return to **equation 3.18** to provide a thermodynamic interpretation of the rate constant. If we identify the equilibrium constant with a Gibbs free energy through the relationship $-RT \ln K^{\ddagger'} = \Delta G^{\ddagger}$ then **equation 3.18** can be rewritten as

$$k(T) = \frac{kT}{h}\exp\left(-\frac{\Delta G^{\ddagger}}{RT}\right), \tag{3.23}$$

where $\Delta G^{\ddagger}$ is the change in Gibbs free energy in going from the reactants to the activated complex. Of course, $\Delta G^{\ddagger}$ can be written in terms of the enthalpy and entropy changes between the reactants and the activated complex:

$$\Delta G^{\ddagger} = \Delta H^{\ddagger} - T\Delta S^{\ddagger}, \tag{3.24}$$

so that **equation 3.23** becomes

$$k(T) = \frac{kT}{h}\exp\left(\frac{\Delta S^{\ddagger}}{R}\right)\exp\left(-\frac{H^{\ddagger}}{RT}\right). \tag{3.25}$$

Recalling that $\Delta H = \Delta(E + pV) = \Delta E + (\Delta n)RT$ for an ideal gas and that for our reaction $A + B \rightleftharpoons (AB)^{\ddagger}$ the value of Δn is -1, we see that $\Delta H^{\ddagger} = \Delta E^{\ddagger} - RT$, where $\Delta E^{\ddagger}$ is the activation energy in going from reactants to activated complexes and can be identified as the activation energy E_a. Thus,

$$k(T) = \exp(1)\frac{kT}{h}\exp\left(\frac{\Delta S^{\ddagger}}{R}\right)\exp\left(-\frac{E_a}{RT}\right). \tag{3.26}$$

When compared to the Arrhenius expression $k = A\exp(-E_a/RT)$, we find that the preexponential factor is identified as

$$A = \exp(1)\frac{kT}{h}\exp\left(\frac{\Delta S^{\ddagger}}{R}\right). \tag{3.27}$$

The point of this exercise is to identify the Arrhenius preexponential factor with an entropy of activation, the entropy change in going from reactants to the activated complex. For a bimolecular reaction, of course, $\Delta S^{\ddagger}$ is always a negative quantity, since two reactants must come together into a necessarily more ordered state before they can react. If the activated complex needs to be even more "ordered" than just two spheres in contact, then the entropy of activation will be correspondingly more negative, and the preexponential factor and reaction rate constant will become smaller. A more "ordered" activation complex is one for which, in collision theory, the steric factor is smaller than 1, or one for which, in activated complex theory, there are few accessible states of the complex.

3.6   SUMMARY

The main point of this chapter was to show how two simple theories, collision theory and activated complex theory, can be used to estimate rate constants. We started by noting that the rate constant could be calculated if we knew the potential energy surface for the reaction. In most cases, however, this surface is known only approximately. Which of its features is most important in determining the rate constant? Simple collision theory concentrates on the difference between the zero-point energies of the activated complex and the reactants, ϵ^*. It assumes that reaction occurs every time a collision provides an energy along the line between the centers of the reactants greater than ϵ^*. The derivation of the collision theory rate constant starts with evaluation of the reaction probability as a function of energy and impact parameter, $P(\epsilon_r, b)$, averages this over impact parameter to obtain the cross section for reaction at a particular energy, $\sigma(\epsilon_r)$, and then averages the energy-dependent rate constant $k(\epsilon_r) = \sigma(\epsilon_r)v$ over the thermal distribution of collision energies. The result is

$$k(T) = \pi b_{max}^2 v_r \exp\left(-\frac{\epsilon^*}{kT}\right), \tag{3.7}$$

where $v_r = (8kT/\pi\mu)^{1/2}$ is the average relative velocity. A hand-waving derivation of this result is that the rate constant is simply the hard-sphere collision rate, $\pi b_{max}^2 v_r$, times the fraction of collisions, $\exp(-\epsilon^*/kT)$, that provide energy greater than ϵ^*.

Simple collision theory overestimates the rate constant because it fails to take into account the fact that the reactants must be oriented properly for reaction even when the collision energy is sufficient to overcome the threshold energy. We can modify the theory by incorporating a steric factor, p, with $p < 1$ to account for the orientation requirement. Then,

$$k(T) = p\pi b_{max}^2 v_r \exp\left(-\frac{\epsilon^*}{kT}\right). \tag{3.8}$$

Consideration of the simple model illustrated in **Figure 3.9** leads to the conclusion that $p = kT/\epsilon'$, where ϵ' describes the way in which the energy required for reaction varies with the angular orientation of the reactants. Thus, this modified collision theory improves on the simple model by incorporating one more feature of the potential energy surface in addition to the threshold energy ϵ^*.

Activated complex theory also improves on the simple model by incorporating information about the "width" of the barrier over the saddle point. The essential assumption of the theory is that the reactants are at equilibrium with activated complexes, so that the equilibrium constant can be used to calculate the overall rate for the reaction, $k_2[AB^\ddagger] = k_2 K^\ddagger[A][B]$. The concepts of statistical mechanics are used to show that

$$k(T) = \frac{kT}{h} \frac{q^{\ddagger\prime}}{q_A q_B} \exp\left(-\frac{\epsilon^*}{kT}\right). \tag{3.17}$$

In this equation, the quantities q are partition functions that describe the number of states accessible at the temperature of interest. They can be evaluated using **Table 3.1.** For a unimolecular reaction taking place at a particular temperature, the corresponding equation is simply

$$k(T) = \frac{kT}{h} \frac{q^{\ddagger\prime}}{q_A} \exp\left(-\frac{\epsilon^*}{kT}\right). \tag{3.20}$$

If all the reactants have the same energy E^*, then the equation simplifies to

$$k_a(E^*) = \frac{W(E^+)}{hN^*(E^*)}, \tag{3.21}$$

where $W(E^+)$ is the number of states of the activated complex and $N^*(E^*)$ is the density of states of the reactant.

By realizing that the equilibrium constant between reactants and activated complexes can be expressed in terms of a free energy of activation, $\Delta G^{\ddagger} = \Delta H^{\ddagger} - T\Delta S^{\ddagger}$, we found that the Arrhenius preexponential factor, A, can be related to the entropy change in proceeding from the reactants to the activated complex.

suggested readings

K. J. Laidler, *Chemical Kinetics* (Harper and Row, New York, 1987).

R. D. Levine, "The Steric Factor in Transition State Theory and in Collision Theory," *Chem. Phys. Lett.* **175,** 331–337 (1990), and references therein.

P. J. Robinson and K. A. Holbrook, *Unimolecular Reactions* (Wiley-Interscience, London, 1972).

Ian W. M. Smith, *Kinetics and Dynamics of Elementary Gas Reactions* (Butterworths, Boston, 1980), pp. 78 ff.

I. W. M. Smith, "A New Collision Theory for Bimolecular Reactions," *J. Chem. Ed.* **59,** 9 (1982).

J. I. Steinfeld, J. S. Francisco, and W. L. Hase, *Chemical Kinetics and Dynamics* (Prentice-Hall, Englewood Cliffs, NJ, 1989), pp. 158 ff.

D. G. Truhlar, B. C. Garrett, and S. J. Klippenstein, "Current Status of Transition-State Theory," *J. Phys. Chem.* **100,** 12771 (1996).

problems

3.1   For a nonlinear system of 5 atoms, on how many coordinates does the potential energy depend?

3.2   For simple collision theory, at what energy is the cross section most dependent on energy?

3.3   The thermal rate constant is the average of (a) the reaction probability over impact parameter, (b) the reaction probability over the thermal energy distribution, (c) the reaction cross section over the thermal energy distribution, or (d) the energy-dependent rate constant over the thermal energy distribution. (Specify which.)

3.4   The steric factor p depends on (a) the angular dependence of the reaction probability, (b) the requirements for the reactants to be in a particular orientation, (c) a ratio of partition functions, (d) the entropy of activation, or (e) more than one of the above. (Specify which.)

3.5   When activated by collision, the molecule $Br-CH_2-CH_2-CH_2-CH_3$ can eliminate either HBr or Br. Which reaction will have the higher Arrhenius A parameter? Why?

3.6   The partition function (a) specifies the number of states below a certain energy, (b) is proportional to a volume in phase space, (c) specifies the number of states accessible at a certain energy, (d) more than one of the above. (Specify which.)

3.7   The condition for reaction, given the modified collision theory assumptions of Section 3.3.2 and **Figure 3.9,** is that $\epsilon_{lc} \geq \epsilon^{*} + \epsilon'(1 - \cos\gamma)$, where ϵ_{lc} is equal to $\epsilon_r(1 - b^2/b_{max}^2)$. For $\epsilon_r > \epsilon^{*}$, this equation defines a maximum angle

γ_{max} for which reaction can occur: $1 - \cos\gamma_{max} = [\epsilon_r(1 - b^2/b_{max}^2) - \epsilon^*/\epsilon'$. The probability for reaction $P(\epsilon_r, b)$ is then just the surface area of a sphere of radius b_{max} for which $\gamma \le \gamma_{max}$:

$$P(\epsilon_r, b) = \frac{\displaystyle\int_0^{\gamma_{max}} 4\pi b_{max}^2 \sin\gamma \, d\gamma}{4\pi b_{max}^2}$$

$$= 1 - \cos\gamma_{max}$$

$$= [\epsilon_r(1 - b^2/b_{max}^2) - \epsilon^*]/\epsilon'.$$

By integrating this value for $P(\epsilon_r, b)$ over $2\pi b \, db$, calculate $\sigma(\epsilon_r)$. Then average this cross section over the thermal distribution to show that $k(T)$ is equal to (kT/ϵ') times the result for simple collision theory. Note that this theory predicts that the steric factor should increase with temperature. The result is reasonable, since the range of angles for which reaction can occur should increase with increasing temperature.[e]

3.8   Recalling that the energy levels for a particle of mass m in a one-dimensional box of length L are given by $\epsilon_n = n^2h^2/(8mL^2)$ and that the partition function is defined as $q = \Sigma \exp(-\epsilon_n/kT)$, show that for this system is $q = (2\pi mkT)^{1/2}L/h$. Hint:

$$\sum_{n=1}^{\infty} \exp(-n^2\alpha^2) \approx \int_0^{\infty} \exp(-n^2\alpha^2) \, dn.$$

3.9   In **equation 3.15,** we assumed that the rate k_2 at which the activated complexes decompose to products was equal to a vibrational frequency $\nu^{\ddagger}$. Although the vibrational frequency cancels in the final analysis, it might appear disturbing to associate a vibrational frequency with motion over a saddle point. In this problem, we consider the motion over the saddle point to be a translation. Let the activated complex be defined as the set of geometries along a length on the reaction coordinate equal to δ. This length, about 0.1 nm, is arbitrary, since we will see that it, like the vibrational frequency, cancels in the final analysis. If the average translational velocity along the reaction coordinate is $<v>$, then the rate of crossing is just $k_2 = <v>/2\delta$, where the factor of 2 is introduced because half of the transition complexes will be moving in the forward direction. For motion in one dimension the average velocity is $<v> = (2kT/\pi m^{\ddagger})^{1/2}$, where $m^{\ddagger}$ is the mass of the activated complex. Starting from **equation 3.14,** substitute for k_2 and write $q^{\ddagger}$ as the product of a translational partition function $q_t^{\ddagger}$ (in one dimension) and the partition function for the remaining degrees of freedom $q^{\ddagger\prime}$ to derive the result in **equation 3.17.**

3.10   The activation energy E_a can be defined from the Arrhenius expression as

$$E_a = kT^2 \frac{\partial \ln k(T)}{\partial T}.$$

[e]For further discussion of this effect, see R. D. Levine, *Chem. Phys. Lett.* **175,** 331 (1990).

a. Use **equation 3.7** to show that

$$E_a = \frac{1}{2} kT + \epsilon^*.$$

b. What is the percent error in taking $E_a = \epsilon^*$ for each of the first three reactions of **Table 2.2** where T is taken to be at the middle of the applicable range?

3.11 a. Use simple collision theory to calculate the Arrhenius A factor for the elementary reaction $NO + O_3 \rightarrow NO_2 + O_2$. Reasonable values for the molecular radii are 0.14 nm for NO and 0.20 nm for O_3. Assume $T = 300$ K.

b. The experimental A factor for this reaction is 1.0×10^8 L mol^{-1} s^{-1}. What is the value of the steric factor p?

3.12 Consider a unimolecular decomposition that proceeds via the Lindemann mechanism:

$$A + A \xrightarrow{k_1} A^* + A,$$

$$A^* \xrightarrow{k_2} 2\,B.$$

The reactant A may also be activated by a collision with a product molecule B:

$$A + B \xrightarrow{k_3} A^* + B$$

Use collision theory to calculate the relative magnitudes of k_1 and k_3. You may assume that the two reactions have the same threshold energy. (*Hint:* What is a reasonable assumption about the relative volumes of B and A?)

3.13 The reaction $F + H_2 \rightarrow H + HF$ is the rate-limiting elementary step in the overall reaction $H_2 + F_2 \rightarrow 2$ HF. This latter process is the key reaction in the HF chemical laser. The reaction proceeds through a transition state:

$$H_2 + F \rightleftharpoons (HHF)^\ddagger \rightarrow H + HF.$$

Suppose that the activated complex is linear. Some properties of the reactants and the activated complex are given in **Table 3.2**. The threshold energy is estimated to be $\epsilon^* = 6.6$ kJ mol^{-1}.

TABLE 3.2	**Properties in the $F + H_2$ Reaction (Problem 3.13)**		
	H$_2$	**F**	**(HHF)‡**
σ (10^{-19} m^2)	2.7	1.8	—
I (10^{-47} kg m^2)	0.46	—	7.43
ν (10^{13} s^{-1})	13.19	—	12.02
			1.19
			1.19
Electronic degeneracy	1	4	4

a. Use collision theory to calculate the preexponential factor and the rate constant for the reaction at 298 K. Compare your answer to the experimental result of $A = 2 \times 10^{11}$ L mol^{-1} s^{-1}.

b. Use activated complex theory to calculate the rate constant for the reaction at 298 K. Assume that the electronic degeneracy for F and for the activated complex is 4.

3.14 Use activated complex theory to determine the temperature dependence of the Arrhenius preexponential factors for the following three gas-phase reactions

a. $O + N_2 \rightarrow NO + N$

b. $OH + H_2 \rightarrow H_2O + H$

c. $CH_3 + CH_3 \rightarrow C_2H_6$

In other words, for $A \propto T^n$, find n for each equation. You may assume that the activated complex is linear for reaction (a) and nonlinear for reactions (b) and (c). Further, you may assume that $h << kT$ and that the electronic degeneracies are unity.

3.15 A theorem by Tolman states that the activation energy for a reaction is the difference between the average energy of those molecules that react and the average energy of all the molecules: $E_a = <E_r> - <E>$.

a. Show for a general reaction of cross section $\sigma(\epsilon_r)$ that the rate constant is

$$k(T) = 2\pi \left(\frac{1}{\pi kT}\right)^{3/2} \left(\frac{2}{\mu}\right)^{1/2} \int_0^\infty \epsilon_r \sigma(\epsilon_r) \exp\left(-\frac{\epsilon_r}{kT}\right) d\epsilon_r.$$

b. Recalling that $E_a = kT^2 \partial \ln k(T)/\partial T$, prove Tolman's theorem.

3.16 The objective of this problem is to see how isotopic substitution might affect the rate of a chemical reaction. For simplicity, we consider the case when the substitution is made at the atom whose motion is most nearly along the reaction coordinate; the effect on the rate constant is called the *primary kinetic isotope effect*. A simple example might be to compare the dissociation RH → R + H to the dissociation RD → R + D. The origin of the effect is due to the fact that different isotopic species have different zero-point vibrational frequencies, both in the ground state and, to a lesser extent, in the excited state. We recall from quantum mechanics that the vibrational frequency in the harmonic oscillator approximation is $\nu = (1/2\pi)(k_{fc}/\mu)^{1/2}$, where k_{fc} is the force constant for the vibration and μ is the reduced mass of the vibrating masses. An approximation to the primary kinetic isotope effect can be calculated by assuming that there is no difference in zero point energies between the two isotopic species in the activated complex and that both species have the same Arrhenius A factor. The justification for this approximation is that the stretching motion, for example, R–H or R–D, is not bound if it is along the reaction coordinate. (To be sure, there will be some zero-point energy difference for the bending motion in the activated complex, but for this crude approximation we will assume it to be negligible. There will also be some difference in the A factor, but assuming no symmetry number difference, this difference will also be small.) Since the zero-point energy for the hydrogen system, $\frac{1}{2}h\nu_H$, lies higher than the zero-point energy for the deuterium system, $\frac{1}{2}h\nu_D$, the former system will have a lower activation energy than the latter by the difference between these two zero points.

a. Given that the higher of the two ground-state vibrational frequencies is known (as ν_H), derive a formula for the ratio of the two rate constants, k_{light}/k_{heavy}.

b. Typical frequencies for hydrogen stretches are given as follows (in cm^{-1}): C–H; 2900; OH; 3300; NH; 3100; SH; 2600. Calculate k_H/k_D at room temperature for reactions involving motion of the hydrogen or deuterium along the reaction coordinate.

4 Chapter Four

Transport Properties

Chapter Outline

4.1   INTRODUCTION

The goal of this chapter is to understand such properties as thermal conductivity, viscosity, and diffusion on a microscopic level. For gases, we can attain this understanding by application of the kinetic theory developed in the last chapter. Although an exact treatment is mathematically cumbersome, simple physical ideas can be used to derive approximate formulas that have the correct dependence on molecular parameters and differ from the exact formulas only by numerical constants of order unity. Thus, our approach focuses on the underlying physics of the process rather than on obtaining exact results.

The outline of the approach is as follows. After briefly discussing the general functional form of the transport equations, we will make four simplifying assumptions that will enable us to easily apply kinetic theory to transport phenomena in gases. The basic theme is that the properties transported, namely, energy, momentum, or concentration, are carried by the motions of molecules. We know something about this motion from our discussions of the Maxwell-Boltzmann distribution. The first step in a general treatment of transport is to calculate the flux of molecules, i.e., the number of molecules that cross an area per unit time. The second step is to calculate how far the molecules travel in a particular direction between collisions. This distance is clearly related to the mean free path, but it is slightly different.

The third step is to combine these two results to calculate a transport equation for an arbitrary property carried by the gas molecules. We will see that the transport always moves the property in a direction opposite to a gradient, or spatial derivative, in the property, and that the proportionality constant is related to the mean velocity of the molecules, the mean free path, and other properties of the molecules. For gradients that are independent of time, it is then relatively straightforward to apply the general equation in turn to thermal conductivity, where energy is transported; to viscosity, where momentum is transported; and to diffusion, where the molecules themselves are transported. For gradients that are not constant in time, the treatment is somewhat more complex but can again be understood using a simple model, as shown in the final section of this chapter.

4.2   THE FUNCTIONAL FORM OF THE TRANSPORT EQUATIONS

The principal features of all transport equations can be appreciated by considering the flow of a liquid through a tube. **Figure 4.1** displays the important parameters. For the liquid under consideration and for a particular choice of diameter, the tube has an inherent conductivity C. Suppose that a pressure differential $\Delta p = p_2 - p_1 > 0$ is placed across the tube so as to force the liquid to flow from left to right. We expect from common experience that the rate of liquid volume that crosses a unit area oriented perpendicular to the flow will depend linearly on both the pressure differential and the conductivity. The flow of a quantity per unit time per unit area is called the *flux* and has dimensions of (quantity) $s^{-1}\,m^{-2}$. In this case, the quantity is the volume of liquid and the linear proportionalities can be expressed by the equation

$$J_z = -C\frac{\partial p}{\partial z},\qquad (4.1)$$

an equation known as Poiseuille's law. In the example above, $\partial p/\partial z$ is simply $-\Delta p/\ell$ and is called the *gradient* of the pressure. Strictly speaking, since the gradient can have different values in different directions, **equation 4.1** should be written in vector form: $\mathbf{J} = -C\nabla p$, where ∇ is the vector $\mathbf{i}\partial/\partial x + \mathbf{j}\partial/\partial y + \mathbf{k}\partial/\partial z$ and $\mathbf{i}$, $\mathbf{j}$, and $\mathbf{k}$ are unit vectors in the x, y, and z directions, respectively. To keep the notation simple, we will focus on the z component of the flux, while remembering that similar equations can be written for the other directions. Note in **equation 4.1** that the gradient $\partial p/\partial z$ is negative since the pressure decreases as z increases, but that the flux is positive because of the negative sign incorporated in **equation 4.1.**

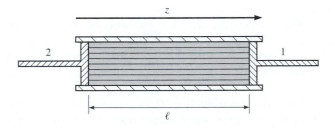

Figure 4.1

The flow of liquid through a tube.

Quantity Transported	Equation	Name
	TABLE 4.1 Transport Equations	
Fluid	$J_z = -C(\partial p/\partial z)$	Poiseuille's law
Heat (thermal conductivity)	$J_z = -\kappa(\partial T/\partial z)$	Fourier's law
Momentum	$J_z = -\eta(\partial v_x/\partial z)$	Viscosity
Particles (diffusion)	$J_z = -D(\partial n^*/\partial z)$	Fick's law
Electrical charge	$J_z = -(1/\rho)(\partial \phi/\partial z)$	Ohm's law

From dimensional analysis we see that the units of the conductivity C are volume $\text{s}^{-1}\,\text{m}^{-1}\,\text{pressure}^{-1}$. While we note here that C is inversely proportional to the viscosity, we defer discussion of the relationship between the conductivity and molecular properties until Section 4.5.

All transport equations have the form of **equation 4.1;** the only differences involve the form of the gradient and the quantity that flows counter to the gradient. The examples of greatest interest are described in **Table 4.1.**

In the case of thermal conductivity, the quantity carried is heat or energy, and it is carried in the direction opposite to the temperature gradient; i.e., heat flows in the positive z direction if the temperature decreases as z increases. The proportionality constant, κ, is called the coefficient of thermal conductivity. Similarly, in the case of diffusion, the quantity carried is the particle itself, and it is carried counter to a density gradient. The proportionality constant, D, is called the diffusion coefficient. Viscosity is at first a bit confusing. The quantity carried is the x component of momentum, but it is carried in the z direction against a gradient of momentum, as discussed in detail in Section 4.5. The proportionality constant is called the viscosity coefficient. Ohm's law concerns the transport of electricity through a conductor against a gradient in electrical potential, as discussed in **Example 4.1.**

example 4.1

Ohm's Law

Objective   Determine Ohm's law for the flux of electrons through a wire, given that its conductivity is $1/\rho$ (ρ is called the resistivity) and that the potential decrease across the wire is V volts per m. Show that the result leads to the common form of Ohm's law: $V = IR$.

Method   The charge flux will have the units of charge per cross-sectional area of the wire per second. It should be proportional to the conductivity of the wire and to the gradient of the electrical potential that pushes the electrons along the wire.

Solution   Since the electrons flow from a region of high potential ϕ to one of low potential, the gradient in the direction of flow is negative; the potential decreases with increasing z, where z is the direction of electron flow. Assuming a linear variation in voltage across the wire, the gradient is thus $\partial\phi/\partial z = -V/\ell$, where ℓ is the length of the wire. The flux of electrons is thus $J_z = -(1/\rho)(\partial\phi/\partial z)$. Writing the flux of electrons as the current, I, per unit area and substituting for the gradient, we obtain $I/A = V/\rho\ell$, or $V = IR$, where $R = \rho\ell/A$ is the resistance of the wire.

Comment   The units of I are amperes (one coulomb of charge per second), while the units of R are ohms. A 1-volt drop in potential across a resistance of one ohm causes a current flow of 1 ampere. The units of the resistivity, ρ, are ohm m.

4.3   THE MICROSCOPIC BASIS FOR THE TRANSPORT LAWS

4.3.1   Simplifying Assumptions

It is clear from **Table 4.1** that the transport laws all have the same basic form, namely that the flux of some quantity is proportional to and in the opposite direction of a gradient. In the case of transport in gases, the explanation of this common form is based on the kinetic theory outlined in Chapter 1. As realized very early by Maxwell and by Boltzmann and later expanded by Enskog and by Chapman,[a] the property transported by the flux must be transported by the individual particles comprising the gas, namely, by molecules subject to the Maxwell-Boltzmann distribution law. In the case of thermal conductivity, the property carried is the energy, $\epsilon = mv^2/2$. In the case of viscosity, the property carried is the momentum mv_x. Diffusion involves the flux of the molecules themselves. While a rigorous theory of transport properties involves both complicated mathematics and physics, the basic form of the answer can be derived from kinetic theory and a few simple assumptions. We will follow this latter route, recognizing that while we expect to capture the basic taste of the argument, the seasoning of our dishes may not be perfect. Most of the equations we derive will show the correct dependence on molecular parameters but will have numerical factors that are not quite correct.

We make the following simplifying assumptions: (1) the molecules behave as rigid spheres with no attractive forces; (2) they all travel with the same speed, equal to the average speed $<v>$, and traverse the same distance, equal to the mean free path λ, between collisions; (3) the molecules taken collectively have an isotropic angular distribution; and (4) each collision results in complete equilibrium with respect to the interchange of the property q which is being transported.

The first assumption is obviously a drastic oversimplification, since we know that it is the forces, both attractive and repulsive, between gas molecules that account for the deviations from ideal gas behavior. The second assumption is more than merely a matter of convenience. It is certainly easier to deal with the average behavior rather than performing each calculation as a function of velocity and finally integrating over the velocity distribution. But this procedure hides the fact that some of the properties we want to transport depend on velocity, $q = q(v)$, so that the rate of transport of this

[a]D. Enskog, *Kungliga Svenska Vetenskapsakademiens Handlingar* **63,** No. 4 (1922) (in German); S. Chapman and T. G. Cowling, *The Mathematical Theory of Non-uniform Gases* (Cambridge University Press, Cambridge, England, 1939).

property is proportional to $vq(v)$. By considering only the average velocity we are in effect replacing $<vq(v)>$ by $<v><q(v)>$, an approximation whose accuracy depends on the exact nature of the distributions. The third assumption turns out to be particularly weak. When molecules collide, they do not completely forget their original direction of motion, so their motion in the presence of a gradient is not likely to be isotropic. The approximation ignores the fact that the gradient affects the velocity distribution. The fourth assumption is likewise a source of error. It may be true for transfer of infinitesimal amounts of the property q per collision, but it will certainly fail when the gradients become large. In view of these approximations, it should be no surprise that the derivations below will introduce incorrect numerical factors. Nonetheless, the essential physical picture is unchanged by these approximations; the property q is carried by molecules whose motions over a wide range of conditions are not too different from those predicted by kinetic theory.

We begin by considering gradients that are stable in time; i.e., gradients that are established by some external means so that the transport of heat, momentum, or concentration does not change the gradient with time. For example, we might hold the ends of a tube of gas at fixed, but different, temperatures by using two large heat baths. Heat would then be transferred through the gas from one bath to another without appreciably changing the gradient.

In the remainder of this section we will first develop an equation for the molecular flux. We will then use this equation to determine two quantities: the flux of a property through a plane and the vertical distance between planes where collisions have occurred. We will finally develop a general flux equation that can be used in subsequent sections to relate the coefficients of thermal conductivity, viscosity, and diffusion to molecular properties such as diameter, speed, heat capacity, and mass.

4.3.2   The Molecular Flux

The first step in a microscopic explanation for transport properties is to recognize that, if molecules are carrying the quantity in question across a unit area in a unit time, we need to know the rate at which the molecules themselves cross the area. **Equation 1.30** and **Figure 4.2** can help in this exercise. How many molecules cross the area A in the indicated plane per unit time? We treat the problem in spherical coordinates,

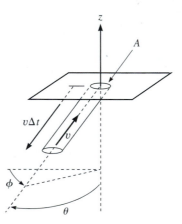

Figure 4.2

The flux of molecules through a plane.

where v ranges from 0 to ∞, θ from 0 to π, and ϕ from 0 to 2π. The relationship between spherical and Cartesian coordinates is discussed in Appendix 1.2. Consider for each possible value of v, θ, and ϕ a cylinder of slant height $v\Delta t$ tilted at angles θ and ϕ with respect to the z axis, where the slant height is chosen so that all molecules within the cylinder with velocities centered on v, θ, and ϕ will cross area A in the time Δt. The volume of the cylinder depends both on the slant height, $v\Delta t$, and on $\cos \theta$: $V = Av\Delta t \cos \theta$. The number of molecules crossing A in Δt is then given simply as the number of molecules in the cylindrical volume times the probability that a molecule will have a velocity v centered on angles θ and ϕ. The number of molecules in the volume is $n^{*}V = n^{*}Av\Delta t \cos \theta$, while the probability of having the given velocity is $(m/2\pi kT)^{3/2} \times \exp(-mv^{2}/2kT)v^{2}\sin \theta\ d\theta\ d\phi\ dv$.[b] Thus,

$$\text{number} = n^{*}Av\Delta t \cos \theta \left(\frac{m}{2\pi kT}\right)^{3/2} \exp\left(-\frac{mv^{2}}{2kT}\right)v^{2}\sin \theta\ d\theta\ d\phi\ dv. \quad \textbf{(4.2)}$$

The number of molecules in the cylinder with velocities centered on (v,θ,ϕ) that cross a unit area of the plane in a unit time is then the flux distribution function:

$$J(v,\theta,\phi)v^{2}\sin \theta\ d\theta\ d\phi\ dv = \frac{\text{number}}{A\Delta t}$$

$$\quad \textbf{(4.3)}$$

$$= n^{*}v^{3}\left(\frac{m}{2\pi kT}\right)^{3/2} \exp\left(-\frac{mv^{2}}{2kT}\right) \cos \theta \sin \theta\ d\theta\ d\phi\ dv.$$

Equation 4.3 is evidently a distribution function giving the probability that a molecule with a speed in the range $v \rightarrow v + dv$ and direction in the range $\theta \rightarrow \theta + d\theta$, $\phi \rightarrow \phi + d\phi$ will pass through the plane in a unit time. We can use it to calculate two important quantities.

We would first like to know the flux of molecules J_{z} that cross the plane from below regardless of their velocity and direction. To find this quantity, we simply need to integrate **equation 4.3** over all the variables, but the range of integration for θ should be from 0 to $\pi/2$ (see **Figure 4.2**) since we want only those molecules moving upward through the plane. Thus,

$$J_{z} = n^{*}\int_{0}^{\infty}\int_{0}^{2\pi}\int_{0}^{\pi/2} v^{3}\left(\frac{m}{2\pi kT}\right)^{3/2} \exp\left(-\frac{mv^{2}}{2kT}\right) \cos \theta \sin \theta\ d\theta\ d\phi\ dv. \quad \textbf{(4.4)}$$

The integration over v (with an additional factor of 4π) was performed in **equation 1.31**: the answer here is simply $1/(4\pi<v>)$. The integration over ϕ gives a factor of 2π. Thus,

$$J_{z} = n^{*}\frac{1}{4\pi}<v>2\pi \int_{0}^{\pi/2} \cos \theta \sin \theta\ d\theta$$

$$\quad \textbf{(4.5)}$$

$$= n^{*}\frac{1}{2}<v>\int_{0}^{\pi/2} \sin \theta\ d(\sin \theta) = n^{*}\frac{1}{2}<v>\left[\frac{\sin^{2}\theta}{2}\right]_{0}^{\pi/2},$$

or

---

[b]Note that integration of the probability over the angles would give a factor of 4π, so that the probability would be identical to that given in **equation 1.31**.

$$J_z = \frac{1}{4} n^* <v>, \tag{4.6}$$

where we recall from Chapter 1 that $<v> = (8kT/\pi m)^{1/2}$, where m is the mass of the gas molecules. This important equation gives the flux of molecules in a particular direction. Of course, for an isotropic gas, the flux of molecules has the same value in any direction, a conclusion that is clear from the fact that the right-hand side of **equation 4.6** does not depend on direction.

Suppose now that the molecules each carry an amount q of some property. Then the flux of that property will be simply the flux of molecules times the amount of the property each carries. In particular, for the $+z$ direction,

$$J_z = \frac{1}{4} <v> n^* q. \tag{4.7}$$

4.3.3   The Vertical Distance between Collisions

The next question we consider is the distance in the z direction traveled by the average molecule between collisions. Although the total average distance is the mean free path λ, the distance in the z direction will be somewhat shorter, since molecules with positive z component velocities move at a variety of angles θ with respect to the z axis. The second result that we will derive from **equation 4.3** is that the average z distance between collision planes is $2\lambda/3$.

Let the slant length of the cylinder in **Figure 4.2** be $v\Delta t = <v>/Z_1 = \lambda$. Then the vertical distance between the indicated plane and the plane in which the molecule last had its collision is $\lambda \cos \theta$. We wish to find the average of this quantity. Since **equation 4.3** gives the flux probability, the average of $\lambda \cos \theta$ will simply be

$$<\lambda \cos \theta> = \frac{\int \lambda \cos \theta \, J(v,\theta,\phi) \, d\tau}{\int J(v,\theta,\phi) \, d\tau} \tag{4.8}$$

$$= \lambda \frac{\int_0^{\pi/2} \cos^2\theta \sin \theta \, d\theta}{\int_0^{\pi/2} \cos \theta \sin \theta \, d\theta} = \lambda \frac{\left[\dfrac{\cos^3\theta}{3}\right]_0^{\pi/2}}{\left[\dfrac{\cos^2\theta}{2}\right]_0^{\pi/2}} = \frac{2}{3}\lambda,$$

where the volume element abbreviated as $d\tau$ is equal to $v^2 \sin \theta \, d\theta \, d\phi \, dv$ and where the integral in the denominator is used for normalization. Consequently, we see that the average distance traveled in the z direction between collisions is $2\lambda/3$.

4.3.4   The General Flux Equation

To calculate the flux of the property q it is convenient to consider a plane located perpendicular to the direction of the gradient. Let the gradient be in the z direction, and let the plane be located at the arbitrary position z_0. As shown schematically in **Figure 4.3,** we calculate the net flux into the plane at z_0 as the flux due to the

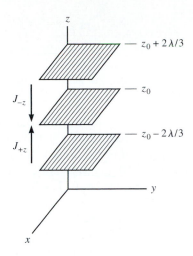

Figure 4.3

Transport between layers separated by the mean free path.

upward motion of molecules that made their last collision in the plane at $z_0 - 2\lambda/3$ and due to the downward motion of molecules that made their last collision in the plane at $z_0 + 2\lambda/3$. From **equation 4.7,** the flux from a plane at z is given by $J_z = \frac{1}{4}n^*q<v>$, where n^*, q, and $<v>$ are evaluated at the position z. While assumption 2 enables us to treat $<v>$ as constant, in principle both n^* and q can vary between planes. Introducing temporarily the notation $\rho_q(z)$ as n^*q evaluated at location z, the upward flux of the property q is then given by $J_{+z} = \frac{1}{4}<v>\rho_q(z_0 - 2\lambda/3)$, where $\rho_q(z_0 - 2\lambda/3)$ is the value of $\rho_q = n^*q$ for upward traveling molecules that had their last collision in the plane at $z_0 - 2\lambda/3$. Similarly, the downward flux of the property is given by $J_{-z} = \frac{1}{4}<v>\rho_q(z_0 + 2\lambda/3)$. The net flux in the upward direction is then

$$J_z = J_{+z} - J_{-z}$$

$$= \frac{1}{4}<v>\left[\rho_q\left(z_0 - \frac{2\lambda}{3}\right) - \rho_q\left(z_0 + \frac{2\lambda}{3}\right)\right].$$

(4.9)

If the gradient is constant or if its change is small over dimensions corresponding to the mean free path, then we may approximate $\rho_q(z_0 \pm 2\lambda/3)$ by the first two terms in a Taylor series expansion about the position z_0:[c]

$$\rho_q\left(z_0 \pm \frac{2\lambda}{3}\right) \approx \rho_q(z_0) \pm \frac{2\lambda}{3}\left(\frac{\partial\rho_q}{\partial z}\right).$$

(4.10)

Substitution of **equation 4.10** into **equation 4.9** and replacement of ρ_q by n^*q yields

[c]The Taylor series expansion for $y(x_0 + \Delta x)$ is given as $y(x_0 + \Delta x) = y(x_0) + \Delta x(dy/dx) + \frac{1}{2}(\Delta x)^2$ $(d^2y/dx^2) + \cdots$, where the derivatives are evaluated at $x = x_0$. If (dy/dx) is nearly constant over the range of Δx, then (d^2y/dx^2) will be small and only the first few terms in the expansion will be needed.

$$J_z = \frac{1}{4}<v>\left[-\frac{4\lambda}{3}\left(\frac{\partial(n^*q)}{\partial z}\right)\right]$$

$$= -\frac{1}{3}<v>\lambda\frac{\partial(n^*q)}{\partial z}. \tag{4.11}$$

Equation 4.11 will form the basis for much of our further discussion. At this point it is worthwhile to make two comments. First, the result does not depend on having a gradient that is independent of position. If the gradient is constant everywhere in space, then **equations 4.10** and **4.11** are exact, but even if the gradient changes as a function of position, **equation 4.11** will give an excellent approximation to the flux through the plane at z_0 as long as the change in the gradient is small over distances within roughly one mean free path of z_0. Second, **equation 4.11** suggests that a nonzero flux will result from either a gradient in the molecular density, n^*, or a gradient in the property q, or both. In our discussions of thermal conductivity and viscosity below, we will assume that there is no net movement of the molecules; that only the property $q = \epsilon$ for thermal conductivity or $q = mv_x$ for viscosity changes with position. In this case, since the number density does not change with position, we see that $\partial(n^*q) = n^*\partial q$. In the case of diffusion, however, the property in flux is the number density itself, so $q = 1$ and $\partial(n^*q) = \partial n^*$.

4.4   THERMAL CONDUCTIVITY

A fundamental observation in the development of the second law of thermodynamics is that heat flows from a hot body to a cold one. The phenomenological description of this flow was discussed in Section 4.2 and is embodied in the equation called Fourier's law: $J_z = -\kappa(\partial T/\partial z)$, where J_z is the flux of heat (energy) in the z direction and κ is the coefficient of thermal conductivity. Since the units of the flux are energy per area per time, we see that Fourier's law has dimensions (J m^{-2} s^{-1}) = κ (K m^{-1}), or that the dimensions of κ are (J m^{-2} s^{-1})/(K m^{-1}) = J m^{-1} s^{-1} K^{-1}.[d] Since 1 J of energy per second is also equal to 1 watt of power, alternative units for κ are W m^{-1} K^{-1}. **Table 4.2** gives some values for κ.

TABLE 4.2	Thermal Conductivity Coefficients, κ, for Various Substances at 273 K and 1 atm
Substance	κ (J m^{-1} s^{-1} K^{-1})
Cu	400
Fe	80
He	0.144
Ar	0.0162
N_2	0.0237
H_2	0.174
O_2	0.0240
CO_2	0.0142
CH_4	0.0300

[d]Kappa, κ, is used here for the thermal conductivity coefficient and should not be confused with the isothermal compressibility coefficient, which sometimes also uses this symbol.

example 4.2

The Heat Flow through Fiberglass Insulation

Objective Calculate the rate of heat loss through a wall insulated with fiberglass. Let the wall be 3 m × 4 m, ignore the conductivity of any other wall materials, and take the thickness of the insulation to be 15 cm, the temperature difference between the inside and outside of the wall to be 10 K, and the coefficient of thermal conductivity for fiberglass to be 5×10^{-2} W m^{-1} K^{-1}.

Method According to Fourier's law, the flux of energy is given by $J_z = -\kappa(\partial T/\partial z)$. The flux is the heat per unit time, so that the total heat loss in watts (joules per second) is the area times the flux: AJ_z.

Solution The gradient is $-(10$ K$)/0.15$ m, so that the total heat loss is $(3$ m × 4 m$)(5 \times 10^{-2}$ W m^{-1} s$^{-1})(10$ K$)/(0.15$ m$) = 40$ W.

Of course, thermal conductivity is not the only method for heat transport. Heat is also transferred by radiation, as from the sun to Earth, or by convection, as in winds that move weather fronts. In our consideration of thermal conductivity, we will separate these processes and analyze the flow of heat (energy) in the absence of net movement of either photons or matter. To be sure, even in conduction the heat is transported by the movement of particles, usually by the motion of molecules, but, in metals, also by the motion of electrons. However, we will assume that there is no *net* molecular motion in conductivity. Thus, the conducted heat moves like the baton in a relay race; it is passed from one particle to another. This view is true, and the macroscopic equations valid, for heat flow through solids, liquids, or gases. In the latter case, however, we can easily come to a microscopic understanding of the coefficient of thermal conductivity.

The kinetic theory that we have developed describes the collisions that provide the opportunity for gases to exchange energy, so that **equation 4.11** should predict the essential features of thermal conductivity in gases, subject to the simplifying assumptions made in the last section. The property transported by the molecules is their energy, ϵ, and by assumption 4, this energy is equilibrated at every collision. If we assume no net motion of the molecules, then $\partial(n^*q) = n^*\partial q$. If the energy per mole is $q = U/N_A = \epsilon$, **equation 4.11** then becomes

$$J_z = -\frac{1}{3}n^*\lambda <v>\left(\frac{\partial \epsilon}{\partial z}\right). \qquad (4.12)$$

Recalling from Chapter 1, Section 1.6, that $(\partial U/\partial T)_v = C_V$, the constant volume molar heat capacity, we write the gradient $(\partial \epsilon/\partial z)$ as

$$\frac{\partial \epsilon}{\partial z} = \frac{1}{N_A}\frac{\partial U}{\partial z}$$

$$= \frac{1}{N_A}\frac{\partial U}{\partial T}\frac{\partial T}{\partial z} \qquad (4.13)$$

$$= \frac{C_V}{N_A}\frac{\partial T}{\partial z}.$$

Thus,

$$J_z = -\frac{1}{3}n^*\lambda <v> \frac{C_V}{N_A}\left(\frac{\partial T}{\partial z}\right).$$

(4.14)

Comparison of **equation 4.14** with Fourier's law yields

$$\kappa = \frac{1}{3}n^*\lambda <v> \frac{C_V}{N_A}.$$

(4.15)

This expression for the thermal conductivity coefficient may be simplified by using **equation 1.47,** repeated here for use with a single component so that $n_2^* = n^*$ and $b_{max} = d$, the molecular diameter:

$$\lambda = \frac{1}{\sqrt{2}\pi d^2 n^*}.$$

(4.16)

Substitution of **equation 4.16** into **equation 4.15** yields

$$\kappa = \frac{<v>C_V}{3\sqrt{2}\pi d^2 N_A}.$$

(4.17)

Note that the result for κ is independent of pressure because the n^* dependence in the general expression for the flux and the $1/n^*$ dependence of the mean free path exactly cancel one another. A qualitative explanation for the cancellation is that, while there are fewer molecules crossing a given area per unit time at low pressure, they travel a longer distance between collisions. Although it is found experimentally that κ is independent of pressure over most pressures of interest, this independence breaks down at very high pressures where the molecules no longer behave like an ideal gas and at very low pressures where the mean free path reaches macroscopic dimensions. In the latter case, truncation of the Taylor expansion used in **equation 4.10** is no longer valid after two terms. At extremely low pressures, the mean free path is limited only by collision at the cold surface or the hot surface, and the thermal conductivity coefficient is then directly proportional to n^*.

It is important to comment that the heat capacity for a real molecule is larger than that for a monatomic ideal gas: $C_V > 3R/2$. The reason, of course, is that real molecules have rotational and vibrational degrees of freedom in addition to translational ones. While many vibrational motions are of high enough frequency not to contribute to the heat capacity, the rotational degrees of freedom contribute R per mole for diatomic molecules and $3R/2$ per mole for polyatomic ones.

example 4.3

The Thermal Conductivity Coefficient of N$_2$ at 273 K and 1 atm

Objective   Estimate the thermal conductivity coefficient of N$_2$ at 1 atm and 273 K, given that the molecular diameter of N$_2$ is 370 pm.

Method   Use **equation 4.17** recalling that $C_V \approx 5R/2$.

Solution   First calculate $<v> = (8kT/\pi m)^{1/2}$:

$$<v> = (8kT/\pi m)^{1/2}$$

$$= \left\{ \frac{8(1.38 \times 10^{-23} \text{ J K}^{-1})(273 \text{ K})(6.02 \times 10^{23} \text{ amu/g})(1000 \text{ g/kg})}{[3.1415 \, (28 \text{ amu})]} \right\}^{1/2}$$

$$= 454 \text{ m/s} \tag{4.18}$$

Then evaluate $C_V = 5R/2 = 5(8.314 \text{ J mol}^{-1} \text{ K}^{-1})/2 = 20.8$ J mol^{-1} K^{-1}.
Finally,

$$\kappa = \frac{(454 \text{ m/s})(20.8 \text{ J/mol})}{3\sqrt{2}\pi(6.02 \times 10^{23} \text{ molec/mol})(370 \times 10^{-12} \text{ m})^2} \tag{4.19}$$

$$= 8.59 \times 10^{-3} \text{ J m}^{-1} \text{ s}^{-1} \text{ K}^{-1}.$$

Comment   Note that the information that the pressure is 1 atm is irrelevant to the solution. Since we used approximations in arriving at **equation 4.17** and in evaluating the heat capacity, the answer is not exactly equal to the measured value listed in **Table 4.2.**

4.5   VISCOSITY

Most people are familiar with the viscous drag of water impeding a swimmer or air impeding a plane. What are the causes of these forces and how can we understand them at a molecular level? Consider two plates of area A separated by a distance in the z direction and immersed in a fluid, as shown in **Figure 4.4.** If the upper plate is drawn through the fluid with a velocity v_x while the lower plate is stationary, then there will be a force exerted on the lower plate in the x direction due to the frictional

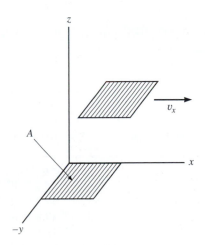

■ **Figure 4.4**

The viscous force on a stationary plate exerted by a moving one.

drag of the fluid; an equivalent force in the negative x direction will have to be applied to hold the lower plate stationary. The force transmitted downward to the stationary plate will be proportional to the area A and is given by Newton's law of viscosity: $F_x = -\eta A(\partial v_x/\partial z)$, where the constant η is called the coefficient of viscosity. From Newton's law we recall that $F = ma = dp/dt$, where p is the momentum, so that the force transferred per unit area is the same as momentum transferred per unit time per unit area, or momentum flux. Thus, the gradient in velocity (or the proportional gradient in momentum) between the two plates causes a flux of momentum that is transmitted by the fluid. Note that, while the momentum and force are in the x direction, the *flux* of momentum is in the z direction: $J_z = F_x/A = -\eta(\partial v_x/\partial z)$.

The units of the flux are momentum per second per area or, equivalently, force per area, so that the flux equation has dimensions (force/area) $= \eta$ (distance/time)/distance. Thus, the units of η are (force/area)/(1/time) or N m^{-2} s. A pascal of pressure is also a N m^{-2}, so that equivalent units for η are Pa s. In older texts, one often encounters the cgs unit for η called a poise; 1 poise = 1 dyne cm^{-2} s = 1 gm cm^{-1} s^{-1} = 0.1 N m^{-2} s.

Table 4.3 provides some viscosity coefficient data for a few materials. Note that both liquids and gases obey the macroscopic viscosity equation. We will focus first on gases and return to the frictional forces in liquids later in this chapter.

In the case of gases and under the assumptions listed in Section 4.3.1, the transfer of momentum must be described by the general flux **equation 4.11,** with $q = mv_x$. We again assume that there is no net transport of molecules, so that $\partial(n^*q) = n^*\partial q$. Substitution of $q = mv_x$ leads to

$$J_z = -\frac{1}{3}n^*<v>\lambda m\left(\frac{\partial v_x}{\partial z}\right),$$ (4.20)

so that

$$\eta = \frac{1}{3}n^*<v>\lambda m.$$ (4.21)

Note that since λ is proportional to $1/n^*$, the viscosity coefficient will be independent of pressure. This prediction was one of the early triumphs of the kinetic theory

TABLE 4.3   Viscosity Coefficients at 273 K for Various Substances

Substance	η (Pa s)
Glycerol	0.95
Olive oil	0.08
Water, liquid, 298 K	0.9×10^{-3}
He	18.8×10^{-6}
H_2	8.4×10^{-6}
Ar	22.2×10^{-6}
O_2	19.2×10^{-6}
CO_2	13.8×10^{-6}
N_2	16.6×10^{-6}
NH_3	9.2×10^{-6}
CH_4	10.3×10^{-6}

of gases. An alternative formulation of **equation 4.21** recognizes that the density ρ is equal to the product of the number density n^* and the mass m, so that

$$\eta = \frac{1}{3}\rho\langle v\rangle\lambda. \tag{4.22}$$

Again, while the numerical factors in these equations are incorrect, the functional form is correct. However, little is gained by using the correct hard-sphere numbers since real molecules do not behave like hard spheres.

Equations **4.21** and **4.22** provide a convenient method for estimation of molecular diameters. Substitution of **equation 4.16** into **equation 4.21,** for example, leads to

$$\eta = \frac{\dfrac{1}{3}n^*\langle v\rangle m}{\sqrt{2}\pi d^2 n^*}, \tag{4.23}$$

or

$$d = \left(\frac{\langle v\rangle m}{3\sqrt{2}\pi\eta}\right)^{1/2}. \tag{4.24}$$

Example 4.4 illustrates this calculation.

example 4.4

Finding Molecular Diameters from Viscosity Coefficients

Objective   Given that the viscosity coefficient for argon at 298 K is 22.2×10^{-6} Pa s, calculate its molecular diameter.

Method   Use **equation 4.24** after calculating $\langle v\rangle$.

Solution   First, calculate $\langle v\rangle = (8kT/\pi m)^{1/2}$:

$$\langle v\rangle = (8kT/\pi m)^{1/2}$$

$$= \left[\frac{8(1.38 \times 10^{-23}\text{ J K}^{-1})(298\text{ K})(6.02 \times 10^{23}\text{ amu/g})(1000\text{ g/kg})}{(\pi 40\text{ amu})}\right]^{1/2}$$

$$= 397\text{ m/s}. \tag{4.25}$$

Then calculate d:

$$d = \left[\frac{\langle v\rangle m}{(3\sqrt{2}\pi\eta)}\right]^{1/2}$$

$$= \left[\frac{(397\text{ m/s})(40\text{ amu})}{3\sqrt{2}\pi(6.02 \times 10^{23}\text{ amu/g})(1000\text{ g/kg})(22.2 \times 10^{-6}\text{ N s m}^{-2})}\right]^{1/2}$$

$$= 299\text{ pm}. \tag{4.26}$$

Although there are other methods for measuring the viscosity coefficient of a fluid, one convenient technique is to determine the volume of the fluid that passes by a unit area of a tubing per unit time; i.e., the volume flux. We have already seen in the opening section of this chapter that this flux is proportional to the product of the pressure gradient and a conductivity coefficient: $J_z = -C(\partial p/\partial z)$, where C depends on the nature of the fluid and the size of the tubing. The dependence on the nature of the fluid comes about because, while the fluid has a finite velocity in the center of the tube, the molecules in contact with the edges of the tube must have zero velocity. Consequently, C should be inversely proportional to the viscosity coefficient of the fluid. A detailed calculation shows that $C = a^2/8\eta$, where a is the radius of the tube. The volume of liquid passing through the tube per unit time is given simply by the volume flux times the area of the tube: $J_z A = dV/dt = -CA(\partial p/\partial z) = -(\pi a^4/8\eta)(\partial p/\partial z)$. This last expression, whose complete derivation is given in Appendix 4.1, is perhaps the most useful form of the *Poiseuille formula* describing the laminar flow of a liquid:

$$\frac{dV}{dt} = -\frac{\pi a^4}{8\eta}\left(\frac{\partial p}{\partial z}\right). \tag{4.27}$$

It enables determination of the viscosity coefficient from a measurement of the rate of volume change. An alternative form of the Poiseuille formula is obtained by multiplying both sides of **equation 4.27** by the liquid's density, ρ:

$$\frac{dm}{dt} = -\frac{\pi a^4 \rho}{8\eta}\left(\frac{\partial p}{\partial z}\right). \tag{4.28}$$

example 4.5

Using the Poiseuille Formula

Objective   Find the viscosity coefficient of a liquid flowing through a tube 0.1 cm in radius and 50 cm in length. When the pressure drop across the tube is 0.1 atm, the volume of liquid emerging from the tube is 1 cm³/s.

Method   Since we know the flow rate and the pressure gradient, we can use Poiseuille's formula, **equation 4.27,** to calculate the viscosity coefficient.

Solution   The flow rate is $dV/dt = 1$ cm³ s⁻¹ = 10⁻⁶ m³ s⁻¹. The pressure gradient is (0.1 atm)/(0.50 m) = 0.2 atm m⁻¹. Then from the Poiseuille formula, η is

$$\eta = \frac{(\pi r^4/8)(\partial p/\partial z)}{(dV/dt)}$$

$$= \frac{\pi(0.001 \text{ m})^4(0.2 \text{ atm/m})(101.3 \times 10^3 \text{ Pa/1 atm})}{8(10^{-6} \text{ m}^3 \text{ s}^{-1})} \tag{4.29}$$

$$= 7.96 \times 10^{-3} \text{ Pa s}.$$

The liquid might very well be olive oil (See **Table 4.3**).

While **equations 4.27** and **4.28** are useful for liquids, where the density is rather insensitive to pressure, for gases the density changes dramatically with pressure: $\rho = Mp/RT$, where M is the molecular weight of the gas. Substitution for ρ and recognition that $n = m/M$ gives

$$\frac{dn}{dt} = -\frac{\pi a^4}{8\eta RT}\left(p\frac{\partial p}{\partial z}\right). \tag{4.30}$$

Because the number of moles of gas crossing any area per unit time, dn/dt, is constant, it must also be true that $p(dp/dz)$ is a constant. If we call the constant B, then $p\,dp = B\,dz$. Integration gives $p^2 = 2Bz + C$. Applying this equation to pressures p_1 at z_1 and p_2 at z_2 yields two equations: $p_1^2 = 2Bz_1 + C$ and $p_2^2 = 2Bz_2 + C$, where C is a constant of integration. Subtraction gives $B = (p_2^2 - p_1^2)/2(z_2 - z_1)$, so that for gases

$$\frac{dn}{dt} = -\frac{\pi r^4}{16\eta RT}\left(\frac{p_2^2 - p_1^2}{z_2 - z_1}\right). \tag{4.31}$$

4.6   DIFFUSION

Anyone whose nose is in working order can attest to the fact that diffusion is an important process. The kitchen smells that woke us in the morning as children or the fragrance from an opened bottle of perfume reach us even if there are no convective currents in a room. The mixing process is spontaneous, but the rate of interdiffusion of two substances has yet to be discussed. Experimental observation shows that diffusion in fluids against a gradient obeys Fick's law, whose form is by now quite familiar: $J_z = -D(\partial n^*/\partial z)$, where J_z is the flux of molecules, $(\partial n^*/\partial z)$ is the gradient in number density, and D is the diffusion coefficient. Dimensionally, the equation is (number time^{-1} area^{-1}) = $D \times$ (number/volume)/distance, so that the dimensions of D are thus distance^2 per unit time, or m^2 s^{-1}. Because the diffusion of one substance into another can depend on the properties of each substance, it will be useful to add subscripts to D. Let D_{12} be the coefficient describing the diffusion of type 1 into molecules of type 2 and let D_{11} be the diffusion coefficient for diffusion of molecules of type 1 into other molecules of the same type. One might well wonder how the latter coefficient could be measured; indeed, it cannot. But D_{11} can be approached quite closely by studying the diffusion of one isotope of a substance in another isotope of the same substance. **Table 4.4** lists some typical diffusion coefficients.

TABLE 4.4	Diffusion Coefficients at 273 K and 1 atm for Various Substances	
Substances	**D_{11} or D_{12} (m^2 s^{-1})**	
H_2–H_2	1.5×10^{-4}	
O_2–O_2	1.9×10^{-5}	
N_2–N_2	1.5×10^{-5}	
CO_2–CO_2	1.0×10^{-5}	
Xe–Xe	5.0×10^{-6}	
O_2–N_2	1.8×10^{-5}	
O_2–CO_2	1.4×10^{-5}	

example 4.6

The Number of O_2 Molecules Crossing an Area per Second While Diffusing through N_2

Objective   Find the number of O_2 molecules diffusing through N_2 molecules and crossing a 0.2 m^2 area at 273 K if the concentration gradient is 40 torr per centimeter and the diffusion coefficient is that given in **Table 4.4**.

Method   Use the diffusion equation, $J_z = -D(\partial n^*/\partial z)$, to calculate the flux. The number crossing the given area is then the flux times the area.

Solution   The diffusion coefficient for O_2–N_2 is $D = 1.8 \times 10^{-5}$ m^2 s^{-1}. Since $p = nRT/V$, we can convert the pressure gradient to a number density gradient by dividing the pressure by RT. The gradient is thus calculated to be

$$\frac{\partial n^*}{\partial z}$$

$$= \frac{(40 \text{ torr/cm})(1 \text{ atm}/760 \text{ torr})(100 \text{ cm/m})(6.02 \times 10^{23} \text{ molecules mole}^{-1})(10^3 \text{ L}/1 \text{ m}^3)}{(0.082 \text{ L atm mole}^{-1} \text{ K}^{-1})(273 \text{ K})}$$

$$= 1.42 \times 10^{26} \text{ (molecules/m}^3) \text{ m}^{-1}. \tag{4.32}$$

Thus, the flux is

$$J_z = -(1.8 \times 10^{-5} \text{ m}^2 \text{ s}^{-1})[1.42 \times 10^{26} \text{ (molecules/m}^3) \text{ m}^{-1}]$$

$$= -2.55 \times 10^{21} \text{ molecules s}^{-1} \text{ m}^{-2}. \tag{4.33}$$

The number crossing the area of 0.2 m^2 per unit time is then the flux times the area, or $(0.2 \text{ m}^2)(2.55 \times 10^{21} \text{ molecules s}^{-1} \text{ m}^{-2}) = 5.10 \times 10^{20}$ molecules/s.

Like thermal conductivity and viscosity, diffusion in gases can be understood by the application of kinetic theory. In this case, however, we must focus on the motion of the molecules themselves. Because their number density changes with position we cannot bring n^* out of the differential $\partial(n^*q)$ in **equation 4.11,** and because it is the molecules themselves that are being transported, $q = 1$ and the flux is the flux of molecules. **Equation 4.11** then becomes

$$J_z = -\frac{1}{3}\langle v \rangle \lambda \left(\frac{\partial n^*}{\partial z}\right), \tag{4.34}$$

and comparison with Fick's first law, $J_z = -D(\partial n^*/\partial z)$, shows that

$$D = \frac{1}{3}\langle v \rangle \lambda. \tag{4.35}$$

As might be expected from the severity of the approximations made in Section 4.3.1, the numerical factor in **equation 4.35** is incorrect, even for hard spheres, but our understanding of the underlying science is enhanced little by correcting it. It is worth noting, however, that the value of the mean free path depends on whether we are considering the diffusion of a molecule of type 1 into other molecules of type 1 or into molecules of another type, 2. For so-called *self-diffusion,* the mean free path is given by **equation 4.16,** since this equation describes how far a molecule travels before colliding with another of the same type. For one molecule of type 1 diffusing through molecules of type 2, however, we must review the derivation of presented in Section 1.7 just prior to **equation 1.47.** If the mean free path for a type 1 molecule in molecules of the same type is $\lambda = \bar{c}/Z_1$, then the mean free path for a type 1 molecule in molecules of type 2 should be $\lambda = \bar{c}/Z_2 = \bar{c}/[\pi b_{max}^2 v_r n_2^*]$. Note that although the calculation of $\bar{c}$ involves the mass of molecules of type 1, the calculation of v_r involves the reduced mass. Furthermore, b_{max} is the average of the diameters of molecules of type 1 and 2. Thus, the mean free path will depend on the properties of both types of molecules. In real systems, molecules of type 1 will diffuse both through others of the same type and through those of type 2, so that the mean free path is somewhat more complicated than that in either of the above calculations; it depends inversely on the total number density, not just on the number density of type 2 molecules.

example 4.7

Calculating the Diffusion Coefficient for N_2

Objective   Approximate the diffusion coefficient of N_2 in N_2 at 300 K and 1 atm given that the molecular diameter is 218 pm (see Example 1.7).

Method   Use **equation 4.35,** noting that under these conditions we have calculated in Example 1.7 the mean free path of N_2 as 3.87×10^{-7} m and the average velocity as 673 m/s.

Solution $D = (1/3)v\lambda = (1/3)(673 \text{ m/s})(3.87 \times 10^{-7} \text{ m}) = 8.68 \times 10^{-5}$ $m^2 \text{ s}^{-1}$.

Comment   That this value is higher than that listed in **Table 4.4** is due only partly to the fact that v and λ are higher at 300 K than at 273 K. Because we have made several simplifying approximations, **equation 4.35** is not expected to be numerically accurate.

4.7 TIME-DEPENDENT TRANSPORT

We have assumed in the preceding sections that the gradient of temperature, momentum, or concentration was steady in time. For example, in the case of diffusion we see from $J_z = -D(\partial n^*/\partial z)$ that if the gradient of concentration is steady in time then the flux of particles will also be steady. We now address the situation in which the gradient changes in time, as it might, for example, if a drop of one material were introduced into another or if heat were momentarily applied to one end of a rod of conductive material. In both cases we see that the gradient is large immediately after the perturbation, but that the diffusion of molecules or the flow of heat tends to cause the gradient to diminish as time progresses.

To be able to describe these processes, we introduce a notation that recognizes that the flux can depend both on position and on time, $J = J(z,t)$. The time dependent flux can be related to the gradient by considering two surfaces of area A separated by a distance Δz, as shown in **Figure 4.5**. Suppose that molecules are diffusing in the positive z direction. What is the change of concentration in the volume $A\Delta z$ per unit time? The concentration is increased by the number of molecules that flow into the volume from below. Because $J(z,t)$ is the number of molecules per unit time per unit area that cross the plane located at z, the change in concentration is given by $J(z,t)$ times A divided by the volume: $\partial n^*(z,t)/\partial t = J(z,t)A/A\Delta z = J(z,t)/\Delta z$, where the dependence of n^* on z and t is made clear by the notation $n^*(z,t)$. Similarly, the concentration is decreased by the molecules that flow out of the volume to regions above; the change is given by $\partial n^*(z,t)/\partial t = -J(z + \Delta z, t)A/A\Delta z = -J(z + \Delta z, t)/\Delta z$. Thus, the net rate of concentration change is

$$\frac{\partial n^*(z,t)}{\partial t} = \frac{J(z,t) - J(z + \Delta z, t)}{\Delta z} \tag{4.36}$$

In the limit when Δz is very small, the quantity on the right-hand side of **equation 4.36** is simply $-\partial J(z,t)/\partial z$, so that

$$\frac{\partial n^*(z,t)}{\partial t} = -\frac{\partial J(z,t)}{\partial z}. \tag{4.37}$$

At any time t, however, the flux is related to the number density gradient, as we have seen in the previous section:

$$J(z,t) = -D\frac{\partial n^*(z,t)}{\partial z}. \tag{4.38}$$

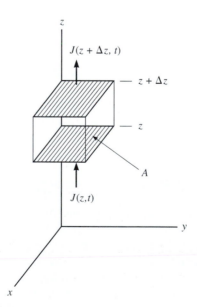

■   **Figure 4.5**

The change in flux with time.

If we take the partial derivative of both sides of **equation 4.38** with respect to z, we obtain

$$\frac{\partial J(z,t)}{\partial z} = -D\frac{\partial^2}{\partial z^2}n^*(z,t). \tag{4.39}$$

Finally, using **equation 4.37** we see that **equation 4.39** can be rewritten as

$$\frac{\partial n^*(z,t)}{\partial t} = D\frac{\partial^2}{\partial z^2}n^*(z,t). \tag{4.40}$$

Equation 4.40 is known as the *time-dependent diffusion equation* or as Fick's second law.

Consider the diffusion of N molecules that start at $z = z_0$ at $t = 0$ in the cross-sectional area A of a tube of infinite length. How will this distribution change in space as a function of time? The solution to **equation 4.40,** as shown in Problem 4.16, is

$$n^*(z,t) = \frac{N}{A}\frac{1}{2\sqrt{\pi Dt}}\ \exp\left[-\frac{(z-z_0)^2}{4Dt}\right]. \tag{4.41}$$

Figure 4.6 displays the concentration profile predicted by **equation 4.41** for different values of Dt. With increasing time, the concentration spreads over larger distances. In fact, if we normalize the right-hand side of **equation 4.41** (which amounts to multiplication by A/N) we will obtain a function that gives the probability that a molecule will be found at a position z at a time t. This function is thus a distribution function for the position at a particular time, and we can use it to calculate average positions. Of course, because the distribution function is symmetric around z_0, the average distance that a molecule has traveled from that position after a time t is zero.

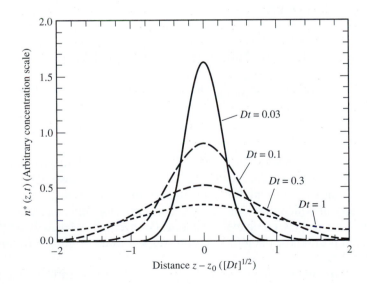

Figure 4.6

Plot of $n^*(z,t)$ following diffusion from a starting condition where all molecules are at $z = 0$ at $t = 0$.

It is useful, however, to calculate the root-mean-squared distance, $z_{rms} = <(z - z_0)^2>^{1/2}$. Using **Table 1.1** to evaluate the integral, this distance is given by

$$[z_{rms}]^2 = \int_{-\infty}^{\infty} (z - z_0)^2 \frac{1}{2\sqrt{\pi Dt}} \exp\left[-\frac{(z - z_0)^2}{4Dt}\right] dz$$

$$= \frac{1}{2\sqrt{\pi Dt}} 2\frac{1}{4} \sqrt{\pi}(4Dt)^{3/2} \qquad (4.42)$$

$$= 2Dt,$$

so that

$$z_{rms} = (2Dt)^{1/2}. \qquad (4.43)$$

We thus see that the root-mean-squared distance that a molecule diffuses is proportional to the square root of the diffusion coefficient and to the square root of the time.

example 4.8

The rms Distance Traveled by a Molecule in a Day

Objective   Find the rms distance that a molecule of naphthalene travels by diffusion in 1 day through the atmosphere assuming the diffusion coefficient is 1.5×10^{-6} m² s⁻¹. Naphthalene is the principal component in moth balls.

Method   Use **example 4.43,** but recognize that this is a three-dimensional problem and not merely a one-dimensional one.

Solution   Note that the square of the distance from the center of a three-dimensional object is $r^2 = x^2 + y^2 + z^2$, so that $<r^2> = <x^2> + <y^2> + <z^2> = 3<z^2>$. One day is (24 hr)(60 min/hr)(60 s/min) $= 8.64 \times 10^4$ s. Thus $(z_{rms})^2 = (2Dt)$ or $(r_{rms})^2 = (6Dt) = [6(1.5 \times 10^{-6}$ m² s⁻¹)(8.64 × 10⁴ s)] $= 0.78$ m² or $r_{rms} = 0.88$ m.

Comment   Note that molecules do not travel far in a day by diffusion. Convection is more often the mode of transport.

We can gain some physical insight into diffusion by considering a process known as the *one-dimensional random walk.* Consider a molecule constrained to move in the z direction in steps of length ℓ, and suppose that after each step the molecule has no memory of which direction it traveled in previous steps; its choice of direction for the next step is completely random. On average, what will be the root-mean-squared position of the molecule with respect to its original position after it has taken N steps?

While this problem can be solved mathematically in closed form, the solution is somewhat complex (see Problem 4.18). It is far easier to write a simple computer program to predict the position. Given a position of z_i after the ith step, the position after the $(i + 1)$th step is given by

$$z_{i+1} = z_i + \ell \ \text{sign}[\text{RND}(\ ) - 0.5], \qquad (4.44)$$

where RND( ) is a random number between 0 and 1 and sign[ ] is a function that is equal to $+1$ if the argument is nonnegative and -1 otherwise.

Figure 4.7 displays the results of six random walks starting at a position z_0. Note that the positions of the particles spread out with increasing number of steps. If we run, say, 1000 trajectories we can compute an accurate average for the root-mean-squared displacement from z_0 as a function of the number of steps. This average for a typical calculation is shown in **Figure 4.8,** which demonstrates that the root-mean-squared displacement in units of ℓ is equal to the square root of the number of steps.

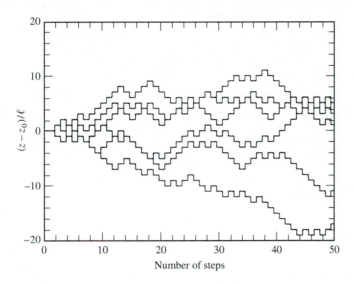

▇ Figure 4.7

Random walks: the position as a function of the number of steps for six one-dimensional random walks.

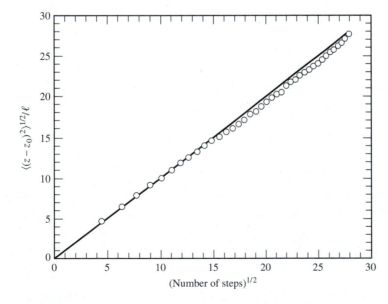

▇ Figure 4.8

Root-mean-squared distance traveled as a function of the square root of the number of steps for a one-dimensional random walk.

What we learn from this computer experiment (or from the more rigorous closed-form solution derived in Problem 4.18) is that molecular diffusion is just like a random walk in one dimension. Let the total time for N steps be equal to N times the average time per step, τ: $t = N\tau$ or $N = t/\tau$. The observation from our computer experiment is that $(z_{rms})^2 = N\ell^2$, or $(z_{rms})^2 = \ell(\ell/\tau)t$. The expression

$$z_{rms} = (\ell^2 t/\tau)^{1/2} \tag{4.45}$$

is known as the *Einstein-Smoluchowski* equation.[e] We can interpret it as follows. Note that $\ell/\tau = <v>$, the average velocity, so that $z_{rms} = [\ell<v>t]^{1/2}$. If we take the step size in the z direction to be that calculated in **equation 4.8**, we find that $z_{rms} = [(2/3)\lambda<v>t]^{1/2}$, or, using **equation 4.35**, $z_{rms} = [2Dt]^{1/2}$. This last equation is exactly what we have calculated in **equation 4.43.**

In retrospect, it should come as no surprise that the one-dimensional random walk agrees with our diffusion calculation. Assumption 4 in Section 4.3.1 made the approximation that complete equilibrium is attained after every collision. When applied to the motion of molecules, this assumption means that there should be no preferential direction for the velocity after any collision. Thus, the assumption that leads to **equation 4.43** and the assumption of a random walk are equivalent. Both are slightly in error when compared to the real situation, but both capture the essential physical situation.

4.8   SUMMARY

By assuming that the motion of molecules is responsible for the transport of properties such as heat, momentum, and concentration in gases, we have found how the constants κ, η, and D in the flux equations for these properties depend on microscopic molecular properties. The relationships were found by making four simplifying assumptions in Section 4.3.1 and by treating the motion of molecules using the kinetic theory developed in Chapter 1. We found that the flux of molecules across a surface is given by

$$J_z = \frac{1}{4} n^* <v> \tag{4.6}$$

and that the average vertical distance between collisions is $2\lambda/3$. Armed with these equations, we showed that when the gradient is constant in time, the flux of a property q in the vertical direction is given by

$$J_z = -\frac{1}{3}<v>\lambda \frac{\partial(n^* q)}{\partial z}. \tag{4.11}$$

Use of this equation with q equal to ϵ, p_x, or 1 gave equations for the following coefficients:

Thermal Conductivity:

$$\kappa = \frac{1}{3} n^* \lambda <v> \frac{C_V}{N_A} = \frac{<v>C_V}{3\sqrt{2}\pi d^2 N_A}, \tag{4.17}$$

[e]A. Einstein, *Ann. d. Physik* **17**, 549 (1905); **19**, 371 (1906); M. v. Smoluchowski, *Ann. d. Physik* **21**, 756 (1906).

Viscosity:

$$\eta = \frac{1}{3} n^* <v> \lambda m, \qquad (4.21)$$

and *Diffusion:*

$$D = \frac{1}{3} <v> \lambda. \qquad (4.35)$$

It is important to remember that, although these equations capture the essential features of transport properties, the numerical coefficients are not quite correct. Those seeking more accurate formulas are referred to one of the texts listed in the reading list at the end of this chapter.

When the gradient is not constant in time we found, using diffusion as an example, that the derivative of the quantity with time was proportional to the second derivative of the quantity in space:

$$\frac{\partial n^*(z,t)}{\partial t} = D \frac{\partial^2}{\partial z^2} n^*(z,t). \qquad (4.40)$$

For a starting condition in which all the molecules have a specified z component at time zero, the root-mean-squared distance traveled as a function of time is given by

$$z_{rms} = (2Dt)^{1/2}. \qquad (4.43)$$

Diffusion of molecules in a gas is analogous to a one-dimensional random walk.

appendix 4.1

The Poiseuille Formula

Consider the flow of a fluid through a cylindrical tube of radius a whose axis is coincident with the x direction and which is subject to a pressure gradient along its length. The velocity of the fluid will be a function of the radial position r from the center of the tube. Molecules at $r = a$ will be in contact with the surface of the tube and will have zero velocity in the x direction, while those in the center of the tube at $r = 0$ will have the largest velocity. The volume V of fluid passing a cross-sectional area of the tube per unit time is given by integrating the area of coaxially concentric shells of thickness dr times the velocity in each shell:

$$\frac{dV}{dt} = \int_0^a v_x(r)(2\pi r)\,dr, \qquad (4.46)$$

where $2\pi r\,dr$ is the area of the shell and $v_x(r)$ is the velocity in the x direction as a function of r. To perform the integration, we first need to determine $v_x(r)$.

To evaluate the radial dependence of the velocity, consider a small cylindrical volume element of the fluid coaxial with the x axis, as shown in **Figure 4.9.** The cross-sectional area of the cylinder is πr^2, and the length is dx. The pressure on the left side of the volume is p, while that on the right side is $p - dp$. When the pressure differential is constant in time, the velocity of the fluid through the cylinder will be constant; its acceleration will be zero. From Newton's law, zero acceleration

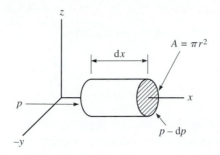

■   **Figure 4.9**

The force due to the pressure differential is equal and opposite to the force due to the viscous drag.

means that the total force on the fluid is zero. A fluid in the volume will thus accelerate its flow in the $+x$ direction until the force in the $-x$ direction due to its viscous drag is exactly equal to the force due to the pressure differential. The force due to the pressure differential is the area times dp: $F_{+x} = \pi r^2\, dp$. The force in the $-x$ direction can be calculated from the flux of momentum in the r direction, $J_r = -\eta(\partial v_x/\partial r)$, so that $F_{-x} = J_r A = -\eta(\partial v_x/\partial r)2\pi r\, dx$, where $A = 2\pi r\, dx$ is the surface area of the outside of the cylinder. Thus

$$\pi r^2\, dp = -\eta 2\pi r dx \frac{\partial v_x}{\partial r},$$

$$\frac{\partial v_x}{\partial r} = -\frac{r}{2\eta}\frac{dp}{dx}. \tag{4.47}$$

This equation can be integrated to give

$$v_x = -\frac{r^2}{4\eta}\frac{dp}{dx} + C, \tag{4.48}$$

where C, the constant of integration, can be evaluated by the boundary condition that the velocity is zero at the wall of the cylinder: $v_x(a) = 0$. The result is

$$v_x = -\frac{a^2 - r^2}{4\eta}\frac{dp}{dx}. \tag{4.49}$$

Figure 4.10 shows the velocity distribution predicted by **equation 4.49.**

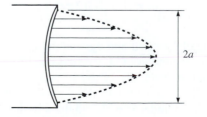

■   **Figure 4.10**

The velocity distribution of a fluid in a cylindrical tube.

We now substitute **equation 4.49** into **equation 4.46** and integrate:

$$\frac{dV}{dt} = \int_0^a \frac{a^2 - r^2}{4\eta} \frac{dp}{dx}(2\pi r)\,dr$$

$$= \frac{2\pi}{4\eta} \frac{dp}{dx} \left[\frac{a^2 r^2}{2} - \frac{r^4}{4} \right]_0^a \qquad \textbf{(4.50)}$$

$$= \frac{\pi a^4}{8\eta} \frac{dp}{dx}.$$

This last equation is simply the Poiseuille formula given in **equation 4.27** with the pressure gradient in the x direction rather than the z direction.

suggested readings

R. S. Berry, S. A. Rice, and J. Ross, *Physical Chemistry* (Wiley, New York, 1980).

J. O. Hirschfelder, C. F. Curtiss, and R. B. Bird, *Molecular Theory of Gases and Liquids* (Wiley, New York, 1954).

W. Kauzmann, *Kinetic Theory of Gases* (W. A. Benjamin, New York, 1967).

E. H. Kennard, *Kinetic Theory of Gases* (McGraw-Hill, New York, 1938).

L. B. Loeb, *The Kinetic Theory of Gases* 3rd ed. (Dover, New York, 1961).

F. R. W. McCourt, J. J. M. Beenakker, W. E. Köhler, and I. Kuscer, *Non-equilibrium Phenomena in Poly-atomic Gases* (Clarendon Press, Oxford, 1990), Chapter 6.

R. D. Present, *Kinetic Theory of Gases* (McGraw-Hill, New York, 1958).

problems

4.1   The coefficient of viscosity does not depend on the number of molecules per unit volume. Explain why not.

4.2   The transport coefficients κ, η, and D all increase as the square root of the temperature, and decrease as the square of the average molecular diameter. Explain why without reference to any formula. Of the three transport coefficients, κ and D vary as $1\sqrt{m}$, whereas η varies as $\sqrt{m}$. Why?

4.3   The rate of a certain surface catalyzed reaction is proportional to the rate at which molecules hit the surface. The rate will increase with an increase in which of the following properties? (a) the mass of the molecules, (b) the velocity of the molecules, (c) the heat capacity of the molecules, (d) the number density of the molecules, (e) the area of the surface.

4.4   Consider a thought experiment in which horses are transported by molecules and suppose that the number of horses is proportional to the number of bushels of oats: $H = kO$. The transport coefficient relating the flux of horses to the gradient of oats depends on which of the following parameters? (a) the weight of the horse, (b) the velocity of the molecule, (c) the proportionality constant k, (d) the mean free path, (e) the speed of the horse.

4.5   Why is the coefficient of thermal conductivity larger for helium than that for argon? Why is the coefficient of thermal conductivity for N_2 larger than that for argon?

4.6   The viscosity coefficient of O_2 is greater than that of CO_2. Which molecule has the greater molecular diameter?

4.7   How does the root-mean-squared distance traveled by a diffusing molecule vary with temperature? (a) not at all, (b) increases (c) decreases. How does it vary with pressure? (a) not at all, (b) increases, (c) decreases.

4.8   Two bugs each execute a one-dimensional random walk with the same step size, but the second bug takes steps twice as often as the first. After a given time the second bug will be (a) twice as far from the origin as the first, (b) $\sqrt{2}$ times as far, (c) the same distance.

4.9   In a tube of infinite length, consider the diffusion of molecules that start at $z = z_0$ at $t = t_0$. The concentration of molecules at a location different than z_0 (a) increases monotonically with time, (b) stays the same, (c) decreases monotonically, (d) increases then decreases, or (e) decreases then increases.

4.10   If thermal conductivity is independent of number density, why is it advantageous to evacuate the region between the walls of a dewar flask?

4.11   The thermal conductivity of silver is about 4 J K^{-1} cm^{-1} s^{-1}. Calculate the heat flow in watts through a silver disk 0.1 cm in thickness and having 2 cm^2 area if the temperature difference between the two sides of the disk is 10 K.

4.12   The heat capacity of N_2 is about 20 J K^{-1} mol^{-1} and its diffusion coefficient is 1.5×10^{-5} m^2 s^{-1}. How much heat will be conducted in 1 s across a 1-cm space between two parallel plates 2 m^2 in area if the plates differ in temperature by 5 K and the space between the plates is filled with nitrogen at 1 atm and 300 K? You may assume that the ideal gas law holds under these conditions.

4.13   a.   Calculate the coefficient of thermal conductivity for nitrogen at 303 K. Assume that πd^2 for N_2 is 7×10^{-20} m^2 and $C_{vm} = (5/2)R$.

          b.   In a double glazed window the panes are separated by 5 cm. What is the rate of heat transfer in watts from a warm room at 323 K to the cold exterior at 283 K through a window of area 1 m^2? Assume that air has the coefficient of thermal conductivity calculated in part (a).

          c.   To approximately what pressure in torr would one have to evacuate the space between the two windows before κ would be decreased appreciably for the value calculated in part (a)?

4.14   The self-diffusion coefficient of CO is $D = 1.75 \times 10^{-5}$ m^2 s^{-1} at 273 K and 1 atm. The density of CO under these conditions is 1.25 kg m^{-3}. Calculate the molecular diameter.

4.15   The heat capacity of N_2 is 20.9 J K^{-1} mol^{-1}, and its viscosity at room temperature is 1.7×10^{-4} poise (1 poise = 1 g cm^{-1} s^{-1}). How much heat will be conducted in 1 s across a 1-mm space between two parallel plates 10 cm $\times$ 10 cm in size if the plates differ in temperature by 5 K and if the space between the plates is filled with N_2 at 1 atm?

4.16   Show by direct differentiation that **equation 4.41** is the solution to **equation 4.40.**

4.17 Write and test a computer program to verify the general result, presented in **Figure 4.8,** that the root-mean-squared distance traveled in a one-dimensional random walk is proportional to the square root of the number of steps.

4.18 The computer experiment on the random walk showed that the root-mean-squared distance traveled in a random walk is proportional to the square root of the number of steps taken. This result can be shown more rigorously by consideration of the following problem.

    a. Suppose a drunken sailor leaves a bar at closing time and executes a one-dimensional random walk in the z direction along the sidewalk. Enumerate all the possible sequences of steps for which, after six steps each of length ℓ, she could be at distances -6ℓ, -4ℓ, -2ℓ, 0, 2ℓ, 4ℓ, or 6ℓ from the doorway of the bar.

    b. Show that the probabilities obtained in part (a) agree with the following formula, which can be used to calculate the absolute value of the sailor's distance from the bar:

$$P(z) = \frac{n!}{[1/2(n + s)]!\,[1/2(n - s)]!\,2^n},$$

    where n is the number of steps, $s = z/\ell$, and $N! = N(N - 1)(N - 2)$ ... (1).

    c. A very accurate approximation to $N!$ for large N is given by *Stirling's approximation:*

$$\ln N! = \left(N + \frac{1}{2}\right)\ln N - N + \ln(2\pi)^{1/2}$$

    Use this approximation to show that

$$P(z) = \left(\frac{2}{\pi n}\right)^{1/2} \exp\left(-\frac{s^2}{2n}\right).$$

    [*Hint:* You will need to approximate $\ln(1 + x) \approx x - x^2/2$.]

    d. Substitute $s = z/\ell$, and let the number of steps n be given by the total time divided by the time per step: $n = t/\tau$, to show that

$$P(z,t) = \left(\frac{2\tau}{\pi t}\right)^{1/2} \exp\left(-\frac{z^2\tau}{2t\ell^2}\right).$$

    e. Finally, compare the above equation with **equation 4.41**, letting $z_0 = 0$, to derive the Einstein-Smoluchowski relationship, **equation 4.45**.

5 Chapter Five

Reactions in Liquid Solutions

Chapter Outline

5.1   INTRODUCTION

Our goal in this chapter is to see what fundamental differences there may be between reactions in the gas phase and reactions in liquid solutions. We will see that, in most cases, the rate for a reaction in solution once the reactants have come together is comparable to that for the same reaction in the gas phase, but that the solution may control the rates at which the reactants come together or products separate. In some cases, the solvent influences the reaction by providing a "cage" around the reactants and products. In other instances, particularly when the reaction involves charge displacement, the solvent may influence the reaction by differentially stabilizing the reactants, products, or the transition state. Finally, some reactions in solution occur on a very short time scale, so we will briefly investigate several experimental approaches to measuring their rate constants.

In comparing a reaction in solution to the same reaction in the gas phase, we often find that the mechanism of the reaction is the same and that the magnitude of the rate constant is quite similar. These facts may at first seem surprising, since gas-phase reactions are based on individual *bimolecular* collisions, whereas the density in solution is so high that an individual reactant is usually in direct contact with more than one other molecule. Why then should the simple bimolecular picture be correct? The answer is that most reactions in solution do not involve the solvent, and that the rate of bimolecular encounters between two reactants is not appreciably different in solution than in the gas phase. A typical concentration in solution, say 4×10^{-2} molar, corresponds to a high but not unreasonable concentration in the gas phase, approximately 1 atm, so that for reactions where the solvent is not one of the reactants, the rate of the reaction in solution is typically within an order of magnitude of the rate in the gas phase. Of course, for reactions where the solvent is a reactant, or for reactions involving solvated ions, the rate is obviously much different in solution. For the most part, however, the solvent is simply something that crowds the reactants.

5.2   THE CAGE EFFECT, FRICTION, AND DIFFUSION CONTROL

5.2.1   The Cage Effect

The difference then between a reaction in the gas phase and one in solution is much like the difference between a romantic encounter on an empty beach and one on a crowded dance floor. The romance of closeness is not appreciably changed by the surrounding solvent of dancers, but it is more difficult to find one another in a crowd, and correspondingly difficult to separate once the dance has ended. The solvent tends to slow the rate of approach of the reactants, so that they must diffuse toward one another through the solution, but it also keeps them together for many "collisions" once they come in contact. This latter phenomenon is often referred to as the *cage effect*.

How difficult is it for products to escape the solvent cage? In an interesting experiment a molecular beam of I_2^- surrounded by a varying number of CO_2 solvent molecules was subjected to a pulse of laser light. The light dissociates the I_2^- to $I + I^-$, but in a large enough cage of CO_2 molecules the I and I^- cannot escape one another. When they recombine, the energy released, about 167 kJ/mol, is dissipated by evaporation of CO_2 molecules from the cluster, so that a smaller cluster with I_2^- at its core is detected. On the other hand, if the cage is small enough, the product I and I^- can separate from one another, so that the cluster of $I_2^-(CO_2)_n$ breaks up into two smaller clusters, one containing I^- and one containing I. By measuring the size of I_2^- and I^- containing product clusters, Papanikolas et al. were able to determine the branching ratio of caged versus uncaged products. **Figure 5.1** displays the fraction of dissociations that underwent recombination due to the cage effect as a function of the size of the starting cluster. Whereas 6 CO_2 molecules do not form a large enough cage to contain the dissociated products, 16 CO_2 molecules cause complete caging. The fraction of caged products increases roughly linearly for cluster sizes between 6 and 16 CO_2 molecules.

If it takes only 16 solvent molecules to cage products having 167 kJ/mol, one can easily imagine that a reactant pair, having much less initial translational energy, will be held by the cage for an appreciable number of bimolecular collisions. How can we describe in simple terms the forces in solution that are responsible for the cage effect?

5.2.2   The Langevin Equation

We consider a particle undergoing collisions in a liquid. Unlike the motion in a gas, the motion of liquid molecules cannot be presented in a closed-form solution. The mean free path, so useful in describing transport properties in gases, is undefined

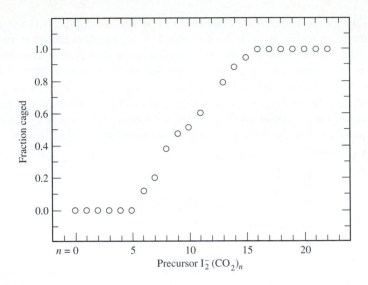

Figure 5.1

The cage effect for I_2^- in CO_2.

From J. M. Papanikolas, J. R. Gord, N. E. Levinger, D. Ray, V. Vorsa, and W. C. Lineberger, *J. Phys. Chem.* **95,** 8028 (1991).

in a liquid. Indeed, the concept of a "collision" loses meaning when solvent molecules are constantly in interaction with their neighbors. Transport properties still follow the macroscopic description outlined in Chapter 4, but the microscopic description cannot be described in terms of binary collisions. However, the situation is not quite as intractable as one might expect, since with so many interactions it becomes relatively easy to describe the average behavior. This is the approach taken at the turn of the century by Paul Langevin for the motion of a Brownian particle in a solution.

Langevin's model for a macroscopic Brownian particle in solution is that, whatever the particle's initial velocity, its average velocity decreases to zero with time because of the frictional forces in the fluid. To be sure, the particle is continually buffeted by collisions with the molecules of the fluid, but on average these collisions are random. The equation of motion for the Brownian particle is thus written as

$$m\frac{dv}{dt} = -\zeta v + f(t),$$ (5.1)

where v is the velocity of the particle, m is its mass, ζ is a coefficient of friction, and $f(t)$ is a function that represents the random forces on the particle due to collisions with the fluid.

Suppose we average over an ensemble of Brownian particles. The ergodic hypothesis of statistical mechanics assures us that such an average is equivalent to an average over time. On average, the forces on the ensemble of Brownian particles must vanish:

$$<f(t)> = 0,$$ (5.2)

so that averaging both sides of **equation 5.1** gives

$$m\frac{d<v>}{dt} = -\zeta <v>.$$ (5.3)

The solution to this equation is found by straightforward integration:

$$<v> = <v>_0 \exp\left(-\frac{\zeta}{m}t\right), \tag{5.4}$$

where $<v>_0$ is the initial average velocity. We see that the velocity decays exponentially with a time constant equal to m/ζ. This process of velocity decay is called *dissipation* and describes how the directed velocity is transferred to the molecules of the fluid.

Appendix 5.1 demonstrates that the Langevin equation, **equation 5.1,** can be solved to show that for times long compared to m/ζ the mean squared displacement of a macroscopic Brownian particle is given by

$$<x^2> = \frac{2kT}{\zeta}t. \tag{5.5}$$

Note that under these conditions, the Brownian particle suffers many collisions with the medium, and its mean squared displacement is determined by the friction coefficient ζ.

We now make a connection with our previous study of diffusion in Chapter 4. We write a version of **equation 4.43** for motion in the x direction as

$$x_{rms} = (2Dt)^{1/2}, \tag{5.6}$$

and then combine **equation 5.5** with **equation 5.6** by noting that $x_{rms} = <x^2>^{1/2}$:

$$\frac{2kT}{\zeta}t = 2Dt,$$

$$D = \frac{kT}{\zeta}. \tag{5.7}$$

It is thus clear that for motion in a liquid the diffusion coefficient is inversely proportional to the friction coefficient.

Further insight comes from the work of Stokes, who showed that the frictional force on a spherical particle of radius a moving through a fluid is

$$F = -\zeta v = -6\pi\eta a v, \tag{5.8}$$

where η is the coefficient of viscosity. Thus, $\zeta = 6\pi\eta a$, and

$$D = \frac{kT}{6\pi\eta a}. \tag{5.9}$$

These equations remind us that, for motion in a liquid, higher viscosity coefficients are equivalent to higher friction coefficients and give rise to lower diffusion coefficients. It thus becomes clear that the forces that prevent particles from moving freely in a fluid are the viscous or frictional forces. It is these forces that are responsible for the cage effect, which we recall can hold reactants together for a number of collisions.

If the rate of reaction between a pair of molecules is high enough, it is possible that the frictional forces responsible for the cage effect will hold the reactant pair together long enough that nearly every encounter between the pair will lead to reaction. Under such circumstances, the rate of the reaction is controlled not by the rate constant for an individual reactive collision between the pair of reactants, but rather by the rate at which they can encounter one another by diffusion. Reactions of this type are called *diffusion-controlled* reactions.

5.2.3   A Simple Model for Diffusion Control

Since the rate of a reaction in solution is controlled partly by how fast the reactants encounter one another and partly by how fast they react once they make their encounter, we consider a mechanism for reaction in solution composed of the following steps:

$$A + B \underset{k_{esc}}{\overset{k_{enc}}{\rightleftharpoons}} AB$$

$$AB \overset{k_r}{\to} P,$$

(5.10)

where A and B are the reactants in solution, P represents the products, k_{enc} is the rate at which A and B encounter one another, k_{esc} is the rate at which the products escape the solvent cage, and k_r is the rate at which they react when within the cage. The steady-state solution to this reaction sequence, which is very similar to the Michaelis-Menten and Lindemann mechanisms (Chapter 2), is simply

$$\frac{d[P]}{dt} = \frac{k_r k_{enc}}{k_{esc} + k_r}[A][B].$$

(5.11)

Note that when the rate of reaction within the cage, k_r, is very slow compared to the escape rate, then the first reaction will basically be at equilibrium, so that the overall rate of the reaction will be given by $(k_{enc}/k_{esc})k_r$. On the other hand, when the reaction within the cage is very fast compared to the escape rate, then the rate of the overall reaction is controlled by the rate at which the reactants encounter one another. Under such conditions, the reaction rate is called diffusion controlled, since the rate is limited by the rate at which reactants can diffuse toward one another. We will see below that for neutral particles this rate is proportional to the sum of the diffusion coefficients for A and B through the solvent and to the radius of the cage, R: $k_{enc} = 4\pi R(D_A + D_B)$. For oppositely charged reactants, say A^+ reacting with B^-, there is an additional multiplicative factor due to the mutual electrostatic attraction.

5.2.4   The Diffusion-Controlled Rate Constant

We have seen in Section 5.2.1 that the solvent cage can have a substantial influence on reaction rates in solution. When the reaction within the cage is very fast compared to the escape rate, then the rate of the overall reaction is controlled by the rate at which the reactants encounter one another. Under such conditions, the reaction rate is called diffusion controlled, and it is now of interest to develop an expression for the overall rate of such reactions. Consider a simple model for this diffusion-controlled limiting rate constant in which species A reacts with species B every time the two approach one another to within their contact distance R, the sum of their two radii, as shown in **Figure 5.2.**

We now briefly generalize to three dimensions the treatment of diffusion in Chapter 4. In three dimensions, the flux is related to the *gradient* of the concentration: $J = -D\nabla c$, where J is a three-dimensional vector, and ∇ is called the gradient operator. In Cartesian coordinates, $\nabla = \mathbf{i}(\partial/\partial x) + \mathbf{j}(\partial/\partial y) + \mathbf{k}(\partial/\partial z)$, with $\mathbf{i}, \mathbf{j}$, and $\mathbf{k}$ as unit vectors in the x, y, and z directions. In spherical coordinates, $\nabla = \mathbf{a}_r(\partial/\partial r) + \mathbf{a}_\theta(1/r)(\partial/\partial\theta) + \mathbf{a}_\phi(1/r\sin\theta)(\partial/\partial\phi)$, with $\mathbf{a}_r, \mathbf{a}_\theta$, and $\mathbf{a}_\phi$ as unit vectors in the r, θ, and ϕ directions.

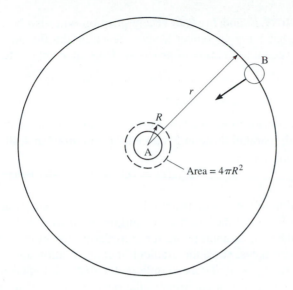

■ **Figure 5.2**

In a diffusion-controlled reaction every reactant B that approaches to within a radius R of A will react. Reaction causes a gradient in [B] that gives rise to a flux of B toward A.

Returning to **Figure 5.2,** since reaction will deplete the concentration of B around each A, the reaction itself will establish a concentration gradient, and this gradient will cause molecules of type B to flow toward those of type A. Let $A(r,\theta,\phi)$ represent the spatially dependent concentration of A, and let $B(r,\theta,\phi)$ represent the spatially dependent concentration of B. The three-dimensional vector representing the flux $\boldsymbol{J}_{\text{A–B}}$ of reactants A and B toward one another is equal to the flux of B toward A due to concentration gradient,

$$-\boldsymbol{J}_{\text{B}} = D_{\text{B}}\nabla B(r,\theta,\phi), \tag{5.12}$$

plus the flux of A toward B due to concentration gradient,

$$-\boldsymbol{J}_{\text{A}} = D_{\text{A}}\nabla A(r,\theta,\phi), \tag{5.13}$$

so that

$$\boldsymbol{J}_{\text{A–B}} = -(\boldsymbol{J}_{\text{A}} + \boldsymbol{J}_{\text{B}}) = (D_{\text{A}} + D_{\text{B}})\nabla B(r,\theta,\phi). \tag{5.14}$$

In this last equation, we have assumed quite reasonably that $\nabla A(r,\theta,\phi) = \nabla B(r,\theta,\phi)$, since the gradients in the two reactants are caused by the same effect, namely, the fact that A molecules around B are depleted by reaction and vice versa. Note that the gradient $\nabla B(r,\theta,\phi)$ is positive, so that our choice of sign gives that the flux $\boldsymbol{J}_{\text{A–B}}$ is also positive, as it should be.

   Suppose that we have a mixture of reactants but that we prevent them from reacting. Their concentrations will then be the bulk, equilibrium concentrations; that is, $B(r,\theta,\phi) = [\text{B}]$ and $A(r,\theta,\phi) = [\text{A}]$, where $B(r,\theta,\phi)$ and $A(r,\theta,\phi)$ are the spatially dependent concentrations and [B] and [A] are the bulk ones, which we will assume to be constants. Now we imagine that the reaction is suddenly turned on. The concentration of A in the vicinity of B will decrease, and vice versa, so that a concentration gradient is formed. But after a short time, steady state will be approached, so that the flux of A and B toward one another will be constant. The concentrations $A(r,\theta,\phi)$ and $B(r,\theta,\phi)$ at any position will also be constant. Under the assumption that every encounter leads to reaction, the flux of A and B toward one another will equal to the flux of products. Thus, at steady state

$$J_{rxn} = J_{A-B} = (D_A + D_B)\nabla B(r,\theta,\phi)$$
$$= \text{constant.} \quad (5.15)$$

Now let us consider the steady-state mathematical solution. Following arguments that exactly parallel those in Section 4.7 (see **equation 4.40**), we find that

$$\frac{\partial B(r,\theta,\phi)}{\partial t} = -\nabla J_{A-B}(t) = (D_A + D_B)\nabla^2 B(r,\theta,\phi). \quad (5.16)$$

We wish a solution to the rate constant for which the concentrations of A and B at any position are fixed in time, so that, at equilibrium, $\partial B(r,\theta,\phi)/\partial t = 0$.

The coordinates most appropriate for considering the motion of the A–B pair are spherical coordinates, since the gradient in concentration and (we assume) the gradient in electrical potential depend only on the distance between the pair. Thus, $B(r,\theta,\phi) = B(r)$. In spherical coordinates the operator ∇^2, called the Laplacian, is given by

$$\nabla^2 = \frac{\partial^2}{\partial r^2} + \frac{2}{r}\frac{\partial}{\partial r} + \frac{1}{r^2}\frac{\partial^2}{\partial^2\theta} + \frac{1}{r^2}\cot(\theta)\frac{\partial}{\partial\theta} + \frac{1}{r^2\sin^2\theta}\frac{\partial^2}{\partial\phi^2}. \quad (5.17)$$

Substituting **equation 5.17** into **equation 5.16** and recognizing that $B(r)$ depends only on r and that $\partial B(r)/\partial t = 0$, we find that the solution for $B(r)$ needs to obey the equation

$$\frac{\partial B}{\partial t} = (D_A + D_B)\left(\frac{\partial^2 B(r)}{\partial r^2} + \frac{2}{r}\frac{\partial B(r)}{\partial r}\right)$$
$$= 0. \quad (5.18)$$

The solution to this equation, as may be readily verified by substitution, is

$$B(r) = \frac{c_1}{2r} + c_2, \quad (5.19)$$

where the constants c_1 and c_2 must still be determined from the boundary conditions. When the distance between A and B is sufficiently large, i.e., as $r \to \infty$, $B(r)$ must approach its bulk concentration, [B]. Thus, we find that $c_2 = $ [B].

To evaluate c_1, let us calculate the concentration of B at the distance R equal to the sum of the two radii. To do this, we examine in more detail the gradient caused by the reaction. Consider the flux of B into of a sphere of radius r centered on a particular reactant of type A, as shown in **Figure 5.2**. Every B flowing into the sphere eventually reacts with A at a distance R, where the concentration of B is $B(R)$. Thus, the rate of the reaction is then just the flux of B, in number per time per area, times the area of the sphere times concentration of A:

$$\frac{d[P]}{dt} = k_r[A]B(R) = J_{rxn}4\pi r^2[A], \quad (5.20)$$

where [P] is the concentration of products, k_r is the phenomenological rate constant we wish to determine (see Section 5.2.3), and $4\pi r^2$ is the area of the sphere around A at a distance r. Combination of **equations 5.15 and 5.20** leads to

$$k_r[A]B(R) = [A]4\pi r^2(D_A + D_B)\frac{dB(r)}{dr}. \quad (5.21)$$

In this equation [A] is constant (equal to the macroscopic concentration), $B(R)$ is constant (to be determined) and $B(r)$ denotes how the microscopic concentration of B varies with distance. Division of both sides of the equation by $[A]r^2$ and multiplication by dr yields

$$k_r B(R) \frac{dr}{r^2} = 4\pi(D_A + D_B) \, dB(r). \qquad (5.22)$$

Finally, integration of both sides over dr from $r = R$ to $r = \infty$ gives

$$k_r B(R) \int_R^\infty \frac{1}{r^2} \, dr = 4\pi(D_A + D_B) \int_R^\infty d[B(r)],$$

$$\frac{k_r B(R)}{R} = 4\pi(D_A + D_B)[B(r = \infty) - B(r = R)],$$

or

$$B(R)\left[\frac{k_r}{R} + 4\pi(D_A + D_B) \right] = 4\pi(D_A + D_B)[B],$$

$$(5.23)$$

$$B(R) = \frac{[B]}{1 + \dfrac{k_r}{4\pi(D_A + D_B)R}}.$$

In this last equation, we have used the fact that $B(r = \infty)$ is just the bulk concentration [B]. Comparison of the last line of this equation with **equation 5.19** leads, after some algebra, to the conclusion that

$$c_1 = \frac{-2R[B]}{1 + \dfrac{4\pi(D_A + D_B)R}{k_r}}, \qquad (5.24)$$

but we will have little use for this equation now that the last line of **equation 5.23** gives us an expression for $B(R)$. Substitution of this solution for $B(R)$ into **equation 5.20** gives

$$\frac{d[P]}{dt} = \frac{k_r}{1 + \dfrac{k_r}{4\pi(D_A + D_B)R}} [A][B]. \qquad (5.25)$$

Thus, when $k_r \gg 4\pi(D_A + D_B)R$, the overall rate constant for the reaction is given by

$$k = 4\pi(D_A + D_B)R. \qquad (5.26)$$

We see that the rate constant, k, is then completely controlled by the encounter rate, k_{enc}, defined in Section 5.2.3; i.e., it is controlled by diffusion. In this limiting case, the reaction occurs instantly when the reactants approach to within their average diameter R. **Example 5.1** illustrates the utility of **equation 5.26**.

example 5.1

Calculating Diffusion-Controlled Rate Constants

Objective Given that the diffusion coefficient of many species in aqueous solution is on the order of 10^{-9} m²/s, calculate the diffusion limited rate constant for a pair of reactants whose average diameter is 2.0 nm.

Method Use **equation 5.26**.

Solution The rate constant should be $k = 4\pi(D_A + D_B)R$, with $D_A \approx D_B$ and $R = 2.0$ nm. Thus, $k = 4\pi(2 \times 10^{-9}$ m²/s$)(2.0 \times 10^{-9}$ m/molecule$)(10^6$ cm³/m³$)(6.02 \times 10^{23}$ molecule/mol$) = 3.0 \times 10^{13}$ cm³ mol⁻¹ s⁻¹ $= (3.0 \times 10^{13}$ cm³ mol⁻¹ s⁻¹$)(1$ L/1000 cm³$) = 3.0 \times 10^{10}$ L mol⁻¹ s⁻¹.

Equation 5.26 provides the diffusion-controlled rate constant in the case when the reactants are uncharged. When the reactants are ionic, the situation is somewhat more complicated because, in addition to the concentration gradient caused by reaction, there is also a concentration gradient caused by the attraction or repulsion of charged particles. A detailed examination, discussed in Appendix 5.2, shows that

$$k = 4\pi(D_A + D_B)\beta, \qquad (5.27)$$

where

$$\beta = \frac{1}{\displaystyle\int_R^\infty \frac{e^{U(r)/kT}}{r^2}\, dr}, \qquad (5.28)$$

and $U(r)$ is the potential of interaction between the ionic reactants. Note that when $U(r) = 0$, the integration in **equation 5.28** can be performed to yield $\beta = R$, and we recover **equation 5.26** from **equation 5.27**. In general, this potential will be given for charged particles by $U(r) = z_A z_B e^2/\epsilon r$, where ϵ here refers to the dielectric constant of the solution, z_A and z_B are the integer charges on the ions, and e is the magnitude of the charge on an electron. The integration in **equation 5.28** then shows that

$$\beta = \frac{-z_A z_B r_0}{1 - \exp(z_A z_B r_0/R)}, \qquad (5.29)$$

where $r_0 = e^2/\epsilon kT$ and is equal to about 0.7 nm in water at 25°C.

5.3 REACTIONS OF CHARGED SPECIES IN SOLUTION: IONIC STRENGTH AND ELECTRON TRANSFER

As discussed in the Introduction to this chapter, most rate constants in solution are similar to those for the corresponding reaction in the gas phase, except when the rate of the reaction is limited by how fast the reactants can diffuse through the solution. Another situation for which the solution-phase rate constant can differ substantially from the gas-phase rate is when the reactants or the activated complex interact strongly with the solvent. An example of such interaction is the electrostatic stabilization of ionic reactants or complexes by the solvent. Two situations will be considered. In the first, we will examine the effect on the rate constant of additional ions in the solution, and we will find that the rate constant is influenced by the *ionic strength* of the solution. In the second situation we will see that even a neutral solvent can influence the reaction rate if the energy of solvation is substantially different for the reactants and products. An example of this second effect is when the reaction involves an electron transfer, either from one molecule to another or between two different sites on the same molecule. As the reaction proceeds, a dielectric solvent must rearrange its structure to attain the minimum energy, and this solvent reorganization will have an influence on the rate constant. How the rate constant varies with the solvent reorganization energy is the subject of Marcus theory, which we will briefly develop.

We consider first the influence on reactions of solutions with high ionic strength.

5.3.1   Reaction Rates and Ionic Strength

It is well known that any ion in solution is stabilized by being surrounded by an ionic "atmosphere" of oppositely charged particles; this effect forms the physical basis of the Debye-Hückel theory.[a] The stabilization increases with the square root of the ionic strength, $I_c = \frac{1}{2}\sum c_i z_i^2$. The question we approach in this section is how to modify activated complex theory to account for the electrostatic interaction between reactants or activated complexes and the ions in solution.

Our discussion of activated complex theory in Section 3.4 has assumed that the equilibrium constant between reactants and products can be written simply in terms of a ratio of concentrations. However, we know from our study of thermodynamics that equilibrium constants are actually related to activities rather than concentrations. To modify the ACT, we return to the fundamental assumptions in ACT embodied in **equation 3.13**, $d[\text{products}]/dt = k_2[AB^\ddagger]$. If we now write the equilibrium constant for $A + B \rightleftharpoons AB^\ddagger$ in terms of activities we obtain

$$K^\ddagger = \frac{a^\ddagger}{a_A a_B} = \frac{\gamma^\ddagger}{\gamma_A \gamma_B}\frac{[AB^\ddagger]}{[A][B]}, \tag{5.30}$$

so that

$$\frac{d[\text{products}]}{dt} = k_2[AB^\ddagger] = k_2\frac{\gamma_A\gamma_B}{\gamma^\ddagger}K^\ddagger[A][B]$$

$$= \frac{\gamma_A\gamma_B}{\gamma^\ddagger}k_2\frac{k_1}{k_{-1}}[A][B]. \tag{5.31}$$

The overall rate constant in **equation 5.31** differs from that in our previous equation **equation 3.13** only by the multiplicative factor $\gamma_A\gamma_B/\gamma^\ddagger$. Let k_0 be the rate constant when all activity coefficients are equal to unity; in other words, let k_0 be the rate constant we have already evaluated by ACT. Then when the activity coefficients are not equal to unity, the rate constant should be given as

$$k = k_0\frac{\gamma_A\gamma_B}{\gamma^\ddagger}, \tag{5.32}$$

---

[a] Peter Debye was awarded the Nobel Prize in Chemistry in 1936 for his contributions to the understanding of molecular structure through his investigations on dipole moments and on the diffraction of X-rays and electrons in gases.

or

$$\log_{10}k = \log_{10}k_0 - \log_{10}\gamma^{\ddagger} + \log_{10}\gamma_A + \log_{10}\gamma_B. \qquad (5.33)$$

For reactions in ionic solutions at low enough concentrations, we might reasonably assume that the Debye-Hückel limiting law is valid:

$$\log_{10}\gamma_i = -Az_i^2 I_c^{1/2}, \qquad (5.34)$$

where z_i is the charge on the species involved in the reaction, I_c is the ionic strength, and A is a constant equal to 0.50 $(L/mol)^{1/2}$ for water at 25°C. Substitution of **equation 5.34** into **equation 5.33** and realization that $z^{\ddagger} = z_A + z_B$ lead to

$$\log_{10}k = \log_{10}k_0 - A[z_A^2 + z_B^2 - (z_A + z_B)^2]I_c^{1/2}$$

$$= \log_{10}k_0 + 2Az_Az_BI_c^{1/2}. \qquad (5.35)$$

Note that **equation 5.35** predicts that the logarithm of the rate constant should vary linearly with the square root of the ionic strength. In addition, reactions between ions of like charge should have rate constants that increase with $I_c^{1/2}$, whereas those between ions of unlike charge should decrease. This behavior is illustrated in **Figure 5.3.**

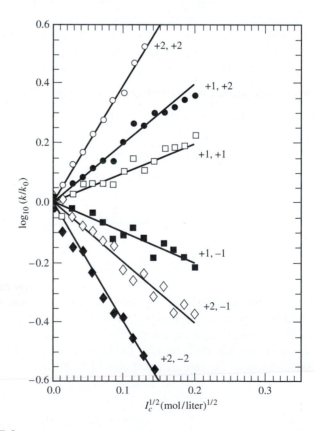

■ **Figure 5.3**

Effect of increasing ionic strength on the rate constant for reaction between ions of varying charges. Values of z_A, z_B are indicated.

The physical basis for these effects can be grasped by considering the change in apparent activation energy with increasing ionic strength. When unlike ions form an activated complex, the complex is less charged than the reactants and is thus less stabilized by increasing ionic strength than the reactants; the apparent activation energy increases so that the rate constant decreases. When like ions form an activated complex, on the other hand, the complex has a higher charge than the reactants and is more stabilized by increasing ionic strength; the apparent activation energy decreases and the rate constant increases. This behavior of the rate constant with ionic strength is often called the *primary salt effect*.

5.3.2   Electron Transfer Reactions: Marcus Theory

Another situation in which the solvent can have a profound influence on the rate constant occurs when the energy of solvation differs substantially for reactants and products. This situation occurs frequently in electron transfer reactions, since as the electron moves from the donor site to the acceptor site the structure of the solvent must adjust to accommodate the new charge distribution. Since all oxidation-reduction reactions in solution involve the transfer of an electron, the determination of the rate constant for electron transfer reactions is an extremely important chemical problem. This problem has been considered in detail by Marcus,[b] and we develop here a simplified derivation of his results.[c] It can be shown that the more complete derivation gives the same answer in the limit when the distance between donor and acceptor sites never becomes too small.

The overall reaction that we would like to consider can be symbolized by the following scheme:

$$D + A \rightleftharpoons (DA) \rightleftharpoons (D^+A^-) \rightleftharpoons D^+ + A^-,$$

where the species in parentheses represent having the donor and acceptor at a distance short enough so that the electron can be transferred, and (DA) and (D^+A^-) represent this "contact pair" before and after transfer. In many oxidation-reduction reactions, the overall rate constant is limited by the rate for the electron transfer, so we will concentrate on this step of the process. In some cases, D and A are different sites on the same molecule, so that, again, the overall rate constant is determined by the rate of the electron transfer.

Consider an electron located on the donor molecule, which itself is surrounded by a number of solvent molecules. The energy of the electron will depend on the nuclear positions of all the atoms in the donor and solvent molecules, so that there will in general be $3N - 6$ nuclear coordinates, where N is the total number of atoms. Under the assumption that the electron moves much more rapidly than the nuclei, the energy of the system can be adequately approximated by calculating the electronic energy for a each possible nuclear configuration. We now imagine how this energy varies along a particular coordinate, one of the $3N - 6$ coordinates. The coordinate we will choose to examine is the reaction coordinate, the one whose nuclear displacements would lead along a minimum energy path from the nuclear configuration of (DA) to that of (D^+A^+).

[b]R. Marcus, *J. Chem. Phys.* **24**, 966 (1956); ibid., **24**, 979 (1956); ibid. **26**, 867 (1957); ibid. **26**, 872 (1957); *Disc. Farad. Soc.* **29**, 21 (1960); *J. Phys. Chem.* **63**, 853 (1963); *J. Chem. Phys.* **38**, 1858 (1963); ibid., **39**, 1734 (1963); ibid., **43**, 679 (1965).

[c]I am grateful to Prof. A. C. Albrecht for providing this derivation and to M. Stimson for bringing it to my attention.

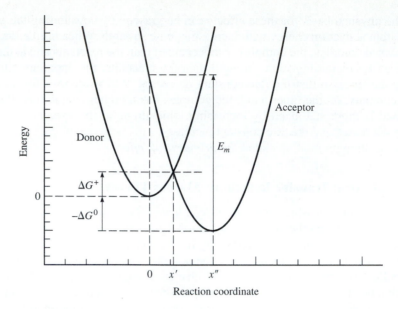

Figure 5.4

Energy dependence as a function of reaction coordinate for electron on donor or acceptor.

When the electron is on the donor, the energy will be a minimum at a particular location along the reaction coordinate; let us arbitrarily label this as position zero along that coordinate, as shown in **Figure 5.4.** The parabolic curve labeled "donor" shows how the energy of the system might vary with displacement when the electron is on the donor. If the electron were on the acceptor, the energy of the system would be different; its minimum will in general be at a different location, say x'', along the reaction coordinate, and the energy of the minimum will differ from that of the donor by ΔG^0, the free energy of the reaction. Note that, as drawn in the figure, ΔG^0 is negative (the products are more stable than the reactants), so that the positive energy difference between the minima of the two parabolas is $-\Delta G^0$. The parabolic curve labeled "acceptor" shows how the energy of the system might vary with displacement when the electron is on the acceptor. If we assume, as did Marcus, that the coupling between the donor and acceptor electronic energy states is weak, then the energy of the transition state will be given by the point of intersection between the two curves. The key to determining the rate constant for the reaction is to find the value of ΔG^+ in the figure. From **equation 3.23,** we know that the rate for the process is given simply by $k_{ET}(T) = (kT/h)\exp(-\Delta G^+/kT)$.

We now suppose that the curves describing how the energy changes with position along the reaction coordinate can be approximated by parabolas, both for the donor and for the acceptor; i.e., we will assume that $E = x^2$ for both parabolas.[d] It can be shown that this approximation is equivalent to the full theory developed by Marcus in the limit when the donor and acceptor sites are not too close together. Let us label by x' the reaction coordinate position where the donor and acceptor parabola intersect. From the point of view of the donor curve, the value of ΔG^+ is simply $\Delta G^+ = x'^2$. Let us also define $E_m = x''^2$ as the value, relative to its minimum, of the

---

[d]Actually, we need only assume that $E \propto x^2$; e.g., $E = Cx^2 = (\sqrt{C}\,x)^2$. The arguments given in the text are then appropriate provided that we then transform variables so that the reaction coordinate, now already plotted in arbitrary units, is plotted in units of $\sqrt{C}$ times the current arbitrary unit.

acceptor parabola at the location of the minimum energy for the donor. This energy, called the *reorganization energy,* is the energy required to reorganize the nuclei of the acceptor and its surrounding solvent into the configuration of the donor and its surrounding solvent in the absence of back transfer of the electron. We now calculate the energy of the intersection point for the two parabola above the minimum energy for the acceptor. As measured from the bottom of the acceptor parabola, this energy is $-\Delta G^0 + \Delta G^+ = (x'' - x')^2 = x''^2 - 2x'x'' + x'^2$. Substituting ΔG^+ for x'^2 and E_m for x''^2, we obtain

$$-\Delta G^0 + \Delta G^+ = E_m - 2x'x'' + \Delta G^+,$$

$$2x'x'' = E_m + \Delta G^0,$$

$$x' = \frac{E_m + \Delta G^0}{2x''}, \tag{5.36}$$

$$\Delta G^+ = x'^2 = \frac{(E_m + \Delta G^0)^2}{4x''^2}.$$

Finally, noting again that $E_m = x''^2$, we obtain the final result for G^+:

$$\Delta G^+ = \frac{(E_m + \Delta G^0)^2}{4E_m}. \tag{5.37}$$

The rate constant for the electron transfer reaction is thus

$$k_{ET}(T) = \frac{kT}{h} \exp\left(-\frac{(E_m + \Delta G^0)^2}{4E_m kT}\right). \tag{5.38}$$

The form of **equation 5.38** makes an interesting prediction about the rate constant for the reaction. If E_m is large and ΔG^0 is positive or just slightly negative, then $(E_m + \Delta G^0)$ will be positive and the rate constant will be relatively small. **Figure 5.5** shows the positions of the parabolic curves for the same value of E_m (the same displacement between the two parabolas) and for values of ΔG^0 ranging from positive in panel (A) to increasingly negative values in panels (B)–(D). In panel (A) $\Delta G^0 > 0$, while in panel (B) $\Delta G^0 < 0$; both panels have $E_m + \Delta G^0 > 0$. Thus, the rate should be small for panel (A) and a little larger for panel (B). As ΔG^0 becomes increasingly negative, eventually $(E_m + \Delta G^0)$ will become zero, and the rate constant will become a maximum; this situation is shown in panel (C) of the figure. If ΔG^0 becomes even more negative, then $E + \Delta G^0$ will be negative and its square will again increase. **Equation 5.38** predicts that the rate constant will then actually *decrease* with increasing free energy change, $-\Delta G^0$. The reason why is shown in panel (D) of the figure, where it can be seen that the activation energy for the electron transfer reaction has now increased. The region of free energy change over which the rate constant decreases as the reaction releases more free energy is called the "Marcus inverted region."

That a rate constant should decrease with increasing exothermicity was quite counterintuitive at the time Marcus put forth his model, and the existence of the Marcus inverted region was in doubt for nearly thirty years until G. Closs, J. R. Miller, and their coworkers reported the confirming set of experiments.[e] The results of their measurements are shown in **Figure 5.6,** which plots the logarithm of the

[e]J. R. Miller, L. T. Calcaterra, G. L. Closs, *J. Am. Chem. Soc.* **106,** 3047 (1984); G. L. Closs, L. T. Calcaterra, N. J. Green, K. W. Penfield, and J. R. Miller, *J. Phys. Chem.* **90,** 3673 (1986); G. L. Closs and J. R. Miller, *Science* **240,** 440 (1988).

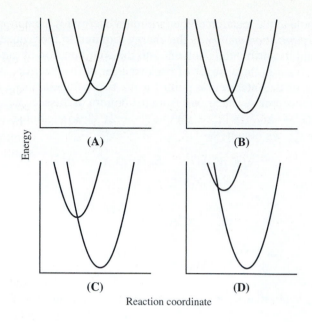

Figure 5.5

Intersecting donor and acceptor parabolas for increasingly negative values of ΔG^0 going from panels (A) to (D). Note that activation energy for the reaction is a minimum in panel (C), but that it increases in going from (C) to (D).

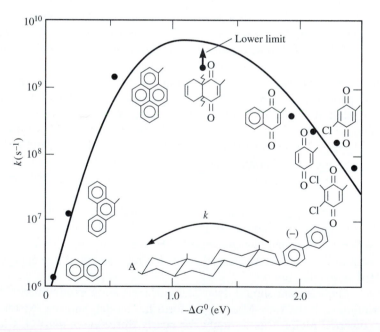

Figure 5.6

Intramolecular electron transfer rate constants as a function of free energy change. The transfer occurs from biphenyl anions to the eight acceptors attached at A in the structure shown.

From G. L. Closs, L. T. Calcaterra, N. J. Green, K. W. Penfield, and J. R. Miller, *J Phys. Chem.* **90**, 3673 (1986). Reprinted with permission from *The Journal of Physical Chemistry.* Copyright 1986 American Chemical Society.

energy transfer rate constant versus $-\Delta G^0$ for a series of molecules consisting of a donor group separated from various acceptor groups by a rigid spacer molecule. By varying the composition of the acceptor and measuring the rate of electron transfer following pulse radiolysis to produce the radical anion, the authors were able to see how the electron transfer rate varied with increasing $-\Delta G^0$. As shown in the figure, the rate did indeed decrease with increasing $-\Delta G^0$ in accordance with the Marcus theory (solid line). Similar experiments were performed by other groups.[f] While the Marcus theory is now the accepted standard for electron transfer reactions,[g] it should be noted that this important area of chemistry is still one of active research.

5.4   EXPERIMENTAL TECHNIQUES

It should come as no surprise that some reactions in solution are very rapid. Reactions involving oppositely charged ions, reactions in which the solvent participates as a reactant, or reactions involving the motion of light particles such as protons or electrons might reasonably be expected to proceed quickly. For such rapid processes, which often occur more rapidly than the reactants can be mixed, special methods are necessary to determine reaction rate constants.

5.4.1   The Temperature Jump Technique

One method pioneered in the 1950s by Manfred Eigen and his coworkers[h] is a relaxation technique. In this method, the reactants start already mixed and in equilibrium with the products. Some property of the system that affects the equilibrium constant, for example the temperature or pressure, is then suddenly changed, and the concentration of a reactant or product is monitored as it changes to achieve its new equilibrium value. **Figure 5.7** displays a typical apparatus for the so-called temperature jump version of the experiment. A power supply (PS) charges a capacitor (C) to a high voltage. A triggered spark gap (G) then discharges the voltage through a cell, simultaneously starting the sweep of an oscilloscope and heating the reactant/product mixture by a few degrees Kelvin. A light source (L) supplies a frequency absorbed by one of the reactants or products, and this frequency is resolved by a monochromator (M), detected by a photomultiplier tube (PMT) and its intensity displayed on the oscilloscope. The resulting waveform is then analyzed to obtain the kinetic information. The best time resolution of such relaxation techniques is typically in the 100-ns regime. **Table 5.1** gives a few protonation and deprotonation rates measured by this technique.

How is the observed signal in this experiment related to the rate constants for the forward and reverse reactions? For simplicity, we first consider an opposing first-order reaction, A $\rightleftharpoons$ B. Immediately after the perturbation, the concentrations of A and B are each displaced from their new equilibrium values by an amount we

[f]M. P. Irvine, R. J. Harrison, M. A. Strahand, and G. S. Beddard, *Ber. Bunsenges. Phys. Chem.* **89,** 226 (1985); M. P. Irvine, R. J. Harrison, G. S. Beddard, P. Leighton, and J. K. M. Sanders, *Chem. Phys.* **104,** 315 (1986).
[g]R. Marcus won the 1992 Nobel Prize in Chemistry for his original work in this area.
[h]M. Eigen, *Disc. Farad. Soc.* **17,** 195 (1954). Eigen shared the 1967 Nobel Prize in Chemistry for his work in this area.

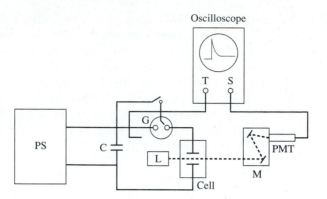

Figure 5.7

Schematic of apparatus used for temperature jump experiment.

TABLE 5.1	Protonation and Deprotonation Rates Measured by Relaxation Techniques	
Reaction	$k_f(M^{-1} s^{-1})$	$k_r(s^{-1})$
$H^+ + OH^- \rightarrow H_2O$	1.3×10^{11}	2.6×10^{-5}
$H^+ + NH_3 \rightarrow NH_4^+$	4.3×10^{10}	24
$OH^- + HCO_3^- \rightarrow H_2O + CO_3^-$	6.1×10^9	1.4×10^2
$OH^- + HPO_4^{-2} \rightarrow H_2O + PO_4^{-2}$	2×10^9	3×10^2

will call x_e. We have already seen in **equation 2.51** of Section 2.4.1 that a system of opposing reactions displaced from equilibrium will relax exponentially with a time dependence given by the sum of the forward and reverse rate constants: $A(t) = A_e + x_e\exp[-(k_1 + k_{-1})t]$. Thus, by measuring the relaxation rate and knowing the equilibrium constant, which is simply the ratio of the forward and reverse rate constants, one can determine the rate constants separately.

For second-order or more complex reactions, the time dependence will be somewhat more complicated than that for the first-order reaction above, but for small displacements the system will relax exponentially toward equilibrium with a time dependence determined by the sum of the forward and reverse rate constants, each perhaps multiplied by a reactant or product equilibrium concentration. For example, Problem 5.3 shows that for the system $A + B \rightleftharpoons C$ the result for the relaxation of A is $A(t) = A_e + x_e\exp\{-[k_f(A_e + B_e) + k_r]t\}$. **Example 5.2** shows how this result can be used to determine the rate constants for $H^+ + OH^- \rightleftharpoons H_2O$.

example 5.2

Calculating Rate Constants from a Relaxation Experiment

Objective   Find the forward and reverse rate constants for the recombination reaction $OH^- + H^+ \rightleftharpoons H_2O$ given that the equilibrium constant at 25°C for the reaction written in this direction is $K_e = [H_2O]/([H^+][OH^-]) = 0.51 \times 10^{16}$ M^{-1}. The OH^- concentration is observed to change from its original value to its new equilibrium value at this temperature exponentially with a time constant of $\tau = 36.8\ \mu s$.

Method

The observed rate is the reciprocal of the time constant, so $1/\tau = k_f([H^+]_e + [OH^-]_e) + k_r$. We also know K_e, so we can both calculate the equilibrium concentrations and relate k_f to k_r: $k_f/k_r = K_e$. With two equations and two unknowns, we should be able to solve for both k_f and k_r.

Solution

First, notice that $k_r = k_f/K_e$ and that $[H^+] = [OH^-] = ([H_2O]/K_e)^{1/2}$. Consequently, $1/\tau = 1/(36.8 \times 10^{-6}\ s) = k_f\{2([H_2O]/K_e)^{1/2} + (1/K_e)\}$; $k_f = 1/\{2[(1000\ g/L)/(18\ g/mol)(0.51 \times 10^{16}\ M^{-1}]^{1/2} + 1/(0.51 \times 10^{16}\ M^{-1})\}\{36.8 \times 10^{-6}\ \mu s\} = 1.30 \times 10^{11}\ M^{-1}\ s^{-1}$. Thus $k_r = k_f/K_e = (1.30 \times 10^{11}\ M^{-1}\ s^{-1})/(0.51 \times 10^{16}\ M^{-1}) = 2.55 \times 10^{-5}\ s^{-1}$.

5.4.2   Ultrafast Laser Techniques

Even the relaxation techniques described above are not fast enough to measure some reaction rates in solution. A more recently developed method for measuring fast processes relies on the extremely short pulse durations available from modern lasers. Although this field is still developing rapidly, the current record for short pulse generation is a few femtoseconds (1 fs $= 10^{-15}$ s), substantially shorter than the time it takes molecules to vibrate. On such time scales, the reactants and solvent molecules stand still. Thus, even the fastest solution reactions occur on somewhat longer time scales, those compatible with laser pulses in the picoseconds (1 ps $= 10^{-12}$ s) and nanosecond region. Such pulses can be used to clock the concentration changes through a method generally known as the pump and probe technique (for more detail on this technique, see Section 7.5.1).

Figure 5.8 shows a schematic diagram for a typical pump-probe apparatus. A short pulse laser generates bursts of radiation. The light beam from this laser is then divided into two beams of different intensity using a partially transmitting

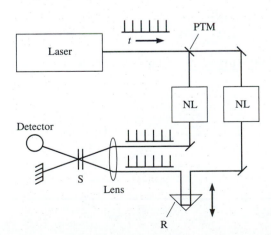

■ Figure 5.8

Schematic diagram of the pump-probe technique using fast laser pulses to probe reaction rates in solution.

mirror (PTM), and each light beam is then converted to the desired wavelength by the use of nonlinear optical processes (NL). Because light travels roughly 0.3 mm in 1 ps, the pathlengths for the two beams can be adjusted so that the pulse of low intensity reaches the sample (S) after the pulse of high intensity. The delay time is scanned by translating the retroreflector (R). The pump pulse is used to initiate the reaction of interest, and the absorption of the probe pulse, proportional to the concentration of a reactant or product, is monitored as a function of the position of the reflector (R). By such a technique one can measure the reaction rate on a picosecond or even femtosecond timescale (see Section 7.5.6).

For example, the pump-probe technique has been used to investigate the isomerization of 1-1'-binaphthyl, shown in **Figure 5.9.** The angle between the two naphthyl groups is changed by about 40° upon excitation from the ground to the first excited singlet state of this molecule. Since the molecule absorbs at a different wavelength in its excited state geometry, the reaction can be followed by exciting the first singlet state with one short laser pulse and observing as a function of time delay the increase in absorption of a second pulse tuned to probe the product. As one might expect, the time it takes to increase the dihedral angle between the two naphthyl groups depends on the surrounding solvent, which must be pushed out of the way as the geometry changes. Because the forces of friction dominate, the time for reaction varies with the viscosity of the solvent. **Table 5.2** shows some typical experimental results and compares these to a theoretical result based on the frictional forces. Experiments such as this allow one to directly evaluate the role of friction in chemical reactions that we considered in Section 5.2.2.

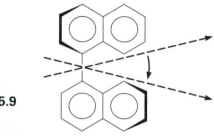

Figure 5.9

1-1'-binapthyl.

| TABLE 5.2 | **Experimental and Theoretical Reaction Times for the Isomerization of 1-1'-binaphthyl in Solvents of Varying Viscosity** |

Solvent	Viscosity (cP)*	Experimental Rise Time (ps)	Theoretical Prediction (ps)
Ethanol	1.08	12.2 ± 1.0	12.2
n-Propanol	2.23	13.7 ± 1.0	13.9
n-Butanol	2.95	15.0 ± 1.0	15.2
n-Pentanol	4.33	17.6 ± 1.0	17.7
n-Hexanol	5.27	19.2 ± 1.0	19.5
n-Heptanol	6.90	24.0 ± 2.0	22.9
n-Octanol	8.95	27.0 ± 2.0	27.4

From D. P. Millar and K. B. Eisenthal, *J. Chem. Phys.* **83,** 5076 (1985).

*1 poise = 0.1 N m^{-2} s

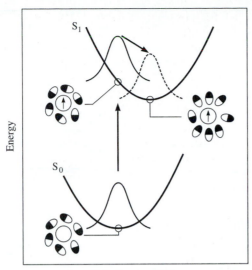

Solvation coordinate

Figure 5.10

The change in electric dipole on excitation induces a change in the solvation. Emission from S_1
back to S_0 can then be used to study the dynamics of the solvation.

From R. M. Stratt and M. Maroncelli, *J. Phys. Chem.* **100,** 12981–12996 (1996). Reprinted with permission
from *The Journal of Physical Chemistry.* Copyright 1996 American Chemical Society.

Another use of ultrafast lasers is to measure the solvation dynamics associated
with the electron transfer reactions we considered in Section 5.3.2. **Figure 5.10**
shows the principle. Suppose optical excitation from the ground singlet S_0 to the
first excited singlet S_1 results in a substantial change in the dipole moment of the
molecule. The solvent cannot adjust to the change in dipole on the time scale of the
excitation, nor is the geometrical arrangement of the solvent likely to be of the lowest energy configuration for the new electronic configuration. With increasing time,
however, the solvent molecules will rearrange their orientation so as to stabilize the
new dipole moment and lower the energy.

Imagine now that one could resolve in time the emission from S_1 back to S_0.
Immediately after the excitation, emission would be at roughly the wavelength of
the excitation, but as time progresses, the emission will occur to longer wavelengths
as the energy on the upper potential surface approaches a minimum. **Figure 5.11**
shows how the emission spectrum shifts to the red as time progresses from 0 to 50
ps. The authors used a fluorescence up-conversion technique to obtain the subpi-
cosecond time resolution. The fluorescence at a particular time delay and wavelength was combined with a probe laser pulse to produce a signal at the sum of the
frequencies of the probe laser and the fluorescence. For a fixed delay between the
probe laser pulse and the initial excitation pulse, the up-converted signal was then
scanned in wavelength to generate a curve in the figure. The time delay between the
probe pulse and the excitation pulse was then changed and the wavelength scan
repeated to obtain the series of curves shown in the figure. Not surprisingly, different solvent molecules respond more or less rapidly depending on their dielectric
constant.

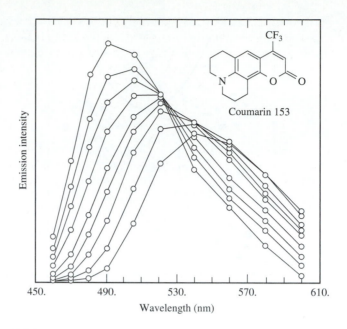

■ **Figure 5.11**

The shift in the fluorescence spectrum of Coumarin 153 in formamide as a function of time following excitation. The times are 0, 0.05, 0.1, 0.2, 0.5, 1, 2, 5, and 50 ps, in order of decreasing peak intensity.

From R. M. Stratt and M. Maroncelli, *J. Phys. Chem.* **100**, 12981–12996 (1996). Reprinted with permission from *The Journal of Physical Chemistry*. Copyright 1996 American Chemical Society.

5.5   SUMMARY

Reactions in solution are found to differ little from their gas phase counterparts when the solvent does not participate as a reactant. The principal difference is that the solvent molecules both prevent the initial approach of reactants and also form a cage around them when they finally encounter one another. The forces responsible for the cage effect are related to the friction coefficient, ζ, in the Langevin equation. Using Langevin's model, the diffusion coefficient D in the fluid can be related to the friction coefficient:

$$D = \frac{kT}{\zeta}. \tag{5.7}$$

If the reaction between reactants is facile when they are caged, then the reaction rate is often controlled by the rate at which the reactants can diffuse toward one another; the reaction is *diffusion controlled*. For such reactions, we found that the rate constant was given by

$$k = 4\pi(D_A + D_B)\beta, \tag{5.27}$$

where, for nonionic species, β is simply equal to R, the average diameter of the reactants. For ionic species,

$$\beta = \frac{-z_A z_B r_0}{1 - \exp(z_A z_B r_0/R)}, \tag{5.29}$$

where $r_0 = e^2/\epsilon kT$, with e here signifying the charge on an electron and ϵ here signifying the dielectric constant of the medium.

When a reaction involves charged species in solution, the solvent can influence the reaction by stabilizing differently the reactants, activated complexes, and products. The influence on the reaction rate of the ionic strength of the solution can be understood by considering how the activity coefficients deviate from unity. When these deviations can be approximated by the Debye-Hückel limiting law, then we obtain the following relationship between the rate constant and the ionic strength:

$$\log_{10}k_0 = \log_{10}k_0 + 2Az_Az_BI_c^{1/2}, \tag{5.35}$$

where I_c is the ionic strength and A is a constant equal to 0.50 $(L/mol)^{1/2}$ for water at 25°C. When unlike ions form an activated complex, the complex is less charged than the reactants and is thus less stabilized by increasing ionic strength than the reactants; the apparent activation energy increases so that the rate constant decreases. When like ions form an activated complex, on the other hand, the complex has a higher charge than the reactants and is more stabilized by increasing ionic strength; the apparent activation energy decreases and the rate constant increases.

A second situation where the solvent can affect the reaction rate constant is during electron transfer. Marcus has addressed the problem of how the activation energy for the reaction depends on ΔG^0, the free energy of the reaction. By using an approximate derivation of his theory, we determined that the rate constant for the electron transfer reaction is given by

$$k_{ET}(T) = \frac{kT}{h}\exp\left(-\frac{(E_m + \Delta G^0)^2}{4E_mkT}\right), \tag{5.38}$$

where E_m, called the reorganization energy, is the energy the donor would have in excess of the acceptor if it and its surrounding solvent were placed at the minimum energy configuration of the acceptor. This equation predicts that the rate constant for electron transfer will first increase with $-\Delta G^0$, that is as the products become more stable with respect to the reactants, but that it will then decrease as $-\Delta G^0$ increases further. This latter region is called the Marcus inverted region.

In the above examples, the solvent influences the stability of reactants, activated complexes and products, but it does not itself react. When the solvent does participate as a reactant, the reaction rates in solution are often large enough to require special experimental techniques for determination of the rate constant. Relaxation and ultrafast laser techniques were briefly described as ways to measure rapid rate constants.

appendix 5.1

The Langevin Equation and the Mean Squared Displacement

In this appendix we solve the Langevin equation, **equation 5.1,** for the root-mean-squared distance traveled by a Brownian particle. Noting that $v = dx/dt$, we multiply both sides of **equation 5.1** by x and rewrite it as

$$mx\frac{d}{dt}\frac{dx}{dt} = -\zeta x\frac{dx}{dt} + xf(t). \tag{5.39}$$

Since $d/dt\,(x\,dx/dt) = (dx/dt)^2 + x\,(d/dt)\,(dx/dt)$, we rewrite the first term:

$$m\frac{d}{dt}\left(x\frac{dx}{dt}\right) - m\left(\frac{dx}{dt}\right)^2 = -\zeta x\frac{dx}{dt} + xf(t). \tag{5.40}$$

We now take the ensemble average of both sides of the above equation. On the right-hand side, $\langle xf(t)\rangle = \langle x\rangle\langle f(t)\rangle$ since we assume that the fluctuating forces are random and independent of position; note that $\langle f(t)\rangle = 0$, as stated in **equation 5.2.** Thus,

$$m\left\langle \frac{d}{dt}\left(x\frac{dx}{dt}\right)\right\rangle = -\zeta\left\langle x\frac{dx}{dt}\right\rangle + m\left\langle \left(\frac{dx}{dt}\right)^2\right\rangle. \tag{5.41}$$

The last term is just $m\langle v^2\rangle$, which is equal to kT for a particle with one degree of freedom (in the x direction, as assumed here). The solution to the resulting differential equation is then simply

$$m\frac{d}{dt}\left\langle x\frac{dx}{dt}\right\rangle = -\zeta\left\langle x\frac{dx}{dt}\right\rangle + kT$$

$$\left\langle x\frac{dx}{dt}\right\rangle = Ae^{-\zeta t/m} + \frac{kT}{\zeta}, \tag{5.42}$$

where the validity of the second step can be verified by differentiation. The constant A must be determined from the initial condition, which we take to be $x = 0$ when $t = 0$ so that $A = -(kT/\zeta)$. We now write $x(dx/dt)$ as $\frac{1}{2}(d/dt)(x^2)$ and obtain the equation

$$\frac{1}{2}\frac{d}{dt}\langle x^2\rangle = \frac{kT}{\zeta}(1 - e^{-\zeta t/m}). \tag{5.43}$$

Integrating both sides of this equation gives the final expression for the mean squared displacement

$$\langle x^2\rangle = \frac{2kT}{\zeta}\left[t - \frac{m}{\zeta}(1 - e^{-\zeta t/m})\right]. \tag{5.44}$$

Let us now investigate two limiting cases for this formula. When t is short enough so that $t/m \ll 1$, then we can approximate the exponential by its first few terms: $\exp(-\zeta t/m) \approx 1 - (\zeta t/m) + \frac{1}{2}(\zeta t/m)^2$, so that

$$\langle x^2\rangle = \frac{kT}{m}t^2. \tag{5.45}$$

This formula just tells us that on such short time scales, the rms displacement goes as $(kT/m)^{1/2}t$, or that the particle just moves freely with its one-dimensional rms speed, $(kT/m)^{1/2}$. On this time scale, when $t \ll m/\zeta$, the particle does not encounter the friction of the medium (note that the friction coefficient does not appear in **equation 5.45**). For a particle of water's density and a radius 100 nm moving in a fluid such as water, the time m/ζ is about 10^{-9} s.

Now consider the case when $t \gg m/\zeta$. Under this condition, **equation 5.44,** reduces to

$$\langle x^2\rangle = \frac{2kT}{\zeta}t. \tag{5.46}$$

This last equation is the long-time mean squared displacement. Note that the mean squared displacement is inversely proportional to the friction coefficient ζ.

Diffusion with an Electrostatic Potential

For neutral species, motion is controlled by the response to a concentration gradient, but for ionic species, motion is also affected by the electrostatic potential between the ions. Before proceeding to calculate the diffusion controlled rate constant, we digress briefly to consider how the diffusion equation, $J_z = -D(\partial c/\partial z)$, would need to be modified if species responded not only to a concentration gradient but also to an electric field.

An ion which is moving in a potential gradient $\nabla\Phi$ caused has a *mobility* $\mu = v/(\nabla\Phi)$. Thus, the mobility is simply the proportionality constant between the velocity v and the potential gradient; the bigger the gradient, the faster the ion will move. If A and B, for example, are oppositely charged, the potential gradient between them will cause them to move toward one another, so that the flux toward one another will be increased. Since v is the velocity toward one another, the flux toward one another is increased by $cv = c\mu\nabla\Phi$, where c is the concentration. The general diffusion equation $J_z = -D(\partial c/\partial z)$ would then need to be modified to give

$$J_z = -\left(D\frac{\partial c}{\partial z} + \frac{z}{|z|}\mu c\frac{\partial\Phi}{\partial z} \right),\qquad(5.47)$$

where z is the charge on the ion and the factor $z/|z|$ is used to take care of the sign of the charge. The first term is the contribution to the flux due to diffusion, whereas the second term is the contribution due to the gradient of the electric field.

At equilibrium in a solution, the overall flux at any point must be zero, so that

$$J_z = 0 = -\left(D\frac{\partial c}{\partial z} + \frac{z}{|z|}\mu c\frac{\partial\Phi}{\partial z} \right).\qquad(5.48)$$

Solution of this equation is relatively straightforward:

$$\frac{dc}{c} = -\frac{z}{|z|}\frac{\mu}{D}\,d\Phi,\qquad(5.49)$$

or

$$\ln\frac{c}{c^0} = -\frac{z}{|z|}\frac{\mu}{D}\Phi,\qquad(5.50)$$

where c^0 is the concentration when the electric field is zero. We now assume that at equilibrium the concentration must obey the Boltzmann law:

$$c = c^0\exp\left(-\frac{U}{kT}\right),\qquad(5.51)$$

where the potential energy U of the ions in the field is given by $U = ze\Phi$.

Substitution of **equation 5.51** into **equation 5.50** and use of $U = ze\Phi$ gives an equation that relates the mobility and the diffusion coefficient:

$$\frac{\mu}{D} = \frac{|z|e}{kT}.\qquad(5.52)$$

Finally, substitution of **equation 5.52** and $U = ze\Phi$ into **equation 5.47** gives

$$J_z = -D\left(\frac{\partial c}{\partial z} + \frac{c}{kT}\frac{\partial U}{\partial z} \right).\qquad(5.53)$$

More generally, we recognize that the flux can have components in each of the three spatial dimensions, so that

$$J = -D\left(\nabla c + \frac{c}{kT}\nabla U\right). \quad (5.54)$$

We now return to the development of Section 5.2.4 and generalize **equation 5.21** for the case of diffusion in an electrostatic field to obtain:

$$k_r[A]B(R) = [A]4\pi r^2(D_A + D_B)\left[\frac{dB(r)}{dr} + \frac{B(r)}{kT}\frac{dU(r)}{dr}\right]. \quad (5.55)$$

In this equation [A] is again constant, $B(R)$ is constant (to be determined), and $B(r)$ denotes how the microscopic concentration of B varies with distance.

We now rewrite the right-hand side of the last equation as

$$k_r[A]B(R) = [A]4\pi r^2(D_A + D_B)e^{-U(r)/kT}\frac{d}{dr}[B(r)e^{U(r)/kT}]. \quad (5.56)$$

Division of both sides of the equation by $[A]r^2\exp[-U(r)/kT]$ and multiplication by dr yields

$$k_r B(R)\frac{e^{U(r)/kT}dr}{r^2} = 4\pi(D_A + D_B)\,dB(r). \quad (5.57)$$

Finally, if we define

$$\beta^{-1} = \int_R^\infty \frac{e^{U(r)/kT}\,dr}{r^2}, \quad (5.58)$$

then integration of both sides of **equation 5.62** over dr from $r = R$ to $r = \infty$ gives

$$k_r B(R)\int_R^\infty \frac{e^{U(r)/kT}}{r^2}\,dr = 4\pi(D_A + D_B)\int_R^\infty d[B(r)e^{U(r)/kT}],$$

$$\frac{k_r B(R)}{\beta} = 4\pi(D_A + D_B)[B(r = \infty)e^{U(r=\infty)/kT} - B(r = R)e^{U(r=R)/kT}]$$

or

$$B(R)\left[\frac{k_r}{\beta} + 4\pi(D_A + D_B)e^{U(R)/kT}\right] = 4\pi(D_A + D_B)[B]$$

$$B(R) = \frac{[B]}{e^{U(R)/kT} + \dfrac{k_r}{4\pi(D_A + D_B)\beta}}, \quad (5.59)$$

where $U(r = \infty)$ has been assumed to be zero. Note that $U(R)$ is very small, since the two ions are in contact, so that $\exp[U(R)/kT] \approx 1$. Substitution of this solution for $B(R)$ into **equation 5.20** gives

$$\frac{d[P]}{dt} \approx \frac{k_r}{1 + \dfrac{k_r}{4\pi(D_A + D_B)\beta}}[A][B]. \quad (5.60)$$

Thus, when $k_r \gg 4\pi(D_A + D_B)\beta$, the overall rate constant for the reaction is given by

$$k = 4\pi(D_A + D_B)\beta. \quad (5.61)$$

This is the equation given as **equation 5.27**.

suggested readings

R. S. Berry, S. A. Rice, and J. Ross, *Physical Chemistry* (Wiley, New York, 1980).

J. H. Espenson, *Chemical Kinetics and Reaction Mechanisms* (McGraw-Hill, New York, 1981).

G. G. Hammes, *Principles of Chemical Kinetics* (Academic Press, New York, 1978).

R. M. Noyes, "Effects of Diffusion Rates on Chemical Kinetics," *Progress in Reaction Kinetics* **1,** 129 (1961).

M. J. Pilling and P. W. Seakins, *Reaction Kinetics* (Oxford University Press, Oxford, 1995).

R. M. Stratt and M. Maroncelli, "Nonreactive Dynamics in Solution: The Emerging Molecular View of Solvation Dynamics and Vibrational Relaxation," *J. Phys. Chem.* **100,** 12981 (1996).

G. A. Voth and R. M. Hochstrasser, "Transition State Dynamics and Relaxation Processes in Solutions: A Frontier of Physical Chemistry," *J. Phys. Chem.* **100,** 13034 (1996).

R. W. Weston, Jr., and H. A. Schwarz, *Chemical Reactions* (Prentice-Hall, Englewood Cliffs, NJ, 1972).

problems

5.1 For a diffusion-controlled reaction the rate constant depends on the size of the reactants both through R and through the diffusion coefficients. In solution the diffusion coefficient can be approximated by the Stokes-Einstein relationship, $D = kT/6\pi\eta r$, where η is the viscosity coefficient of the solvent and r is the radius of the diffusing species. For reactants of the same size, will the diffusion-controlled rate constant depend on the size?

5.2 The reaction rate between species of charge $+1$ and one of charge -2 will (a) increase, (b) decrease, or (c) remain constant with increasing ionic strength?

5.3 For the mechanism

$$A + B \underset{k_r}{\overset{k_f}{\rightleftharpoons}} C$$

show that in the limit of small perturbations the concentration of A relaxes to its new equilibrium value A_e according to the equation $A(t) = A_e + x_e \exp\{-[k_f(A_e + B_e) + k_r]t\}$, where x_e is the initial concentration change between [A] and its equilibrium value; $x_e = A(t = 0) - A_e = B(t = 0) - B_e = C_e - C(t = 0)$.

5.4 The reaction between hydrogen peroxide and iodide has been investigated by F. Bell, R. Gill, D. Holden, and W. F. K. Wynne-Jones in *J. Phys. Chem.* **55,** 874 (1951): $H_2O_2 + 2I^- + 2H^+ \rightarrow 2H_2O + I_2$. The reaction proceeds by two parallel mechanisms and leads to rate law

$$\frac{-d[H_2O_2]}{dt} = \frac{d[I_2]}{dt} = k_1[H_2O_2][I^-] + k_2[H_2O_2][I^-][H^+].$$

The rate constants k_1 and k_2 are observed to vary with ionic strength in the fashion shown in the table. Are these data qualitatively and quantitatively consistent with the primary salt effect in **equation 5.35?**

I_c	$k_1(\text{M}^{-1}\,\text{s}^{-1})$	$k_2(\text{M}^{-2}\,\text{s}^{-1})$
0.000	0.658	19.0
0.0207	0.663	15.0
0.0525	0.670	12.2
0.0925	0.679	11.3
0.1575	0.694	9.7
0.2025	0.705	9.2

5.5   Suppose that the critical distance for reaction of iodine with CCl_4 is 2×10^{-10} m and that the diffusion coefficient of iodine atoms in CCl_4 is 3.0×10^{-9} m^2 s^{-1} at 25°C. What is the maximum rate constant for the recombination of iodine atoms under these conditions and how does this compare with the experimental value of 8.2×10^9 M^{-1} s^{-1} measured by R. M. Noyes in *J. Am. Chem. Soc.* **86,** 4529 (1964)?

5.6   Suppose that three ions, A with charge z_A, B with charge z_B, and C with charge z_C react in a termolecular process. Derive an equation similar to **equation 5.35** to describe the primary salt effect on this reaction.

6

Reactions at Solid Surfaces

Chapter Outline

6.1 INTRODUCTION

Systems of high surface-to-volume ratio play extremely important roles in life processes, industrial manufacture, and even geological change. The exchange of oxygen and carbon dioxide in our lungs or in green plants, the conversion of straight-chain hydrocarbons to aromatic compounds on a platinum catalyst, the formation of $ClONO_2$ on ice crystals responsible for the "ozone hole," and the reactions responsible both for the formation of soil and for the use of its mineral content by living organisms are all due to chemical reactions that occur at the surface of a substance.

In this chapter we examine the special kinetic behavior of systems reacting at surfaces. Our goal is to see how the kinetic principles we have developed can be used to understand the kinetics at surfaces. While we will focus on the solid-gas interface, much of what we will discuss is applicable as well at other important interfaces. We will start by considering the nature of the surface separating two heterogeneous phases. Processes that occur at this interface are characterized by several steps: adsorption and desorption onto/from the surface, reaction at the surface, and diffusion along the surface. After considering a model of adsorption due to Langmuir, we will then investigate unimolecular and bimolecular reactions at surfaces. As we will see, there are many parallels between the catalytic action of surfaces and the catalytic action of enzymes, already studied in Chapter 2. We will next comment briefly on the nature of catalytic sites on surfaces and then discuss surface diffusion in some detail. We end the chapter with a description of two important techniques for surface study: temperature programmed desorption and modulated molecular beam methods.

Of course, the special feature of the gas-solid interface, or indeed of any surface, is that molecules can be exchanged there between two heterogeneous phases. A naive view of the surface dividing two phases is simply a flat layer of atoms, but detailed experiments have demonstrated that surfaces themselves have structure. **Figure 6.1** shows a schematic model of a solid surface. Most of the atoms (or molecules) on the surface are arranged in layers called *terraces,* but the terraces often have either *adatoms* or *vacancies.* Between the terraces are *steps,* often, but not always, of 1 atom. The steps themselves are not straight, but rather have *kinks* and *step-adatoms.* An image of an actual surface, taken by a technique called scanning-tunneling microscopy, is shown in **Figure 6.2.** This structure of surfaces is extremely important to kinetics, since the rate of a surface chemical reaction may vary by several orders of magnitude, depending on the number of steps or on the detailed arrangement of the atoms in the terraces.

The arrangement of atoms on the terraces is, of course, related to the arrangement of atoms in the bulk. While it is beyond our current interest to investigate in detail the crystallographic designations for atomic arrangements in the terraces, it is important to note that different arrangements are possible and that the chemistry that occurs may be dramatically influenced by the structure of the surface. **Figure 6.3** shows a few commonly observed structures.

This chapter will be concerned principally with the characterization of the kinetics of reactions at the gas-solid interface. Such reactions typically involve five main processes: flux of reactants to the surface, adsorption onto the surface, diffusion to reactive sites, reaction, and desorption of the products into the gas phase.

The first process, the flux of reactants to the surface, has already been discussed in Section 4.3.2, where we saw that the flux of molecules through a plane was given by **equation 4.6:** $J_z = \frac{1}{4} n^* <v>$, where n^* is the concentration of the reactant and $<v>$ is its average velocity. This flux times the surface area gives the number of reactants that strike the surface per unit time; it is an upper limit on the rate of the reaction. **Example 6.1** shows that during an exposure of 1 langmuir (1 L = 10^{-6} torr-s) about as many molecules hit the surface per unit area as there are atoms on the surface.

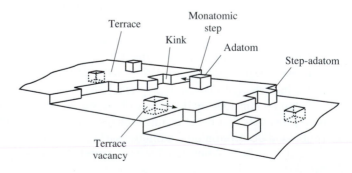

■ **Figure 6.1**

Model of a solid surface depicting different surface sites.

From G. A. Somorjai, *Chemistry in Two Dimensions: Surface* (Cornell University Press, Ithaca, NY, 1981). Reprinted with permission of Dr. G. A. Somorjai.

Figure 6.2

A scanning-tunneling microscope image of an Si(001) miscut 0.5° toward [100]. The surface steps down from left to right.

Courtesy of M. G. Lagally; for a description of a similar figure see B. S. Swartzentuber, Y.-W. Mo, R. Kariotis, M. G. Lagally, and M. B. Webb, *Phys. Rev. Lett.* **65,** 1913 (1990).

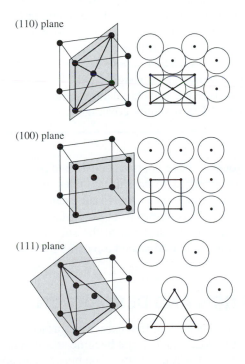

(110) plane

(100) plane

(111) plane

Figure 6.3

The surfaces of single-crystal metals correspond to planes drawn through the solid. A few simple planes for body-centered cubic crystals are given in this figure.

From R. Gomer, "Surface Diffusion," *Scientific American,* August, 1982. Reprinted by permission of Jerome Kuhl.

example 6.1

The Number of Molecules Striking the Surface during a 1-Langmuir Exposure Compared to the Number of Surface Sites

Objective   A 1-langmuir (1-L) exposure is defined as an exposure of 10^{-6} torr-s. Assuming that the surface of interest is a square array of atoms separated by 4 Å and that every atom provides an adsorption site, calculate the fractional coverage of the surface after an exposure to 1 L of a gas that adsorbs on the surface at each collision.

Method   The flux of the adsorbate to the surface is given by $J = \frac{1}{4} <v>n^*$. We need to calculate how many adsorbates hit the surface given a 1-L exposure and to compare this to the number of adsorption sites. The latter number can be calculated from the number of atoms on the surface given the geometry of the surface layer and the spacing.

Solution   We are given that the exposure is $1\,L = 10^{-6}$ torr-s $= n^*\Delta t$, where Δt is the duration of the exposure; thus $n^* = (10^{-6}$ torr-s$)/\Delta t$ and $J\Delta t = \frac{1}{4} <v>(10^{-6}$ torr-s$)$. Supposing $<v>$ to be around 500 m/s gives the number of molecules striking the surface during this exposure: $J\Delta t = (0.25)(500$ m/s$)(10^2$ cm/m$)(10^{-6}$ torr-s$)(3.22 \times 10^{16}$ molec cm^{-3} torr$^{-1}) = 4.0 \times 10^{14}$ molec cm^{-2}, where the last number in parentheses is the conversion factor from torr (1/760 atm) to molecules cm^{-3} at 300 K. Now we need to see how many surface sites per cm^2 are on the surface. The spacing between atoms is 4 Å, so there are $1/(4 \times 10^{-8})^2 = 6.25 \times 10^{14}$ sites cm^{-2}. Thus the coverage is roughly $(4 \times 10^{14})/(6.25 \times 10^{14}) = 0.64$. For order-of-magnitude calculations, one often assumes that an exposure of 1 L gives unity coverage.

The remaining processes have not been considered previously. Molecules in the gas phase must become adsorbed on the surface. They then typically migrate by diffusion to sites where they can react or where they can encounter another reactant. Following reaction, the products desorb from the surface into the gas phase. We will examine each of these processes in detail in the subsequent sections and end this chapter with a discussion of a few advanced topics for the study of surface reactions.

6.2   ADSORPTION AND DESORPTION

An early and important discovery in the history of surface catalysis was the observation by Faraday that molecules must first become attached to, or *adsorb* on, a surface before they can react.[a] The mutual attraction between an approaching molecule and a surface can be attributed to two types of interactions. In the first, called

---

[a]M. Faraday, *Philos. Trans.* **124**, 55 (1834).

physisorption, the attraction is due to the weak *van der Waals* or dispersion forces. Although, in the general case neither the molecule nor the surface atom will have an average dipole moment, the instantaneous positions of the nuclei and electrons will give rise at any time to dipole moments on the molecule and on the surface atom. The attraction between these instantaneous dipole moments leads to the force responsible for physisorption. The strength of the physisorption bond is typically less than 20 kJ mole^{-1}.

In the second type of adsorption, called *chemisorption,* the molecule is attracted to the surface by the same forces that are present in a normal chemical bond, with strengths of 300–500 kJ mole^{-1}. Just as in a normal chemical reaction, a barrier to chemisorption is frequently observed.

Figure 6.4 shows a simplified view based on the ideas of Lennard-Jones[b] of the potential energy $V(R)$ between an approaching molecule and a surface atom as a function of the molecule-surface distance R. Curves are shown for both the physisorption and chemisorption bonds. At infinite separation, the physisorbed molecule is free of the influence of the surface; it is customary to define this state as the zero of potential energy. Note that the physisorption bond is rather weak and that its minimum occurs at larger separation compared to the chemisorption bond. Note also that the chemisorption curve is shown as having its energy at infinite separation somewhat higher than the zero of potential energy. This is because chemisorption

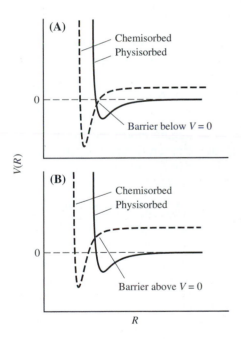

■ **Figure 6.4**

Potential energy curves for chemisorption and physisorption. In (A) the transition from chemisorption to physisorption occurs over a barrier which lies below the energy of infinite separation, while in (B) the barrier lies above this energy. Chemisorption in case (B) is said to be *activated.* The depth of the physisorbed well is exaggerated in the figure compared to the depth of the chemisorbed well.

[b]J. E. Lennard-Jones, *Trans. Farad. Soc.* **28,** 333 (1932).

typically involves a rearrangement of the structure of the adsorbed molecule. If the adsorbed molecule is pulled away from the surface but frozen in structure, its potential energy will generally be higher than that of the free molecule.

It is typical that to pass from the physisorbed state to the chemisorbed state, the system must overcome a potential barrier. As shown in the figure, this barrier corresponds to the crossing between the physisorption and chemisorption potential energy curves. If the barrier is above the energy corresponding to infinite separation (taken as the zero of energy in the figure), then the chemisorption is said to be *activated*; energy must supplied for the chemisorption reaction to take place. If the barrier is below the energy corresponding to infinite separation, the reaction can proceed even if the molecule approaches the surface with nearly zero kinetic energy. Of course, this one-dimensional picture is highly simplified, since the potential of interaction will depend not only on R but also on the orientation of the molecule and on the specific site on the surface which it approaches.

6.2.1   The Langmuir Isotherm

It was Irving Langmuir who first studied the adsorption process quantitatively.[c] In the simplest model, we suppose that molecules can adsorb only at specific sites on the surface, and that once a site is occupied by one molecule, it cannot adsorb a second molecule. The adsorption process can then be represented as

$$A + S \rightarrow A\text{--}S,$$

where A is the adsorbing molecule, S is the surface site, and A–S represents an A molecule bound to the surface site. In a similar way, the reverse desorption process can be represented as

$$A\text{--}S \rightarrow A + S.$$

If [A] is the concentration of molecules and if θ is the fraction of the surface sites covered by A, then in the Langmuir model the rate of adsorption is proportional to $k_a[A](1-\theta)$, while the rate of desorption is proportional to $k_d\theta$, where k_a and k_d are the rate constants for the two processes. At equilibrium we know that the rates of the two processes are equal, so that

$$k_a[A](1-\theta) = k_d\theta,$$

or

$$\frac{\theta}{1-\theta} = \frac{k_a}{k_d}[A]. \tag{6.1}$$

Denoting $K \equiv k_a/k_d$ and solving for the fraction of occupied sites, we obtain the Langmuir adsorption isotherm:

$$\theta = \frac{K[A]}{1 + K[A]}. \tag{6.2}$$

Note that **equation 6.2** predicts that the fraction of occupied sites should increase linearly with [A] when [A] is low enough so that $K[A] << 1$, while it should approach unity when [A] is large enough so that $K[A] >> 1$. This behavior is demonstrated in the plot of θ versus [A] shown in **Figure 6.5.**

[c]I. Langmuir, *J. Am. Chem. Soc.* **38,** 221 (1916); **40,** 1361 (1918). Langmuir won the Nobel Prize in Chemistry in 1932 for this work.

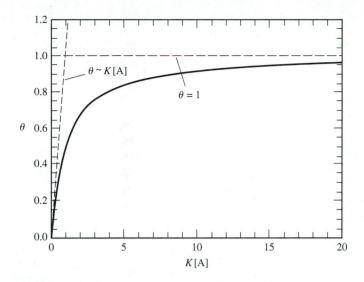

■ **Figure 6.5**

The surface coverage as a function of adsorbate concentration for the simple Langmuir model.

A somewhat different situation occurs if, as is often observed, the adsorbing molecule dissociates when it chemisorbs. In this case, each dissociated molecule occupies two sites rather than one. A detailed consideration of this situation is outlined in Problem 6.6.

6.2.2   Competitive Adsorption

When two or more species can occupy the same sites on a surface the situation becomes more interesting. Let one species A have fractional surface coverage θ_A and let a second species B have coverage θ_B. Then the rates for adsorption of species A and B are $k_a^A(1 - \theta_A - \theta_B)[A]$ and $k_a^B(1 - \theta_A - \theta_B)[B]$, respectively, since in each case the rate of adsorption is proportional to the fraction of free sites $(1 - \theta_A - \theta_B)$. Similarly, the rate of desorption in each case is proportional to the number of sites occupied by each species. For A and B these rates are $k_d^A\theta_A$ and $k_d^B\theta_B$, respectively. At equilibrium the rates of adsorption and desorption for each species must be the same, so that, denoting $K_A \equiv k_a^A/k_d^A$ and $K_B \equiv k_a^B/k_d^B$, we find

$$\frac{\theta_A}{1 - \theta_A - \theta_B} = K_A[A], \tag{6.3}$$

and

$$\frac{\theta_B}{1 - \theta_A - \theta_B} = K_B[B]. \tag{6.4}$$

Equations 6.3 and **6.4** are two simultaneous equations in the variables θ_A and θ_B that we would like to know. After a little algebra, we find that

$$\theta_A = \frac{K_A[A]}{1 + K_A[A] + K_B[B]} \tag{6.5}$$

and that

$$\theta_B = \frac{K_B[B]}{1 + K_A[A] + K_B[B]}. \tag{6.6}$$

Note that if the product of K and the concentration for one of the species, say A, is much higher than for the other, then the surface coverage of the other, B in this case, will be reduced. This is because both species are competing for the same sites on the surface. In the extreme limit when $K_A[A] \gg K_B[B]$, B will be almost totally excluded from the surface, while the surface coverage of A will approach the value it would have in the absence of B; in this limit **equation 6.5** reduces to **equation 6.2.** The situation is analogous to the competitive inhibition in enzyme catalysis that we encountered in Section 2.5.2.

6.2.3   Heats of Adsorption

It is instructive to consider the equilibrium between adsorption and desorption in light of thermodynamics. From the Clausius-Clapeyron equation we know that the quantity $[d \ln(P_A/P_A^0)/d(1/T)]_{\theta=\text{const}}$ gives a measure of the heat of adsorption:

$$\frac{d \ln(P_A/P_A^0)}{d(1/T)}\Bigg|_{\theta=\text{const}} = -\frac{\Delta H_{ad}}{R}, \tag{6.7}$$

where P_A^0 is a reference pressure (normally one atmosphere) and ΔH_{ad} is the molar heat of adsorption. Because $\Delta H_{ad} = \Delta(E + pV)_{ad}$, we find $\Delta E_{ad} = \Delta H_{ad} - \Delta(pV)_{ad}$. For an ideal gas, $\Delta(pV)_{ad} = \Delta nRT = -RT$, so $\Delta E_{ad} = \Delta H_{ad} + RT$.

Figure 6.6 shows the heat of adsorption in kcal/mole for CO on various polycrystalline transition-metal surfaces.

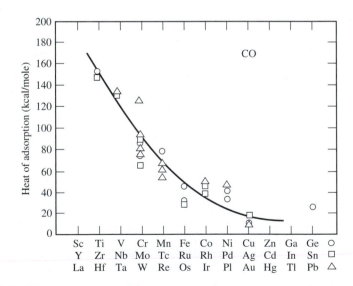

■ Figure 6.6

Heats of adsorption of CO on polycrystalline transition-metal surfaces.

From G. A. Somorjai, *Chemistry in Two Dimensions: Surfaces* (Cornell University Press, Ithaca, NY, 1981). Reprinted with permission of Dr. G. A. Somorjai.

6.3   REACTIONS AT SURFACES: CATALYSIS

Having considered in detail the processes of adsorption and desorption, we now turn to overall reaction mechanisms for surface reactions. In what follows, we assume that the overall mechanism consists of adsorption and desorption of the reactant(s) followed by reaction on the surface. We will assume for the moment that diffusion on the surface and desorption of the products are rapid enough not to be rate limiting. Diffusion on the surface will be treated in more detail in Section 6.4.

Overall reactions at surfaces, like those in the gas phase, can be categorized by the order of the reaction, and, as in the gas phase, the order for an elementary step is equal to the molecularity.

6.3.1   Unimolecular Surface Reactions

A unimolecular elementary step in a reaction at a surface is one whose rate is first order in the surface coverage of the reactant, θ:

$$\frac{d[P]}{dt} = k\theta = \frac{kK[A]}{1 + K[A]}, \tag{6.8}$$

where P is the product of the surface reaction and the last equality follows from **equation 6.2.** Note that the order of the overall reaction depends on the conditions. Thus, for low values of $K[A]$, the reaction rate increases linearly with [A] because adsorption is the rate limiting step; the reaction is first order in [A]. However, at high values of $K[A]$ where the surface is saturated with A, the rate of the reaction becomes zero order in [A] and depends only on the rate constant k. This situation is analogous to the enzyme catalyzed reaction studied in Section 2.5.2.

An example of a unimolecular surface reaction is the decomposition of ammonia on platinum to give nitrogen and hydrogen at temperatures where both products desorb rapidly. (Under conditions where N_2 desorbs rapidly but H_2 does not, the reaction rate is still first order in NH_3 but is inversely proportional to $[H_2]$; see Problem 6.7.) **Example 6.2** shows that dehydrogenation of alcohols at surfaces is also a unimolecular reaction.

example 6.2

Dehydrogenation of Alcohols

Objective   The dehydrogenation of gaseous alcohols, RCH_2CH_2OH, to give an olefin, $RCH_2{=}CH_2$, plus water on many metal oxides proceeds rapidly at 300°C. If the rate of olefin production is 10^{13} molec s^{-1} cm^{-2} of surface area at an alcohol pressure of 1 torr, and the rate constant for the reaction is $k = 10^{14}$ molec s^{-1} cm^{-2}, what would the rate of the reaction be for an alcohol pressure of 2 torr?

Method   From **equation 6.8,** we see that $d[P]/dt = k\theta = k\{K[A]/(1 + K[A])\}$, where [A] represents the concentration of alcohol. We are given $d[P]/dt$ as 10^{13} molec s^{-1} cm^{-2} and k as 10^{14} molec s^{-1} cm^{-2}, so we need to see how $\{K[A]/(1 + K[A])\}$ changes as [A] goes from 1 to 2 torr.

Solution   At 1 torr $\{K[A]/(1 + K[A])\}$ must be 0.1 from the fact that $d[P]/dt$ is 10 times smaller than k, so $K[A] = 0.1(1 + K[A])$, or $0.9K[A] = 0.1$, or $K[A] = \frac{1}{9}$ for $[A] = 1$ torr. If we double $[A]$ to 2 torr, then $K[A] = \frac{2}{9}$, or $\{K[A]/(1 + K[A])\} = \frac{2}{9}/(1 + \frac{2}{9}) = 0.182$, whereas for 1 torr this value was 0.1. Thus the rate increases by a factor of 1.82 to 1.82×10^{14} molec s^{-1} cm^{-2}.

6.3.2   Bimolecular Surface Reactions

For most bimolecular surface reactions, the reaction rate is proportional to the product of the surface coverages of the two reactants:[d]

$$\frac{d[P]}{dt} = k\theta_A\theta_B$$

$$= \frac{kK_AK_B[A][B]}{(1 + K_A[A] + K_B[B])^2},$$

(6.9)

where the last equality follows from **equations 6.5** and **6.6.** For a constant gas-phase concentration of [B], the rate of the reaction will at first increase linearly with [A]. In this region, the adsorption of A is the rate-limiting step in the reaction. As [A] is increased further, the reaction rate will reach a maximum and then finally decrease as 1/[A]. In this last region, the rapid adsorption of A prevents B from reaching the surface. **Figure 6.7** contrasts the overall rate of the reaction as a function of [A] for a unimolecular and bimolecular reaction process.

An example of a bimolecular surface reaction is the addition of hydrogen to conjugated hydrocarbons, such as ethylene. Another example is the exchange between D_2 and NH_3, as described in **Example 6.3.**

example 6.3

H/D Isotope Exchange in Ammonia

Objective   The exchange between D_2 and NH_3 on an iron catalyst at 150°C has been found to exhibit a peak in the rate of NH_2D production as the concentration of NH_3 is increased. Suggest why. Ammonia is known to bind to iron more strongly than deuterium, and deuterium is adsorbed as atomic D.

Method   Consider **Figure 6.7** and **equation 6.9** in the limits of low and high $[NH_3]$.

[d]The situation described in the text is called the Langmuir-Hinshelwood model for a bimolecular reaction, for which both species must be adsorbed on the surface before they can react with one another. An alternative bimolecular model known as the Eley-Rideal mechanism involves the direct reaction of a gas-phase molecule with an adsorbed one, for which $d[P]/dt = k\theta_A[B]$. Only a very few reactions are thought to proceed by the Eley-Rideal mechanism (see Problem 6.12). In reality, the Langmuir-Hinshelwood and Eley-Rideal models are two limiting cases; most reactions will have a mechanism somewhere between these two extremes.

Solution   If $K_{NH_3}[NH_3]$ is larger than $K_D[D]$, then the numerator of the right-hand side of **equation 6.9** will be proportional to $[NH_3]$, whereas the denominator will be proportional to $\{1 + K_{NH_3}\{[NH_3]\}^2$. Thus, $d[NH_2D]/dt \propto [NH_3]/\{1 + K_{NH_3}\{[NH_3]\}^2$. This function is proportional to $[NH_3]$ at low concentrations (where $K_{NH_3}[NH_3] \ll 1$), but inversely proportional to $[NH_3]$ at high concentrations, so the function must have a maximum at some intermediate ammonia concentration.

6.3.3   Activated Complex Theory of Surface Reactions

Activated complex theory, discussed in Chapter 3 in connection with gas-phase reactions, is equally applicable to surface reactions. The only reinterpretation necessary is that the partition functions for species on the surface are now partition functions per unit surface area rather than per unit volume. Thus, following **equation 3.17,** the rate constant for adsorption of a gas-phase molecule on a surface, $G + S \rightleftharpoons [G\text{–}S]^{\ddagger} \rightarrow G\text{–}S$, is

$$k = \frac{kT}{h} \frac{q^{\ddagger\prime}}{q_g q_s} \exp\left(-\frac{\epsilon^*}{kT}\right),   \text{(6.10)}$$

where q_g is the partition function (per unit volume) of the free gas, q_s is the partition function (per unit area) of the unoccupied surface site, and $q^{\ddagger\prime}$ is the partition

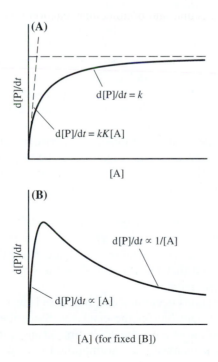

Figure 6.7

Overall rate of (A) first-order and (B) second-order surface reactions as a function of the gas-phase concentration of one of the reactants. In (B) the second reactant has been held at fixed concentration.

function (per unit area) of the activated complex, omitting the degree of freedom along the reaction coordinate. It is usually assumed that the partition function for the unoccupied surface site (involving only internal motions of the solid surface) is unchanged by the adsorption, so that the ratio of $q^{\ddagger\prime}$ to q_s depends only on the internal degrees of freedom of the activated complex. Problem 6.9 shows that **equation 6.10** reduces to $k = \frac{1}{4} <v>$ under certain limiting assumptions about the activated complex, confirming that the rate of collisions of molecules with the surface, $kn^* = \frac{1}{4} n^* <v>$, is an upper limit on the rate of adsorption.

In a similar manner, activated complex theory gives the rate constant for a unimolecular elementary step in a surface reaction, $A \rightleftharpoons A^{\ddagger} \rightarrow$ with A and $A^{\ddagger}$ as both adsorbed species:

$$k = \frac{kT}{h} \frac{q^{\ddagger\prime}}{q_A} \exp\left(-\frac{\epsilon^*}{kT}\right),$$ (6.11)

where q_A is the partition function per unit area of the adsorbate A and $q^{\ddagger\prime}$ is the partition function per unit area of the activated complex, omitting the degree of freedom along the reaction coordinate.

For a bimolecular elementary step between two adsorbed species in a surface process we likewise obtain

$$k = \frac{kT}{h} \frac{q^{\ddagger\prime}}{q_A q_B} \exp\left(-\frac{\epsilon^*}{kT}\right),$$ (6.12)

with similar interpretations for the partition functions.

6.3.4   The Nature of Surface Catalytic Sites

It has long been assumed that the catalytic nature of solid surfaces is due to "active sites" where reaction proceeds preferentially due to a structural feature that can help in some way to lower the barrier between reactants and products. H. S. Taylor[e] was the first to propose that defect sites (steps, kinks, adatoms, etc.), where low-coordinated surface atoms have unsaturated valencies, are most effective in the catalytic process. However, it is only recently that direct evidence for this thesis has been obtained. Zambelli et al.[f] examined the dissociation of nitric oxide on a ruthenium(0001) surface using scanning-tunneling microscopy and found from the distribution of nitrogen atoms after the reaction that the steps between terraces were responsible for the catalysis. **Figure 6.8** shows an image of the Ru(0001) surface with two terraces separated by a step. The left terrace is the upper terrace, and the step is along the dark line in the center of the diagram. The gray spots are nitrogen atoms, and it is noticeable in the image that the density of these atoms decreases with increasing distance from the step. This density distribution is exactly what would be expected if reaction occurs at the step edge. More detailed analysis of this and other images showed that the Ru atoms on the upper level of the step were responsible for the dissociation.

Not all surface reactions occur at steps, but there is usually some feature of the surface that is most responsible for the catalysis. Since the "active sites" are not likely to be at the location where a molecule first adsorbs, it is clear that diffusion to and from the active sites must play an important role. We thus examine surface diffusion next.

[e]H. S. Taylor, *Proc. R. Soc. London Ser. A* **108**, 105 (1925).
[f]T. Zambelli, J. Wintterlin, J. Trost, and G. Ertl, *Science* **273**, 1688–1690 (1996).

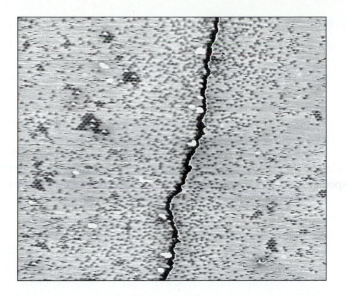

■ **Figure 6.8**

A scanning-tunneling microscopy image of a Ru(0001) surface after the dissociative adsorption of 0.3 langmuirs of NO at 315 K. The dark line in the center of the image is a step separating two terraces, with the upper one on the left. The gray dots are nitrogen atoms, while the darker and broader spots are islands of oxygen atoms. Note that the density of nitrogen atoms decreases with distance from the step.

Reprinted with permission from T. Zambelli, et al., *Science,* **273,** 1688–1690. Copyright © 1996 American Association for the Advancement of Science.

6.4   SURFACE DIFFUSION

Most surface reactions involve diffusion following adsorption. In the case of uni-molecular reactions, the surface adsorbate might need to diffuse to an active site—a kink or step, for example. In the case of a bimolecular reaction, the two adsorbates must diffuse to a common location where they can react to give products.

From the macroscopic point of view, a minor reinterpretation of **equation 4.40** (see Section 4.7) gives us the connection between the diffusion coefficient, *D,* and the change in concentration with time and space:

$$\frac{\partial n^*(x,t)}{\partial t} = D\,\frac{\partial^2}{\partial x^2}\,n^*(x,t), \tag{6.13}$$

where $n^*(x,t)$ is now the number of adsorbates per unit area as a function of position and time, and D is the diffusion coefficient in units of area per time. Thus, if one can measure the surface concentration in the presence of a gradient as a function of position and time, one can deduce the diffusion coefficient. In the absence of a gradient, molecules on a surface execute a random walk, so that the diffusion coefficient can be determined by the root-mean-squared distance the adsorbate travels in a given time. Following **equation 4.43,** this distance is given as

$$x_{\text{rms}} = (2Dt)^{1/2}. \tag{6.14}$$

Several techniques have been developed for determining the rates of surface diffusion. Perhaps the most dramatic is the field emission microscope, developed in

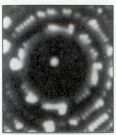

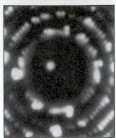

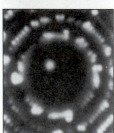

Figure 6.9

Field ion images showing the diffusion of an Ir adatom on an Ir surface. The dot near the center of each photograph is the image of the Ir adatom.

Reprinted with permission from T. T. Tsong, *Physics Today,* **46,** 24 (1993). Copyright American Institute of Physics, 1993.

the 1930s by E. W. Müller.[g] In this technique, a highly magnified image of a pointed metallic sample is formed on a fluorescence screen by field ionization of (typically) helium atoms. The ionization is most efficient at regions of high curvature on the tip surface, which, like lightning rods, have the highest field when the tip is held at a negative potential with respect to the fluorescence screen. Since atoms adsorbed onto the tip provide a highly curved surface, ions are very effectively produced at these locations; the adatoms show up as an intense spot on the screen.

Motion of the adatoms can be followed by examination of images taken at different times. For example, **Figure 6.9** shows several images of an Ir adatom on an Ir surface. By measuring the root-mean-squared displacement of the adatom as a function of time, it is then possible to determine the diffusion coefficient through **equation 6.14.** Diffusion coefficients measured in such studies of single atoms on metal surfaces range from about 1×10^{-8} to 3×10^{-5} in units of $m^2 \ s^{-1}$.

It is interesting to compare the magnitude of diffusion coefficients on a surface to those of molecules in the gas phase. From **Table 4.4** we see that coefficients for gas-phase molecules have magnitudes of 1×10^{-5} to $2 \times 10^{-4} \ m^2 \ s^{-1}$. Thus, diffusion is much slower on a surface than in the gas phase. A key to why this is true comes from the temperature dependence. Whereas diffusion in gases has a very weak temperature dependence, diffusion on surfaces increases dramatically with temperature. From the microscopic point of view, the reason for this behavior lies in a potential energy barrier between adjacent preferred adsorption sites, as shown in **Figure 6.10.** Experimentally it is found that the temperature-dependent diffusion coefficient is given by $D(T) = D_0 \exp(-\Delta E_m / kT)$, where ΔE_m is the barrier to diffusion and D_0 is a constant. Typically, ΔE_m is found to be about one quarter of the energy required for desorption. From a microscopic point of view, D_0 may be interpreted from a rearrangement of **equation 6.14:** $D = \frac{1}{2} x_{rms}^2 / t$. Thus, the diffusion constant is given as $\frac{1}{2}$ times the product of the rate of hopping attempts, $1/t$, times the square of the average distance per hop, x_{rms}^2.

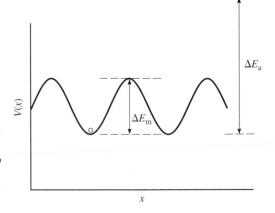

Figure 6.10

Potential energy $V(x)$ as a function of lateral surface displacement x for a molecule diffusing on a surface. The barrier to migration due to diffusion, ΔE_m, is typically one quarter of the barrier to desorption, ΔE_a.

[g]E. W. Müller, *Ergeb. Exakten Naturwiss.* **27,** 290 (1953); R. H. Good, Jr., and E. W. Müller, *Handbuch der Physik,* S. Flügge, ed., **21,** 176 (1956).

example 6.4

Estimating the Surface Diffusion Coefficient of CO on Iron

Objective Estimate the diffusion coefficient for CO on iron at 300 K given the heat of adsorption from **Figure 6.6** and the knowledge that adjacent sites are separated by 1.0 nm.

Method The diffusion coefficient should be one half the product of a hopping attempt frequency times the square of the hop length times a Boltzmann factor accounting for ΔE_m. As an estimate, we may take ΔE_m to be about $0.25\Delta E_a$, which, from the figure is about $(0.25)(50$ kcal/mol). The hopping attempt frequency may be approximated by a low vibrational frequency, about 1.5×10^{12} s^{-1}.

Solution The diffusion coefficient is then given by $D \approx (0.5)(1.5 \times 10^{12}$ s$^{-1})(1.0 \times 10^{-9}$ m$)^2\exp\{-[(0.25)(50 \times 10^3$ cal/mol$)(4.184$ J/cal$)/(8.314$ J K^{-1} mol$^{-1})(300$ K$)]\} = 0.7 \times 10^{-6}$ m^2 s^{-1}.

So far we have considered molecular diffusion on a bare surface, i.e., in the limit of zero coverage. When molecules of either the same or another type partially cover the surface, the diffusion coefficient will be different than that on a bare surface. First of all, fewer sites are available to which the molecules can jump. Second, the lateral interactions between molecules at adjacent sites may affect the rate at which the migration takes place. Measurement of diffusion coefficients as a function of coverage is a field that is currently under investigation by a variety of new techniques. One of these is described in Problem 6.10.

6.5   ADVANCED TOPICS IN SURFACE REACTIONS

6.5.1   Temperature-Programmed Desorption

We consider in this section a method, called temperature-programmed desorption (TPD), that makes it possible to estimate the energy of desorption or reaction as well as the Arrhenius preexponential factor.[h] Consider molecules that are irreversibly adsorbed on the surface indicated in **Figure 6.11** at some low temperature T_0. The leak valve is then closed, the valve to the pump is opened, and the density of product molecules is monitored with a mass spectrometer as the crystal is heated in time at a rate β degrees per second (typically $\beta \approx 10$ K s^{-1}). The crystal temperature is thus $T = T_0 + \beta t$.

The density of desorbed molecules measured by the mass spectrometer will be given by a balance between the rate of desorption and the pumping speed. When the pumping speed is high enough so that no readsorption takes place, the density of desorbed molecules will be directly proportional to the rate of desorption. Of

[h]G. Ehrlich, *J. Appl. Phys.* **32,** 4 (1961); *Adv. Catal. Relat. Subj.* **14,** 271 (1963); P. A. Redhead, *Vacuum* **12,** 203 (1962).

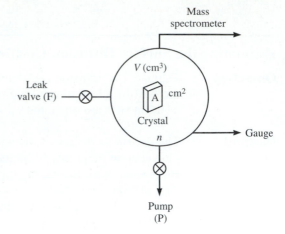

Figure 6.11

Schematic diagram of a temperature-programmed desorption apparatus.

From M. Boudart and G. Djéga-Mariadas-sou, *Kinetics of Heterogeneous Catalytic Reactions* (Princeton University Press, Princeton, NJ, 1984). Copyright © 1984 by Princeton University Press. Reprinted by permission of Princeton University Press.

course, the rate of desorption depends strongly on temperature, so that when the temperature of the crystal reaches a high enough value so that the rate of desorption is appreciable, the mass spectrometer will begin to record a rise in density. At higher crystal temperatures, the surface will eventually become depleted of desorbing molecules, so that the mass spectrometer signal will decrease. The shape and position of the peak in the mass spectrometer signal can be used to learn about the activation energy for desorption and the Arrhenius preexponential factor.

Consider a first-order desorption process: A–S $\rightarrow$ S + A with a rate constant $k_d = A \exp(-\Delta E_a/RT)$. If Γ represents the number of surface adsorbates per unit area,[i] the rate of desorption will be given by

$$-\frac{d\Gamma}{dt} = k_d \Gamma$$
$$= \Gamma A \exp\left(-\frac{\Delta E_a}{RT}\right). \tag{6.15}$$

However, since the heating rate is $dT/dt = \beta$, we see that

$$\frac{1}{dt} = \frac{\beta}{dT}. \tag{6.16}$$

Multiplying by $-d\Gamma$ gives

$$-\frac{d\Gamma}{dt} = -\beta \frac{d\Gamma}{dT}, \tag{6.17}$$

and combination of this last equation for $-d\Gamma/dt$ with **equation 6.15** gives

$$-\frac{d\Gamma}{dT} = \frac{\Gamma A}{\beta} \exp\left(-\frac{\Delta E_a}{RT}\right). \tag{6.18}$$

---

[i]Note that Γ and θ are related. Γ is the number of adsorbates per area, whereas θ is the number of adsorbates per number of sites. Thus Γ times the number of sites per area gives θ.

At the peak of the mass spectrometer signal, the increase in the desorption rate is matched by the decrease in surface concentration per unit area so that the change in $d\Gamma/dT$ with temperature is zero:

$$\frac{d}{dT}\frac{d\Gamma}{dT}\bigg|_{T=T_M} = 0, \tag{6.19}$$

or

$$\frac{d}{dT}\left[\frac{\Gamma A}{\beta}\exp\left(-\frac{\Delta E_a}{RT}\right)\right] = 0,$$

$$\left[\frac{d\Gamma}{dT} + \frac{\Delta E_a}{RT_M^2}\Gamma\right]\frac{A}{\beta}\exp\left(-\frac{\Delta E_a}{RT_M}\right) = 0, \tag{6.20}$$

$$\frac{\Delta E_a}{RT_M^2} = -\frac{1}{\Gamma}\frac{d\Gamma}{dT}.$$

Finally, substitution of **equation 6.18** into this last equation yields

$$\frac{\Delta E_a}{RT_M^2} = \frac{A}{\beta}\exp\left(-\frac{\Delta E_a}{RT_M}\right), \tag{6.21}$$

or

$$2\ln T_M - \ln\beta = \frac{\Delta E_a}{RT_M} + \ln\frac{\Delta E_a}{RA}. \tag{6.22}$$

If different heating rates β are used and the left-hand side of this last equation is plotted as a function of $1/T_M$, we see that a straight line should be obtained whose slope is $\Delta E_a/R$ and whose intercept is $\ln(\Delta E_a/RA)$. Note also that the value of T_M does not depend on the initial coverage.

The situation is a bit more complicated for a second-order desorption process such as $2\,A{-}S \rightarrow A_2 + 2\,S$. Problem 6.11 shows that the resulting equation for T_M is

$$2\ln T_M - \ln\beta = \frac{\Delta E_a}{RT_M} + \ln\frac{\Delta E_a}{AR\Gamma_0}, \tag{6.23}$$

where Γ_0 is the initial surface coverage. Again, a plot of the left-hand side of the equation against $1/T_M$ provides information about ΔE_a and A_d. Note, however, that the temperature of the peak will change with initial surface coverage. This variation is the hallmark of a second-order desorption. For example, note how the D_2 desorption peak at 380 K in **Figure 6.12** shifts to lower temperature with increasing coverage.

It frequently occurs that molecules decompose before they desorb. For example, formic acid decomposes on a nickel surface to H_2, CO, and CO_2 at a temperature of about 388 K. **Figure 6.13** shows the temperature-programmed desorption spectrum of this molecule (the lower-temperature desorption of water is not shown in the figure). If adsorbed separately on this surface, H_2 and CO would desorb at 353 and 438 K, respectively. The fact that the principal peaks for H_2 and CO occur at 388 K indicates that these species are initially produced by decomposition of the acid. Some of the CO escapes from the surface due to the exothermicity of the reaction, while some is readsorbed and then desorbs later at 438 K.

6.5.2 Modulated Molecular Beam Methods

We consider in this section the application of molecular beam methods to surface reactions. These methods provide insight into the kinetics of surface processes as

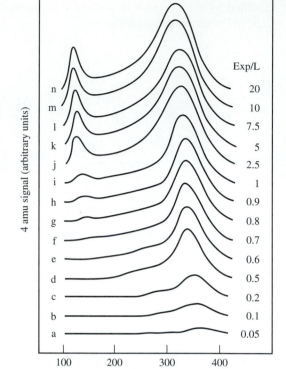

Figure 6.12

Thermal desorption spectra of D_2 from Rh(100) for different exposures in langmuirs (1 langmuir = 10^{-6} torr-s). Note how the peak shifts to lower temperature with increasing coverage.

From Y. Kim, H. C. Peebles, and J. M. White, *Surface Science* **114,** 363 (1982). Reprinted from *Surface Science,* copyright 1982, with permission from Elsevier Science.

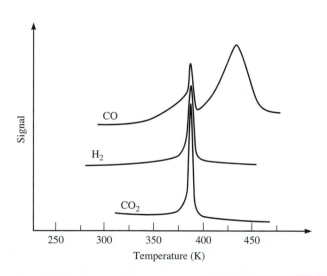

Figure 6.13

TPD of formic acid adsorbed at 325 on a nickel surface.

From M. Boudart and G. Djéga-Mariadassou, *Kinetics of Heterogeneous Catalytic Reactions* (Princeton University Press, Princeton, NJ, 1984) based on original data of J. McCarthy, J. Falconer, and R. Madix, *J. Catal.* **30,** 235 (1973). Copyright © 1984 by Princeton University Press. Reprinted by permission of Princeton University Press.

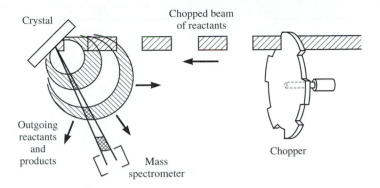

■ **Figure 6.14**

Schematic diagram of the molecular beam-surface scattering technique.

From G. A. Somorjai, *Chemistry in Two Dimensions: Surfaces* (Cornell University Press, Ithaca, NY, 1981). Reprinted with permission of Dr. G. A. Somorjai.

well as detail concerning the dynamics of the gas-surface interaction. Molecular beam methods are also important to our understanding of gas-phase reactions and dynamics, but we will postpone our examination of this field until Chapter 8.

Figure 6.14 shows a typical molecular beam apparatus for studies of surface processes. A molecular beam of reactants is formed using methods to be described in Chapter 8. The beam is chopped by a rotating, slotted disk before it impinges on a solid surface. Unreacted reactants and newly formed products are detected by a rotatable mass spectrometer as a function of time. Two important quantities can be determined using this technique: the time dependence of the product flux and the angular distribution of the scattered species. We consider the angular distribution first.

6.5.2.1   *Angular Distributions*

The angular distribution of the scattered molecules provides information on the exchange (or "accommodation") of momentum during the surface collision. Consider two possible limits. In one, the molecule colliding with the surface "sticks" to the surface long enough to lose all memory of its incoming direction and momentum. After such a strong energy exchange, we can regard the surface and gas to be at equilibrium at some specific temperature. If the molecule subsequently escapes from the surface attractive potential, we might wonder what angular distribution its velocity would have with respect to a line normal to the surface. The answer is shown for CO scattering from Pt(111) in **Figure 6.15,** where the distance of a point from the origin in the plot is proportional to the probability of finding the product velocity at the given angle with respect to the surface normal. For the highest coverage of preadsorbed CO on the surface the scattering is given approximately by $\cos \theta$. In fact, we already know that the $\cos \theta$ distribution, given by the dashed circle in the plot, should be the answer to the equilibrium problem! Consider the velocity distribution function for molecules leaving an area A of the surface. If the surface and gas temperatures are the same, i.e., if there is equilibrium, this distribution will be the same as the velocity distribution for molecules which strike that area of the surface, except that the velocity vectors will be reversed in sign. We have already seen in Section 4.3.2 (see **Figure 4.2** and **equation 4.3**) that the velocity

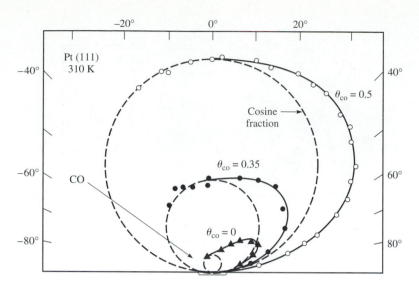

■ **Figure 6.15**

The angular distribution of CO scattered from Pt(111) at 310 K and for three different coverages. The CO beam was incident at 55°, as shown by the arrow. The dashed circles indicate the
distribution expected for purely cosine scattering.

From C. T. Campbell, G. Ertl, H. Kuipers, and J. Segner, *Surface Science* **107,** 207 (1981). Reprinted from
Surface Science, copyright 1981, with permission from Elsevier Science.

distribution for molecules striking the surface starting from the three-dimensional
"volume" element $v^2\sin\theta\,d\theta\,d\phi\,dv$ is proportional to $\cos\theta$. Thus, the angular distribution for molecules leaving the surface in equilibrium with the surface temperature will also be proportional to $\cos\theta$, an observation known as the Knudsen
cosine law.

Now consider another limit for the angular distribution, one in which the molecules simply bounce off an unmoving surface in the same way light might be
reflected from a mirror. In this case the outgoing angle of deflection with respect to
the surface normal will simply be the negative of the incoming angle of incidence.
Such scattering is termed *specular scattering* or *elastic scattering*. **Figure 6.15**
shows that CO scatters nearly specularly when the surface is bare ($\Theta_{CO} = 0$), but
that as the coverage increases the scattering follows more and more the cosine law.
At the highest coverage there is still a remnant of the specular scattering, but the
distribution nearly follows the Knudsen law.

Most angular scattering distributions lie somewhere in between these two limits, as shown in the CO/Pt(111) case for intermediate coverages of CO. Furthermore, even in the specular limit, the peak of the scattering distribution is often
shifted from the exact specular direction. **Figure 6.16** shows such a shift for the
scattering of Ar from Pt. The reason for the shift is that the collision with the surface is rarely elastic. Movement of the surface atoms either imparts energy to or
takes energy from the component of velocity perpendicular to the surface. In this
case, Ar atoms on average gain perpendicular momentum by collision with the
energetic surface atoms.

Occasionally one finds a dramatic departure from either the specular or cosine
scattering. **Figure 6.17** shows the angular distribution of H_2 found after passing H_2
through a thin iron membrane. The H_2 dissociates as it passes through the membrane

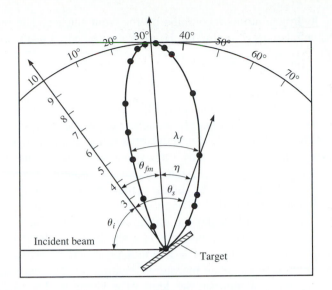

Figure 6.16

Polar plot of the angular distribution of Ar scattered from Pt with a surface temperature of 1173 K and an incident angle of 55°.

From F. O. Goodman and H. Y. Wachman, *Dynamics of Gas-Surface Scattering* (Academic Press, New York, 1976) based on data from A. R. Rudnicki and H. Y. Wachman, *Surf. Sci.* **34,** 679 (1973). Reprinted from *Surface Science,* Copyright 1973, with permission from Elsevier Science.

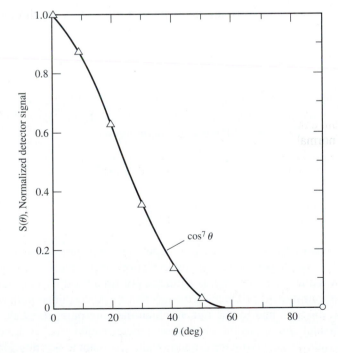

Figure 6.17

Angular distribution of H_2 desorbed from polycrystalline iron at 1140 K.

From T. L. Bradley and R. E. Stickney, *Surf. Sci.* **38,** 313 (1973). Reprinted from *Surface Science,* copyright 1973, with permission from Elsevier Science.

and then the hydrogen atoms recombine on the front surface. The recombination releases substantial energy, forcing the H_2 off the surface in a peak centered on the surface normal.

6.5.2.2   Kinetic Parameters

Modulated molecular beam methods also allow the determination of rate constants. To gain an intuitive understanding of how such an experiment might work, consider the analogy of the sun heating the Earth during a yearly cycle. The average amount of sunlight varies nearly sinusoidally throughout the year, so that if the heating were instantaneous, the peak in the Earth's average monthly temperature should follow the peak in the exposure to sunlight; the warmest month would be June, while the coldest month would be December. However, the rate of heating of the Earth's surface and atmosphere depend on the heat capacity. The result is that the temperature varies sinusoidally, but there is a phase lag of about 1 month. Turning this around, the observation of a 1-month phase lag between exposure and response can be used to calculate the rate of heating, or the heat capacity in this case. Another observation concerns the depth of modulation. The average monthly temperatures over the ocean vary less with season than those over land, because the heat capacity of salt water is higher than that of land. By analogy, measurement of the phase lag and depth of modulation of reactant products following periodic exposure of a surface to reactants provides information on the rate constant for the reaction.

Now let's put these concepts on a firmer mathematical foundation. Consider the simple mechanism outlined below in which a modulated beam of intensity $I(t)$ interacts with the surface:

$$A + S \xrightarrow{s_0 I(t)} A\text{-}S,$$

$$A\text{-}S \xrightarrow{k} A + P, \tag{6.24}$$

where s_0 is the sticking coefficient of A on the surface and k represents the rate constant for reaction of the surface-bound species to products. The differential equations for the intermediate and the products are given by

$$\frac{dC(t)}{dt} = s_0 I(t) - kC(t),$$

$$\frac{dP(t)}{dt} = kC(t), \tag{6.25}$$

where $C(t)$ and $P(t)$ represent the time-dependent concentrations [A–S] and [P], respectively. In the experiment, the concentration of products in the gas phase is measured by a mass spectrometer as a function of time for different incident fluxes $I(t)$. The concentration of products in the gas phase is determined both by their production rate, $dP(t)/dt$, and by the rate at which they are pumped away. Under normal experimental conditions the mass spectrometer signal, $S(t)$, is directly proportional to $dP(t)/dt$. Thus, (taking the proportionality constant to be unity for simplicity) we have $S(t) = dP(t)/dt = kC(t)$.

In the general case, $I(t)$ may vary on a time scale comparable to the reaction, so that it would be inappropriate to use the steady-state principle to solve this problem. Instead, we use the Laplace transformation method of solution outlined in

Appendix 6.1. Consider the transform of the equation for $dC(t)/dt$, where the transformation variable is the pure complex variable $i\omega$. Using **Table 6.1** in the appendix with $s = i\omega$, we find that the transformed equation is

$$i\omega c(\omega) = s_0 I(\omega) - kc(\omega),$$

$$c(\omega) = \frac{s_0 I(\omega)}{k + i\omega}. \tag{6.26}$$

From the equation $S(t) = kC(t)$, we obtain $S(\omega) = kC(\omega)$, so

$$t(\omega) \equiv \frac{s(\omega)}{I(\omega)} = \frac{ks_0}{k + i\omega},$$

$$= \frac{ks_0}{(k + i\omega)}\frac{(k - i\omega)}{(k - i\omega)}, \tag{6.27}$$

$$= \frac{s_0(k^2 - ik\omega)}{k^2 + \omega^2},$$

where $t(\omega)$ is called the *transfer function*. This function indicates how the amplitude and phase of the mass spectrometer signal will depend on the frequency of the input signal. Specifically, writing $t(\omega)$ in vector notation in a complex plane, as shown in **Figure 6.18,** we obtain

$$t(\omega) = a(\omega)\exp[-i\phi(\omega)], \tag{6.28}$$

where the real part, $a(\omega)$, indicates how the amplitude of the signal changes with frequency, while the phase indicates how the output waveform is shifted from the input one. Comparison of **equation 6.28** with **equation 6.27** yields

$$a(\omega) = \frac{s_0}{\sqrt{1 + (\omega/k)^2}},$$

$$\tan \phi(\omega) = \frac{\omega}{k}. \tag{6.29}$$

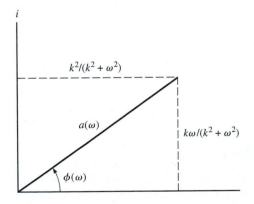

Figure 6.18

The transfer function can be expressed in terms of amplitude $a(\omega)$ and phase $\phi(\omega)$ in the complex plane.

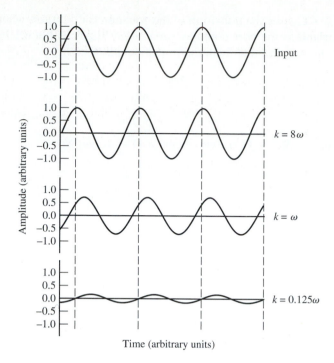

■ **Figure 6.19**

The modulated molecular beam (here assumed to be sinusoidal) and the mass spectrometer signal for the kinetic scheme of the text and for different values of the rate constant k in compared to the frequency of modulation ω.

Figure 6.19 shows an input waveform (assumed to be a sine wave rather than a square wave) and the signal waveforms for various cases of k compared to ω. Note that the signal $s(\omega) = t(\omega)I(\omega) = a(\omega)\exp[-i\phi(\omega)]I(\omega)$. When the modulation of the input signal is slow on the time scale of the reaction kinetics, the output waveform is just like the input waveform, but when the modulation is fast on the time scale of the kinetics, the output waveform is reduced in amplitude and shifted in phase. A measurement of the phase shift as a function of modulation frequency can be used to determine the rate constant. For example, if the rate constant were equal to the modulation frequency, then the phase shift would be 45° and the amplitude would be the original amplitude divided by the square root of 2.

In practice, the input waveform is often a square wave rather than a sine wave. Since a square wave contains frequency components at odd multiples of the fundamental, the transfer function actually can provide information on several frequencies simultaneously. Problem 6.13 outlines a method for recovering the information.

6.6   SUMMARY

Our study of reactions at solid surfaces has taken us from the structure of the surfaces, involving terraces, steps, kinks, and adatoms, to the kinetics of surface processes, involving adsorption, desorption, diffusion, and reaction. The flux of atoms or molecules to the surface is given by

$$J_z = \frac{1}{4}n^*<v>.$$ **(4.6)**

The adsorption of molecules striking the surface can occur through both *physisorption* and *chemisorption,* and the latter process will be activated if the barrier between the physisorption well and the chemisorption well lies above the energy corresponding to infinite separation of the molecule from the surface.

The Langmuir model for adsorption assumes that molecules can adsorb only at specific sites on the surface and that once a site is occupied by one molecule it cannot adsorb a second molecule. The model leads to the following equation for the surface coverage:

$$\theta = \frac{K[A]}{1 + K[A]},$$ (6.2)

where K is the equilibrium constant for adsorption: $K = k_a/k_d$. When two molecules of different types compete for the same adsorption sites, the coverage of one is given by

$$\theta_A = \frac{K_A[A]}{1 + K_A[A] + K_B[B]},$$ (6.5)

with a similar equation for θ_B.

From application of the Clausius-Clapeyron equation we find that

$$\left. \frac{d \ln(P_A/P_A^0)}{d(1/T)} \right|_{\theta=\text{const}} = -\frac{\Delta H_{ad}}{R}.$$ (6.7)

Reactions at surfaces are unimolecular, bimolecular, etc. just as in the gas phase or in solution. For unimolecular processes we found that

$$\frac{d[P]}{dt} = k\theta = \frac{kK[A]}{1 + K[A]},$$ (6.8)

where P is the product of the reaction. For a bimolecular process

$$\frac{d[P]}{dt} = k\theta_A\theta_B$$

$$= \frac{kK_AK_B[A][B]}{(1 + K_A[A] + K_B[B])^2}.$$ (6.9)

The two processes have remarkable differences in their pressure dependence, since in the latter case one of the reactants can "poison" the surface toward adsorption of the second. Activated complex theory can be applied to surface reactions to yield estimates of rate constants.

Surface diffusion plays an important role in surface reactions. The barrier to diffusion on the surface is less than the barrier to desorption, and diffusion constant on a surface can be obtained by measuring how far a molecule moves in a given time:

$$\frac{\partial n^*(x,t)}{\partial t} = D\frac{\partial^2}{\partial x^2}n^*(x,t).$$ (6.13)

A convenient method for estimating the energy of desorption or reaction is by temperature programmed desorption. For a first-order process the temperature T_M at which the signal has a maximum is related to the activation energy by the equation

$$2 \ln T_M - \ln \beta = \frac{\Delta E_a}{RT_M} + \ln \frac{\Delta E_a}{RA}.$$ (6.22)

For a second-order process the temperature at which the signal is a maximum also depends on the surface coverage:

$$2 \ln T_{\text{M}} - \ln \beta = \frac{\Delta E_{\text{a}}}{RT_{\text{M}}} + \ln \frac{\Delta E_{\text{a}}}{AR\Gamma_0}. \tag{6.23}$$

Modulated molecular beam methods allow one to measure both the angular distribution and the reaction rate constant. Under conditions of equilibrium adsorption and desorption, the angular distribution varies as $\cos \theta$ (where θ is the angle measured from the surface normal). Under nonequilibrium conditions, the scattering is often more strongly directed—toward the specular angle in the case of a repulsive interaction at the surface and toward the surface normal if a reaction takes place releasing a large amount of energy.

The rate constant is related to the phase of the modulated product signal and the frequency of modulation:

$$\tan \phi(\omega) = \frac{\omega}{k}. \tag{6.29}$$

appendix 6.1

Integral Transforms

The Laplace transform provides a convenient method for solution of differential equations. Consider the function $F(t)$. Let the Laplace transform of $F(t)$ be defined as

$$f(s) = \mathcal{L}[F(t)] = \int_0^\infty e^{-st} F(t)\, dt, \tag{6.30}$$

where t is a real variable, $F(t)$ is a real function of t whose value when $t < 0$ is zero, $f(s)$ is a function of s, and s is a complex variable. The function $f(s)$ is called the Laplace transform of $F(t)$, while $F(t)$ is called the inverse Laplace transform of $f(s)$.

The usefulness of the Laplace transform for differential equations comes from the observation that the Laplace transformation of derivatives like $dF(t)/dt$ converts them to simple algebraic expressions of s.

Consider the Laplace transform of $dF(t)/dt$:

$$
\begin{aligned}
\mathcal{L}[dF(t)/dt] &= \int_0^\infty \frac{dF(t)}{dt} e^{-st}\, dt, \\[2mm]
&= F(t)e^{-st}\Big|_0^\infty - \int_0^\infty F(t)\, d(e^{-st}), \\[2mm]
&= -F(t = 0) + s\int_0^\infty F(t)e^{-st}\, dt, \\[2mm]
&= -F(t = 0) + sf(s),
\end{aligned}
\tag{6.31}
$$

where we have used integration by parts in going from the first line to the second.

In a similar way, it is possible to build up a table of Laplace transforms of particular functions. **Table 6.1** provides the transforms of some of the more common functions.

TABLE 6.1	Table of Laplace Transforms

$F(t)$	$f(s)$
$F(t)$	$\displaystyle\int_0^\infty e^{-st}\,F(t)\,dt$
$\dfrac{dF(t)}{dt}$	$sf(s) - F(t=0)$
1	$\dfrac{1}{s}$
e^{at}	$\dfrac{1}{s-a}$
te^{at}	$\dfrac{1}{(s-a)^2}$
t	$\dfrac{1}{s^2}$
$\sin bt$	$\dfrac{b}{s^2+b^2}$
$\cos bt$	$\dfrac{s}{s^2+b^2}$
$\dfrac{1}{(a-b)}(e^{at}-e^{bt})$	$\dfrac{1}{(s-a)(s-b)}$

As an example of the usefulness of Laplace transforms, consider the differential equation we encountered (Section 2.4.3, Consecutive Reactions) in **equation 2.63:**

$$\frac{d[B]}{dt} = k_1 A(0)e^{-k_1 t} - k_2[B]. \tag{6.32}$$

The solution can be obtained by taking the Laplace transform of both sides, solving the resulting algebraic equation, and then taking the inverse Laplace transform. First we take the Laplace transform:

$$sb(s) - B(t=0) = \frac{k_1 A(0)}{s+k_1} - k_2 b(s). \tag{6.33}$$

Next, note that $B(t=0) = 0$ from the initial condition that all population in the $A \rightarrow B \rightarrow C$ reaction is initially in A. Solution of the algebraic equation gives

$$sb(s) + k_2 b(s) = \frac{k_1 A(0)}{s+k_1},$$

$$\tag{6.34}$$

$$b(s) = k_1 A(0)\frac{1}{s+k_2}\frac{1}{s+k_1}.$$

If we now take the inverse Laplace transform of both sides of the last equation we obtain

$$[B] = \frac{k_1}{k_2 - k_1} A(0)(e^{-k_1 t} - e^{-k_2 t}). \tag{6.35}$$

This last equation is just **equation 2.64.**

It should be noted that the Laplace transform method is appropriate only for *linear* differential equations; that is, ones that involve in every term only one variable concentration rather than the product of two or more. Transformation then gives a set of linear algebraic equations. The solutions can then be inverse transformed to obtained the concentrations as a function of time.

suggested readings

M. Boudart and G. Djéga-Mariadassou, *Kinetics of Heterogeneous Catalytic Reactions* (Princeton University Press, Princeton, NJ, 1984).

G. Ehrlich and K. Stolt, "Surface Diffusion," *Ann. Rev. Phys. Chem* **31,** 603 (1980).

K. B. Eisenthal, "Photochemistry and Photophysics of Liquid Interfaces by Second Harmonic Spectroscopy," *J. Phys. Chem.* **100,** 12997 (1996).

R. Gomer, "Surface Diffusion," *Scientific American,* **247,** 98–106, August (1982).

F. O. Goodman and H. Y. Wachman, *Dynamics of Gas-Surface Scattering* (Academic Press, New York, 1976).

R. J. Hamers, "Scanned Probe Microscopies in Chemistry," *J. Phys. Chem.* **100,** 13103 (1996).

W. Ho, "Reactions at Metal Surfaces Induced by Femtosecond Lasers, Tunneling Electrons, and Heating," *J. Phys. Chem.* **100,** 13050 (1996).

K. J. Laidler, *Chemical Kinetics,* 3rd ed. (Harper and Row, New York, 1987).

Richard I. Masel, *Principles of Adsorption and Reaction at Solid Surfaces* (Wiley, New York, 1996).

G. M. Nathanson, P. Davidovits, D. R. Worsnop, and C. E. Kolb, "Dynamics and Kinetics at the Gas-Liquid Interface," *J. Phys. Chem.* **100,** 13007 (1996).

C. T. Rettner, D. J. Auerbach, J. C. Tully, and A. W. Kleyn, "Chemical Dynamics at the Gas-Surface Interface," *J. Phys. Chem.* **100,** 13021 (1996).

R. Smith, ed., *Atomic and Ion Collisions in Solids and at Surfaces: Theory, Simulation and Applications* Cambridge University Press, 1997).

G. A. Somorjai, *Chemistry in Two Dimensions: Surfaces* (Cornell University Press, Ithaca, NY, 1981).

J. I. Steinfeld, J. S. Francisco, and W. L. Has, *Chemical Kinetics and Dynamics* (Prentice-Hall, Englewood Cliffs, NJ, 1989).

T. T. Tsong, "Atom-Probe Field Ion Microscopy," *Physics Today* **46,** 24 (1993).

problems

6.1   The analysis of adsorption by Langmuir predicts that with increasing exposure the coverage of a species on a surface (a) increases monotonically, (b) decreases monotonically, (c) increases then decreases, or (d) decreases then increases.

6.2   Under Langmuir's assumptions, if two species can adsorb on a surface, then the equilibrium coverage of one of them will (a) increase, (b) decrease, or (c) remain constant as the exposure of the surface to the other species increases.

6.3   The barrier to diffusion is typically (a) a bit larger than, (b) a bit smaller than, (c) a lot larger than, or (d) a lot smaller than the barrier to desorption.

6.4 For a first-order desorption process, the temperature at which a peak might be expected in the temperature-programmed desorption spectrum depends on which of the following: (a) the Arrhenius A parameter, (b) the Arrhenius activation energy, (c) the rate of surface heating, (d) the coverage, (e) the heat of adsorption, or (f) the diffusion coefficient?

6.5 Suppose a molecule is completely accommodated on the surface before desorbing. To detect the desorbed species most sensitively, we would place a detector so that it accepted molecules (a) moving perpendicular to the surface normal, (b) moving at the specular angle, or (c) moving along the surface normal. (Specify which.)

6.6 Consider adsorption with dissociation: $A_2 + S + S \rightarrow A–S + A–S$. Show from an analysis of the equilibrium between adsorption and desorption that the surface coverage θ is given as a function of $[A_2]$ by

$$\theta = \frac{K^{1/2} [A_2]^{1/2}}{1 + K^{1/2} [A_2]^{1/2}}.$$

6.7 As might be expected, the decomposition of ammonia on platinum is first order in the gas-phase concentration of ammonia. However, at certain temperatures, where the N_2 coverage is nearly zero but the H_2 coverage is not, the reaction is inversely proportional to the H_2 pressure. Suggest a mechanism for this observation, evaluate the rate expression for the mechanism, and show under what limit the mechanism gives the observed result.

6.8 A second-order surface reaction involves two gas-phase species A and B, which are adsorbing and desorbing from the surface. For a fixed concentration of B denoted as $[B]_0$ in the gas phase, it is observed that the overall rate of the reaction has a maximum at a particular concentration of A denoted as $[A]_{max}$. What is the relationship between $[A]_{max}$ and $[B]_0$?

6.9 Show that (6.10) reduces to $k = \frac{1}{4} <v>$ for atomic adsorption under the limiting assumption where the activated complex is mobile in two dimensions and there is no activation energy.

6.10 One method for measuring the surface diffusion coefficient at arbitrary initial coverages is to thermally desorb molecules from the surface using two pulsed laser beams coincident in time and crossed at a slight angle. The interference between the light waves from the two beams creates a concentration grating of adsorbed molecules on the surface. As diffusion takes place, the concentration grating decreases in time. A convenient method for measuring the decay of the grating is to use it to scatter light from a third laser pulse, delayed in time from the first two. For reasonable intensities of the probe beam, second-harmonic generation can be obtained at each of the diffraction peaks. The method, which was developed by Shen and his coworkers,[j] has recently been used by Rosenzweig, Farbman, and Asscher [*J. Chem. Phys.* **98**, 8277 (1993)] to study the diffusion of ammonia on Re. A theoretical analysis shows that the second-harmonic intensity for the nth order diffraction peak should decay as

[j]X.-D. Xiao, D. D. Zhu, W. Daum, and Y. R. Shen, *Phys. Rev.* **B46**, 9732 (1992); *Surf. Sci.* **271**, 295 (1992); X. D. Zhu, Th. Rasing, and Y. R. Shen, *Phys. Rev. Lett.* **61**, 2883 (1988).

$$I_n^{2\omega}(t) = I_n^{2\omega}(0)\exp\left(-\frac{8\pi^2 n^2 Dt}{s^2}\right),$$

where s is the spacing of the grating and D is the diffusion coefficient. If the second-harmonic signal at the first-order diffraction peak falls to $1/e$ of its original value in 200 s and the grating spacing is $s = 8.0$ μm, what is the diffusion coefficient for NH_3 on Re at the temperature and coverage of the experiment (110 K and $\theta = 0.25$)?

6.11   Show that for a second-order desorption process the temperature at the peak maximum in the temperature programmed desorption experiment is given by **equation 6.23.** (*Hint*: The surface coverage at the temperature corresponding to the peak has been shown by Redhead to be $\Gamma_0/2$.)

6.12   A few reactions are thought to proceed by the Eley-Rideal mechanism. For example, the reaction $D_{(g)} + H_{(ad)} \rightarrow HD_{(g)}$ is thought not to require adsorption of the H atom before reaction; that is, the attacking D picks up the H on a single bounce. Evidence that this is the case is shown in **Figure 6.20,** which plots the HD product angular distribution for an incidence D-atom angle of $-60°$ [from C. T. Rettner, *Phys. Rev. Lett.* **69,** 383 (1992)]. The solid circles are for an incident energy of 0.06 eV, whereas the open squares are for an incident energy of 0.33 eV.
a.   Point out two features of this data that support the hypothesis that the reaction proceeds by the Eley-Rideal mechanism.
b.   Suggest another possible experiment that might test this hypothesis.

6.13   We saw in Section 6.5.2.2 that modulated molecular beam techniques could be used to determine kinetic parameters. Normally, the molecular beam is modulated in a square-wave fashion rather than in a sine-wave fashion. The

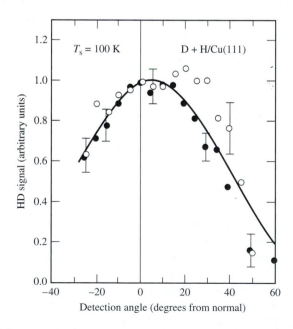

Figure 6.20

Plot of the HD product angular distribution for an incidence D-atom angle of $-60°$.
From C. T. Rettner, *Phys. Rev. Lett.* **69,** 383 (1992). Copyright 1992 by the American Physical Society.

Laplace transform of a square wave contains amplitude at all odd multiples of the fundamental frequency (though with decreasing amplitude as the frequency increases), so that the transfer function, given in **equation 6.28**, also has such frequency components. If the amplitude and phase of the transfer function are plotted in polar coordinates, the resulting figure can be used to determine many aspects of the mechanism of the surface reaction. For example, consider the kinetic scheme

$$A + S \xrightarrow{ps_0 I(t)} C_1,$$

$$\xrightarrow{(1-p)s_0 I(t)} C_2,$$

$$C_1 \xrightarrow{k_1} A + P,$$

$$C_2 \xrightarrow{k_2} A + P,$$

where p and $(1 - p)$ are the fraction adsorbing to the two different species C_1 and C_2, respectively, and both species can react to give the product P. Determine the transfer function for this system of parallel reactions when $p = 0.5$ and plot the transfer function in polar coordinates for the following three values of k_1/k_2: 1, 10, 100.

6.14 Molecular beam techniques can also be used to study processes at the gas-liquid interface. **Figure 6.21** shows the schematic of an apparatus used by Saecker and Nathanson for investigation of the energy exchange between a

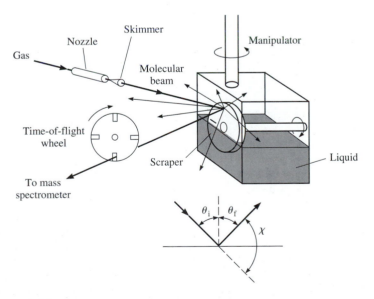

Figure 6.21

Schematic diagram of an apparatus for studying gas-liquid interactions. A rotating wheel and a scraper provide a fresh surface for interaction with the molecular beam. The speed of the scattered beam is measured by a time-of-flight method using a rotating chopper.

From M. E. Saecker and G. M. Nathanson, *J. Chem. Phys.* **99**, 7056 (1993). Reprinted with permission from the *Journal of Chemical Physics.* Copyright, American Institute of Physics, 1993.

variety of atoms and molecules with glycerol [HO–CH$_2$–CH(OH)–CH$_2$–OH]. The speed distribution of the scattered species was determined by timing the flight between the slotted chopper wheel and the ionization region of a quadrupole mass spectrometer.

Figure 6.22 displays the resulting time-of-flight distributions for Ne, CH$_4$, NH$_3$, and D$_2$O. The distributions are characterized by two components, an inelastic component at short times and a trapping-desorption component at longer times. The "inelastic" component corresponds to molecules that collide once with the surface of the liquid and, while losing some energy, scatter impulsively from the surface (see Section 8.4.4 for more discussion of inelastic processes). The trapping-desorption component corresponds to molecules that lose enough energy to bind momentarily to the surface or dissolve in the liquid. These molecules have sufficient contact with the surface to then desorb with a speed characteristic of the surface temperature. Explain why the ratio of the trapping-desorption component to the inelastic component might increase dramatically as shown.

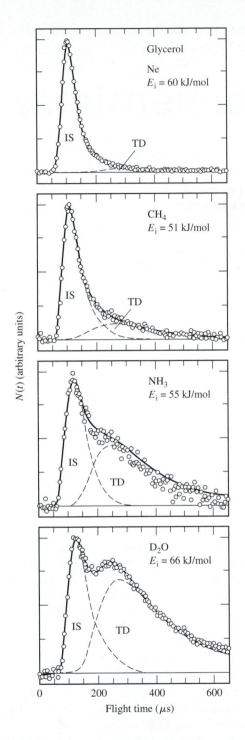

Figure 6.22

Time-of-flight distributions for collisions of various gases with glycerol.

From M. E. Saecker and G. M. Nathanson, *J. Chem. Phys.* **99,** 7056 (1993). Reprinted with permission from the *Journal of Chemical Physics.* Copyright, American Institute of Physics, 1993.

7

Chapter Seven

Photochemistry

Chapter Outline

7.1   INTRODUCTION

The field of photochemistry combines our knowledge of the quantum mechanical interaction of light and matter with the kinetics of chemical processes. It is a field important for both the fundamental information it yields about molecular structure and dynamics and the practical understanding it provides for processes ranging from the chemistry of the atmosphere to the mechanism of vision. In this chapter we explore several aspects of molecular photochemistry. We will concentrate largely on examples in isolated or gas-phase molecules, because it is in these systems that experiments have led to new theoretical concepts. After reviewing fundamental issues concerning the absorption and emission of light, we illustrate some of the fates of photoexcited molecules, including fluorescence, quenching, intramolecular vibrational energy redistribution, internal conversion, intersystem crossing, phosphorescence, and photodissociation. As a practical application of photochemistry and kinetics, we turn to processes affecting the concentration of stratospheric ozone, touching briefly on the origin of the Antarctic "ozone hole" and outlining both the basic Chapman mechanism for ozone production and destruction as well as the catalytic cycles which modify this simple mechanism. A major section of this chapter deals with the dynamics of photochemical processes. It is here that we see how measurement of the dynamics can provide fundamental information about how molecules interact, how they are held together, and what happens when they fall apart. The section begins with a description of the "pump-probe" technique for measuring dynamics and moves quickly to a description of two important probe methods, laser-induced fluorescence and multiphoton ionization. It then turns to a consideration of the rates for unimolecular processes and introduces the principles of the RRKM theory (see Section 7.5.4). It closes with discussions of photochemical angular distributions, photochemistry on very short time scales, and the relationship between photodissociation dynamics, potential energy surfaces, and absorption spectra. A summary section emphasizes the major points of the chapter.

7.2 ABSORPTION AND EMISSION OF LIGHT

Because any photochemical reaction involves the absorption of light, we begin by reviewing the fundamentals of the interaction of light with matter. The intensity of absorption is characterized by the fraction of light that is absorbed when it traverses a thickness ℓ of sample of concentration c or pressure p. This fraction is $(I_0 - I)/I_0$, where I/I_0, the fraction transmitted, is given by the Beer-Lambert law,

$$\frac{I}{I_0} = \exp(-\alpha \ell p), \tag{7.1}$$

where α is called the absorption coefficient. This exponential form of the Beer-Lambert law is usually used for gases, in which case α has the units of $\text{torr}^{-1}\,\text{cm}^{-1}$, and p and ℓ are expressed in torr and cm, respectively. In solutions an equivalent form of the law is more often used,

$$\frac{I}{I_0} = 10^{-\epsilon \ell c}, \tag{7.2}$$

where ϵ has the units of liters $\text{mole}^{-1}\,\text{cm}^{-1}$, and c and ℓ are expressed in moles/liter and cm, respectively. Another form of the law expresses the transmission in terms of a frequency-dependent cross section, $\sigma(\nu)$, for absorption:

$$\frac{I}{I_0} = \exp[-\sigma(\nu)\ell \rho], \tag{7.3}$$

where the absorption cross section might be given in cm^2, with ℓ in cm and the density ρ in molecules cm^{-3}.

The absorption coefficients and cross section are related to three more fundamental quantities, the Einstein coefficients: A_{21} for spontaneous emission from state 2 to 1, B_{21} for stimulated emission from state 2 to 1, and B_{12} for absorption from state 1 to 2. For this two-level system, illustrated in **Figure 7.1**, if $\rho(\nu)$ is the energy density per unit frequency of light at frequency ν, then the change in N_1, the population of level 1, is given as

$$\frac{dN_1}{dt} = -N_1 B_{12}\rho(\nu) + N_2[B_{21}\rho(\nu) + A_{21}]. \tag{7.4}$$

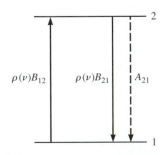

■ **Figure 7.1**

Einstein absorption (B_{12}), stimulated emission (B_{21}), and spontaneous emission (A_{21}) coefficients for a two-level system. The absorption and stimulated emission rates are proportional to the density of radiation at the resonant frequency.

The first term in this expression is the loss of N_1 due to absorption to level 2, while the second term accounts for the production of N_1 due to stimulated and spontaneous emission from level 2. In these equations, $\rho(\nu)$ has units of energy length^{-3} (time^{-1})$^{-1}$ (energy density per unit frequency), and B_{12} and B_{21} have units of energy^{-1} length^3 time^{-2}. At equilibrium for this two level system we have $dN_1/dt = 0$, or

$$\frac{N_2}{N_1} = \frac{B_{12}\rho(\nu)}{B_{21}\rho(\nu) + A_{21}}. \tag{7.5}$$

Comparison of this last equation with the Boltzmann distribution and use of $\rho(\nu)$ for a blackbody distribution provides relationships between the three coefficients (see Problem 7.13):

$$B_{12} = B_{21}\frac{g_2}{g_1},$$

$$A_{21} = \frac{8\pi h\nu^3}{c^3}B_{21}. \tag{7.6}$$

Here, g_1 and g_2 are the degeneracies of levels 1 and 2, respectively.

An important consequence of the Einstein analysis of absorption and emission is the prediction of *laser* action (light amplification by stimulated emission of radiation). **Equation 7.4** shows that emission of light will dominate if

$$N_2[B_{21}\rho(\nu) + A_{21}] > N_1 B_{12}\rho(\nu). \tag{7.7}$$

Thus, provided that the density of radiation is high enough so that $B_{21}\rho(\nu) >> A_{21}$, stimulated emission will occur when

$$\frac{N_2}{N_1} > \frac{B_{12}}{B_{21}} = \frac{g_2}{g_1}, \tag{7.8}$$

where the last equality follows from **equation 7.6.** We will see in Section 8.2 that chemical reactions sometimes produce products with $N_2 > N_1$, an "inverted" population.

A relationship between B_{12} and the absorption cross section $\sigma(\nu)$ can be developed by comparing the differential form of **equation 7.3,**

$$-dI(\nu) = I(\nu)\sigma(\nu)N_1 d\ell, \tag{7.9}$$

and the total decrease in population of N_1 given by the negative of **equation 7.4:**

$$-dN_1 = N_1 B_{12}\rho(\nu)\,dt - N_2[B_{21}\rho(\nu) + A_{21}]\,dt. \tag{7.10}$$

In the first of these equations we use $I(\nu)$ to signify the intensity per unit frequency [so that its units are energy length^{-2} time^{-1} (time^{-1})$^{-1}$]. Thus, $I(\nu)\,d\nu$ is the intensity (in energy length^{-2} time^{-1}) of radiation in the region from ν to $\nu + d\nu$. The density of light times its speed, c, is the intensity: $\rho(\nu)c = I(\nu)$. Thus, division of the intensity decrease $-I(\nu)\,d\nu$, by the energy of the photon, $h\nu$, and by the speed of light gives the decrease in the number density of photons, $-dI(\nu)\,d\nu/(ch\nu)$. Assuming that the transition is of sufficient energy that the population of level 2 is negligible (i.e., that $N_2 << N_1$), the number density of photons absorbed must be equal to the number density of molecules excited from level 1; i.e., $-dI(\nu)\,d\nu/(ch\nu) = -dN_1$. Combining this last equation with **equation 7.10** gives

$$\frac{I(\nu)\,d\nu\sigma(\nu)N_1\,d\ell}{ch\nu} = N_1 B_{12}\rho(\nu)\,dt,$$

or

$$\frac{\sigma(\nu)\,d\nu}{h\nu}\frac{d\ell}{dt} = B_{12}\frac{c\rho(\nu)}{I(\nu)}. \tag{7.11}$$

On the left-hand side of this equation, $d\ell/dt$ can be replaced by the speed of light, c, while on the right-hand side $c\rho(\nu)$ is equal to $I(\nu)$. Consequently,

$$B_{12} = \frac{c\sigma(\nu)\,d\nu}{h\nu}. \tag{7.12}$$

This equation can be used in either of two ways. If we integrate over frequency, we obtain the Einstein coefficient for the entire absorption band. If we multiply the frequency-dependent cross section by the bandwidth of a narrow light source, we obtain the Einstein coefficient for absorption at that particular frequency.

The Einstein coefficients are also related to the transition dipole moment for the transition:

$$A_{21} = \frac{64\pi^2\nu^3}{3hc^3}|\mu_{21}|^2, \tag{7.13}$$

where

$$\mu_{21} = -e\int \psi_{e_2 v_2}(\mathbf{r},\mathbf{R})\mathbf{r}\psi_{e_1 v_1}(\mathbf{r},\mathbf{R})\,d\tau_{el}\,d\tau_{nuc}. \tag{7.14}$$

Under the Born-Oppenheimer approximation, the electronic-vibrational wave functions, $\psi_{ev}(\mathbf{r},\mathbf{R})$ can be expressed as products, $\psi_e(\mathbf{r})\psi_v(\mathbf{R})$, so that

$$\mu_{21} = -e\int \psi_{e_2}(\mathbf{r})\mathbf{r}\psi_{e_1}(\mathbf{r})\,d\tau_{el}\int \psi_{v_2}\psi_{v_1}\,d\tau_{nuc}. \tag{7.15}$$

Since the second of these integrals is the overlap between the vibrational wave functions in the upper and lower levels, the absorption or emission of light will occur preferentially to states for which this overlap is greatest.

Since the electrons move much faster than the nuclei, absorption or emission is not accompanied by a change in the nuclear coordinates; the transition is *vertical* in a diagram such as **Figure 7.2,** plotting potential energy as a function of nuclear coordinates. Note in this figure that the transition is drawn as a vertical arrow and that the preferentially populated upper vibrational level, here shown as $v = 3$, is the

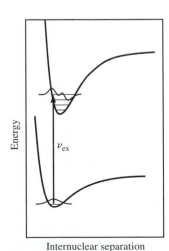

Figure 7.2

According to the Franck-Condon principle, absorption occurs preferentially to the state whose wave function has the maximum overlap with the initial wave function.

one whose vibrational wave function has the maximum overlap with the wave function of the lower level, here taken as $v = 0$. This important consequence of **equation 7.15** is known as the *Franck-Condon principle.*

While the absorption depicted in **Figure 7.2** often results in a line spectrum, where each line corresponds to a transition between not only individual vibrational but also individual rotational levels of the upper and lower electronic states, the lines are never infinitely sharp. The widths of the lines are caused by *Doppler broadening,* where the motion of the molecule relative to the observer or source of light causes emission or absorption of light at different frequencies, and by *lifetime broadening,* where the finite lifetime of the molecule due to emission, dissociation, or other processes gives rise to a width in accordance with the uncertainty principle, $\Delta E \Delta t \approx \hbar$. For molecules with a Maxwell-Boltzmann distribution of velocities, the Doppler-broadened line shape is given by the formula

$$\sigma(\nu) = \sigma_0 \exp\left[-\frac{mc^2}{2kT}\frac{(\nu - \nu_0)^2}{\nu_0^2}\right], \tag{7.16}$$

where ν_0 is the frequency at line center. The constant σ_0 is a normalization factor

$$\sigma_0 = \frac{c}{\nu_0}\sqrt{\frac{m}{2\pi kT}}, \tag{7.17}$$

and the Doppler width (full width at half maximum) is given by

$$\Delta \nu = 2\frac{\nu_0}{c}\sqrt{\frac{2kT\ln 2}{m}}. \tag{7.18}$$

A Doppler profile is shown as the solid line of **Figure 7.3.**

For molecules whose lifetime is given by τ, the Lorentz-broadened line shape is given by

$$\sigma(\nu) = \frac{\Delta \nu}{2\pi}\frac{1}{(\nu - \nu_0)^2 + (\Delta \nu/2)^2}, \tag{7.19}$$

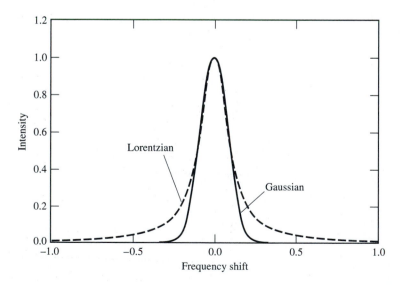

Figure 7.3

Doppler- and Lorentz-broadened line shapes.

where ν_0 is the frequency at line center and $\Delta\nu = 1/(\pi\tau)$ is the full width at half maximum. A Lorentz-broadened profile is shown as the dashed line of **Figure 7.3.**

We can define a *quantum yield* for a photochemical process as simply the number of events of interest divided by the number of photons of light absorbed by the system. For example, if the process of interest were the production of a photochemical product P, then the "quantum yield of P" would be given by the number of product molecules produced during a given time, N_P, divided by the number of photons absorbed by the system during that same time, $N_{h\nu}$: $\Phi_P = N_P/N_{h\nu}$. Alternatively, the quantum yield could be expressed as the ratio of the rate of production of P divided by the rate of absorption of light.

In determining the quantum yield, it will not always be the case that the process of interest is production of a product molecule. We next examine the possible fates of molecules excited by radiation.

7.3   PHOTOPHYSICAL PROCESSES

Absorption of light at visible or shorter wavelengths typically results in a combination of vibrational, rotational, and electronic excitation. We examine in this section the fate of the energy deposited in the molecule. In many cases, the energy is lost by a radiative process, either fluorescence—radiation to a lower electronic state of similar multiplicity as the upper one—or phosphorescence—radiation to a state of different multiplicity. But many nonradiative processes can occur as well, either as the result of collision or, particularly for larger molecules, as collisionless, intramolecular processes. These include intramolecular vibrational energy redistribution, internal conversion, and intersystem crossing. Of more interest from the chemical point of view, the energy can also be used to break a bond in the molecule. Such dissociation can be caused by a direct process, such as absorption to a dissociative electronic state, or by an indirect process, such as absorption to a level which, while bound at the level of the Born-Oppenheimer approximation, is coupled to the continuum of dissociative states by higher-order terms in the Hamiltonian.

7.3.1   Fluorescence and Quenching

Perhaps the simplest photochemical processes are absorption and re-emission of light. According to the Franck-Condon principle, in the absence of collisions that change the initially excited vibrational level, light will be emitted preferentially at two frequencies. One frequency is the same as the frequency that caused the excitation, since at this frequency the overlap between the ground and excited vibrational wave functions is strong. Another frequency, where strong emission will occur corresponds to a downward vertical transition from a location near the outer turning point of vibrational motion on the upper surface, since at this location the wave function also has maximum amplitude.

We now consider the situation when collisions with other molecules cause vibrational relaxation in the upper electronic state. The situation is shown schematically in **Figure 7.4.** If vibrational relaxation were to take place rapidly enough, nearly all the subsequent fluorescence would occur from $v = 0$, as shown in the figure. In reality, of course, the rate of vibrational relaxation is not infinitely fast compared to the rate of fluorescence, particularly when the pressure of the collision partner is low, so that emission from intermediate levels will occur.

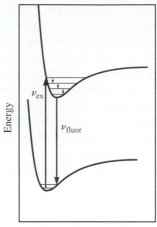

Internuclear separation

■ **Figure 7.4**

Vibrational relaxation in the upper electronic state prior to fluorescence back to the ground electronic state.

Figure 7.5 displays the fluorescence spectrum of I_2 following excitation of the $v = 25$, $J = 34$ level of the $B\,^3\Pi_{0u}$ electronic state with a filtered mercury lamp. The fluorescence is resolved by a monochromator in the absence and presence of a quenching gas, He. In the upper panel, resonance fluorescence is observed only from the initially excited level; two lines are observed corresponding to $\Delta J = \pm 1$ transitions. In the lower panel, emission occurs from other vibrational and rotational levels produced in collision with He.

The way in which the intensity of emission from the initial and intermediate levels varies as a function of pressure can be used to provide a measure of the relaxation. Suppose a molecule M is excited by radiation at an intensity that is constant in time (sometimes called *cw* [continuous wave] radiation). The excited species, M^*, can then emit radiation or suffer a quenching collision that takes it to a state, $M^\dagger$, that does not emit at the frequency being detected. The state $M^\dagger$ might be another vibrational level of the excited electronic state, as in **Figure 7.4,** or it might be another electronic state, as will be discussed in later sections. In either case, the kinetic scheme is

$$M + h\nu_a \xrightarrow{I_a\sigma} M^*,$$

$$M^* + Q \xrightarrow{k_q} M^\dagger, \tag{7.20}$$

$$M^* \xrightarrow{k_f} M + h\nu_f.$$

The rate for excitation of M to M^* is equal to the intensity of radiation at the excitation frequency, I_a, times the cross section for absorption, σ, whereas the intensity of fluorescence, I_f, is proportional to $k_f[M^*]$. Now assume that the initial concentration of M, M_0, is not appreciably perturbed by the excitation and that the assumptions implicit in the steady-state approximation are valid. Solution of the steady-state equations leads (Problem 7.6) to the equation

$$\frac{1}{I_f} = \frac{1}{I_a\sigma M_0}\left(\frac{k_q}{k_f}[Q] + 1\right). \tag{7.21}$$

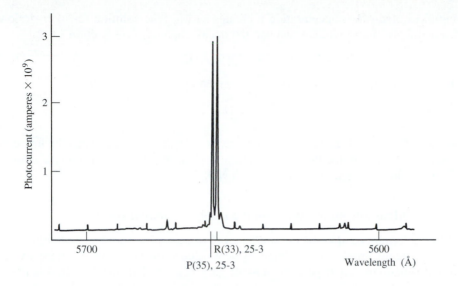

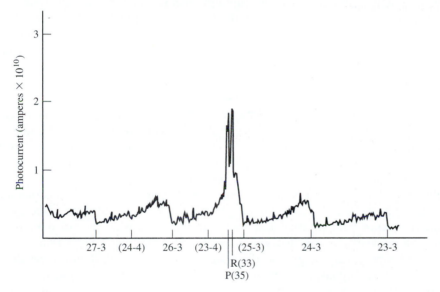

■ **Figure 7.5**

Fluorescence spectrum of I_2 without (*upper panel*) and with (*lower panel*) added helium.

From J. I. Steinfeld and W. Klemperer, *J. Chem. Phys.* **42**, 3475 (1965). Reprinted with permission from the *Journal of Chemical Physics*. Copyright, American Institute of Physics, 1965.

Thus, a plot of $1/I_f$ versus [Q] should yield a straight line whose slope is proportional to the ratio between the quenching and fluorescence rate constants. This equation is known as the *Stern-Volmer* equation, and systems obeying the relationship are said to follow Stern-Volmer kinetics. Problem 7.14 provides an example.

An alternative experimental procedure, which yields somewhat more information, is to excite the molecule M with pulsed radiation rather than with radiation of constant intensity and to detect the decay time of the fluorescence as a function of quencher concentration rather than the fluorescence intensity. Let the pulse of radiation be short compared to the time scale for fluorescence or quenching, and let the initial excited-state concentration it creates be M_0^*. The solution for $M^*(t)$ follows immediately since there are two parallel decay channels (see Section 2.4.2):

$$M^*(t) = M_0^* \exp[-(k_q[Q] + k_f)t]. \tag{7.22}$$

Thus, M^* decays with a time constant τ given by the relationship

$$\frac{1}{\tau} = (k_q[Q] + k_f). \tag{7.23}$$

A plot of τ^{-1} as a function of [Q] thus gives k_q as the slope and k_f as the intercept. Example 2.3 illustrated the use of lifetime measurements to determine the quenching of I^* fluorescence by collisions with NO.

7.3.2   Intramolecular Vibrational Energy Redistribution

While intermolecular collisions are required to effect vibrational energy redistribution in diatomic molecules, in larger systems, due to the coupling of different vibrational modes by small perturbations usually neglected in the Hamiltonian (e.g., anharmonicity, vibration-rotation coupling), energy can become redistributed in a collisionless process known as intramolecular vibrational energy redistribution (IVR). Indeed, as we will see in Section 7.5.4, rapid, collisionless exchange of energy between vibrational modes at high levels of excitation is one of the assumptions of the RRKM theory of unimolecular dissociation.

The observation that intramolecular vibrational energy redistribution takes place only at high levels of excitation provides a key to understanding the basic chemistry of this phenomenon. First, consider some data. **Figure 7.6** shows the fluorescence spectrum of anthracene excited in a jet-cooled molecular beam.[a] When this molecule is excited to a vibrational level only 766 cm^{-1} above the ground vibrational level of the first excited singlet state (S_1), the fluorescence shows the expected line spectrum. Each line corresponds to fluorescence from the selected excited level to a different vibrational level of the ground electronic state (S_0), and the intensity distribution is governed by the Franck-Condon principle. The fluorescence lifetime of the spectral feature marked with an asterisk is shown in the uppermost trace of the left-hand panel; a single exponential decay is observed with a lifetime of about 18 ns. At 1420 cm^{-1} of vibrational energy, the fluorescence spectrum still shows some distinct lines, but it has become much more congested. It appears that fluorescence from S_1 is occurring from many vibrational levels besides the one that was initially excited. In addition, the time-resolved fluorescence from the corresponding spectral feature now shows oscillations (called quantum beats) superimposed on an exponential decay. At 1792 cm^{-1} of vibrational energy, the fluorescence spectrum is nearly continuous and the time-resolved fluorescence shows a very complicated behavior, with most of the fluorescence disappearing in 75 ps. The experiment suggests an interesting phenomenon. It appears from the spectra that the initially excited state redistributes its vibrational energy to other vibrational levels and that the amount of redistribution increases with increasing vibrational excitation.

A detailed theoretical treatment[b] of this phenomenon was provided nearly 20 years before the experimental observation of **Figure 7.6.** Vibrations approximated as harmonic normal modes at low levels of vibrational excitation are actually coupled

[a]A brief description of the supersonic expansion technique needed for such cooling is given in Section 8.4.1.

[b]M. Bixon and J. Jortner, *J. Chem. Phys.* **48**, 715 (1968); **50**, 3284 (1969); **50**, 4061 (1969).

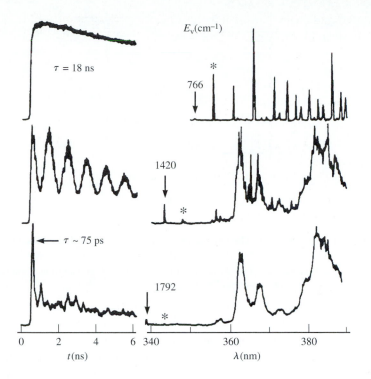

$E_v(cm^{-1})$

$\tau = 18$ ns

766

$\tau \sim 75$ ps

1420

1792

0   2   4   6   340   360   380

t(ns) λ(nm)

■ **Figure 7.6**

Fluorescence spectrum of anthracene at various levels of vibrational excitation. (*left*) Time-resolved fluorescence from corresponding levels of excitation.

From P. M. Felker and A. H. Zewail, *Adv. Chem. Phys.* **70**, 265 (1998). Copyright © 1988 by John Wiley & Sons, Inc. Reprinted by permission of John Wiley & Sons, Inc.

to one another by weak anharmonic, Coriolis, or other forces at higher levels of excitation, but we normally leave out these terms in a zeroth-order approximation to the Hamiltonian. Let the coupling be described by an interaction potential V. When this interaction is evaluated between two zero-order states $|a>$ and $|b>$,[c] the resulting matrix element, $V_{ab} = <b|V|a>$, tells us, roughly speaking, how close in energy two states would have to be before the coupling between them would cause perturbation of the energy levels. Because the density of vibrational levels increases dramatically with energy for large polyatomic molecules, at some level of excitation the spacing between vibrational levels will become smaller than the value of the coupling matrix element. Above this level of energy, the coupling will be observed as an apparent vibrational energy redistribution. Excitation of one "normal mode" will produce fluorescence from another.

To see the effect of the redistribution on the time-resolved fluorescence signal, consider the intramolecular transfer of energy between just two zero-order modes, $|a>$ and $|b>$, shown schematically in **Figure 7.7.** Because of the coupling element V_{ab}, the modes $|a>$ and $|b>$ are not eigenstates of the true Hamiltonian. In the limit where the true Hamiltonian couples only these two modes, the actual eigenstates,

[c]We use the notation $|n>$ here as a shorthand for $\Phi_n(q)$, the vibrational wave function. This is the so-called "bra-ket" notation in which $<m|$ denotes the complex conjugate of $\Phi_m(q)$ and $<m|V|n>$ is an integral, $\int \Phi_m(q)^* V \Phi_n(q) \, dq$.

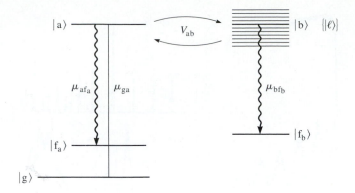

Figure 7.7

Schematic level structure diagram. The two states la> and lb> are coupled by the interaction potential V_{ab}.

From P. M. Felker and A. H. Zewail, *Chem. Phys. Lett.* **102,** 113 (1983). Reprinted from *Chemical Physical Letters,* copyright 1983, with permission from Elsevier Science.

$le_1>$ and $le_2>$, will be linear combinations of la> and lb>: $le_1> = \alpha la> + \beta lb>$ and $le_2> = \beta la> + \alpha lb>$, where $\alpha^2 + \beta^2 = 1$. It is often the case that the mixing is weak, so that we can think, for example, of $le_1>$ as being mostly la> with a little bit of lb> and of $le_2>$ as being mostly lb> with a little bit of la>. For this example $\alpha^2 > \beta^2$, but the specific values of α and β will depend on both V_{ab} and the energy separation between the levels. Because this separation is usually quite small, most excitation mechanisms will excite states $le_1>$ and $le_2>$ coherently (simultaneously and in phase). However, it often occurs that only one of these states has appreciable transition strength from the ground level. In **Figure 7.7** this state is shown as the one composed mostly of the zero-order mode la>.

An analogy can be made to the excitation of two pendulums of nearly the same frequency, coupled by a weak spring. The excitation makes both pendulums swing, but one has a large amplitude of motion and the other a small one. As time progresses, the coupling causes the amplitude of one pendulum to decrease while that of the other pendulum increases. Vibrational amplitude flows from one pendulum to the other and back again at a rate that depends on the difference in frequencies.

Returning to the radiation problem, when the coupled system fluoresces back to different levels of the ground state, a modulation on the fluorescence will be observed corresponding to the oscillatory flow of amplitude between the two coupled levels.

Figure 7.8 shows the fluorescence from two levels of anthracene. The upper trace gives the signal from the level composed mostly of the mode, la>, which is optically coupled to the ground level, while the lower trace gives the signal from the level composed mostly of mode lb>. Note that the oscillations have exactly the same period but they are 180° out of phase, just as we would expect from the pendulum analogy.

Now suppose that not just two vibrational levels are coupled by the interaction, but rather that the density of vibrational levels, ρ_v, is so high that $V\rho_v >> 1$ (the so-called statistical limit). Now the optically connected level, consisting mostly of la>, will be coupled to a very large number of other levels, so that the probability that the energy will return to that level will be small. The fluorescence from the

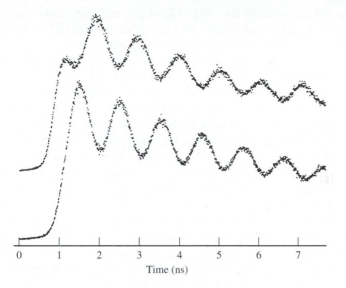

■ **Figure 7.8**

Experimental fluorescence decay curves showing quantum beats.

From P. M. Felker and A. H. Zewail, *Chem. Phys. Lett.* **102**, 113 (1983). Reprinted from *Chemical Physical Letters,* copyright 1983, with permission from Elsevier Science.

optically coupled level will be observed for only a short time, a time characterizing the redistribution of energy to the manifold of other vibrational levels. In the lower left-hand panel of **Figure 7.6,** this time for anthracene with 1792 cm^{-1} of vibrational energy is seen to be about 75 ps.

Intramolecular vibrational energy redistribution is critical to a wide range of chemical problems. It is particularly important to our understanding of unimolecular reactions, to be discussed in more detail in Section 7.5.4.

7.3.3   Internal Conversion, Intersystem Crossing, and Phosphorescence

The intramolecular vibrational energy redistribution we have just examined takes place between the excited level and other densely spaced vibrational levels of the same electronic state. Energy redistribution may also occur from vibrational levels of one electronic state to those of another, a process generally called a *radiationless transition.* When the final electronic state is of the same spin multiplicity as the initial state, this transition is called *internal conversion,* whereas when it is of a different multiplicity the transition is called *intersystem crossing.* Internal conversion and intersystem crossing either can take place as a collisionless, intramolecular processes or can be induced by collisions. The chemistry of these processes is not, in principle, very different from that of intramolecular vibrational energy redistribution; it again involves coupling between the excited state and the dense manifold of highly excited vibrational levels of the lower electronic state. The difference is that the interaction potential involves not only vibrational coupling but also an electronic coupling. Interactions that couple different zeroth-order electronic states are those neglected in making the Born-Oppenheimer approximation; i.e., terms involving the coupling of electronic and nuclear motions. In addition, for intersystem crossing, the coupling involves the spin-orbit interaction.

Figure 7.9 illustrates the process of internal conversion from an initially excited vibrational level of S_1 to high vibrational levels of S_0. It is important to remember that, in most cases, there will be several modes of vibration, so that the potential energy surfaces for S_1 and S_0 will depend on many internuclear coordinates rather than just the one plotted as the abscissa in **Figure 7.9.**

Figure 7.10 illustrates the process of intersystem crossing from an initially excited vibrational level of S_1 to high vibrational levels of the lowest triplet state, T_1. In the presence of collisions, intersystem crossing will typically be followed by vibrational relaxation and phosphorescence back to the ground singlet state.

If the internal conversion or intersystem crossing is caused by collisions, the process can be studied experimentally by observing the loss of fluorescence from the initially excited level of S_1 as the pressure of the collision partner is increased. The intensity of this fluorescence, or in a time-resolved experiment, the lifetime of the fluorescence, will decrease with increasing pressure, following the Stern-Volmer kinetics already discussed in Section 7.3.1.

If the internal conversion or intersystem crossing takes place as an intramolecular process, it is experimentally more difficult to observe. One way to tell that the conversion or crossing has occurred is to compare the lifetime of the fluorescence with the integrated absorption cross section. **Equations 7.12** and **7.6** show, respectively, that the integrated absorption cross section is related to the Einstein B_{12} coefficient and that the Einstein B_{12} coefficient is related to the A_{12} coefficient, the reciprocal of the radiative lifetime. Thus, if the observed lifetime is due solely to radiative processes, it will be equal to the lifetime calculated from the integrated absorption cross section using **equations 7.12** and **7.6.** On the other hand, if internal

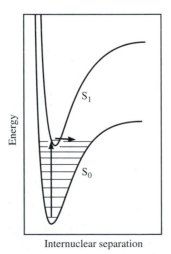

Figure 7.9

Internal conversion from low vibrational levels of the first excited singlet state (S_1) to high vibrational levels of the ground singlet state (S_0).

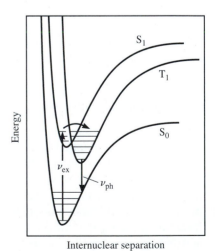

Figure 7.10

Intersystem crossing from S_1 to T_1. Following vibrational relaxation on T_1, phosphorescence occurs to S_0.

conversion or intersystem crossing compete with fluorescence as decay channels for the initially excited state, the observed fluorescence lifetime will typically be shorter than that calculated from the integrated absorption cross section.

A second method for learning that intramolecular internal conversion or intersystem crossing has taken place is to observe the low-pressure fluorescence intensity or fluorescence lifetime as a function of vibrational excitation in the S_1 manifold of states. If fluorescence is the only decay channel, its intensity or lifetime will vary irregularly and only slowly with increasing excitation energy, reflecting primarily the change in Franck-Condon factors for fluorescence as different initial levels are selected. On the other hand, because intersystem crossing and internal conversion depend on the density of vibrational levels in the final electronic state, and because this density increases dramatically with energy, the fluorescence intensity and lifetime will decrease dramatically with increasing excitation energy if either of these intramolecular processes is occurring.

An example of how the fluorescence lifetime varies with excitation energy in a system undergoing internal conversion or intersystem crossing is provided by the benzene molecule. **Figure 7.11** shows the lifetime of benzene fluorescence under collisionless conditions as a function of the vibrational energy above the origin of the first singlet state ($^1B_{2u}$). The experimental points are labeled with their vibrational mode assignment in the excited state; for example, $6^1 1^1$ means one quantum of ν_6 (a carbon–carbon bending mode) and one quantum of ν_1 (the symmetric stretching mode) with zero quanta in all other modes. The important point to notice from the figure is that the lifetime decreases rapidly with increasing vibrational energy. Some more subtle points of the data are worth remarking on as well. Note that the lifetime decreases dramatically with the number of quanta of ν_{16}, an out of plane bending mode. This mode is a so-called "promoting mode" for the radiationless

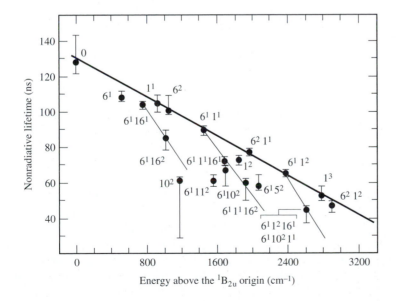

■ **Figure 7.11**

The lifetime of fluorescence from benzene vapor as a function of energy above the origin of the first singlet state.

From K. G. Spears and S. A. Rice, *J. Chem. Phys.* **55**, 5561 (1971). Reprinted with permission from the *Journal of Chemical Physics.* Copyright, American Institute of Physics, 1971.

transition since it is particularly effective in causing an interaction between the excited and ground states of the molecule, states that would not mix if the Born-Oppenheimer approximation were rigorously obeyed. The behavior illustrated for benzene is typical of molecules undergoing internal conversion.

example 7.1

Radiative and Nonradiative Lifetimes

Objective   Integrated absorption data shows that the quantum yield for fluorescence from the 0^0 level of benzene is 0.25. What would be the fluorescence lifetime of benzene excited to this state if there were no internal conversion?

Method   In the absence of collisions the total rate of decay of the 0^0 level is given by the sum of the radiative and nonradiative contributions: $k_{tot} = k_r + k_{nr}$, where $k_{tot} = 1/\tau_{obs}$. In addition, the quantum yield of fluorescence is given by $\Phi_f = k_r/(k_r + k_{nr})$. Thus, the value of $\Phi_f = 0.25$ and $\tau_{obs} \approx 128$ ns (from **Figure 7.11**), can be used to calculate k_r, which is simply $1/\tau_r$, the reciprocal of the fluorescence lifetime in the absence of internal conversion.

Solution   Taking the reciprocal of both sides of the equation $\Phi_f = 0.25$ gives $1 + (k_{nr}/k_r) = 4.0$, or $k_{nr} = 3.0 \, k_r$. We know that $k_{obs} = k_r + k_{nr} = (1 + 3.0)k_r = 1/\tau_{obs} = 1/(128 \text{ ns})$, so that $4.0k_r = 1/(128 \text{ ns})$. Since $k_r = 1/\tau_r$, we find that $\tau_r = (4.0)(128 \text{ ns}) = 512$ ns.

Comment   Note that k_r is called the *radiative* decay rate, while k_{nr} is called the *nonradiative* decay rate.

7.3.4   Photodissociation

Excitation of some molecules results in dissociation into two or more fragments. With the aid of **Figure 7.12** we can distinguish between two types of such photodissociation processes, direct and indirect. In all panels of this figure, the total energy of the molecule following absorption of the photon is fixed and given by the level of the vertical arrow. As time progresses, potential energy is converted into kinetic energy, and the internuclear separation between the fragments increases. In a direct process, all motion takes place on a single excited potential energy surface and the dissociation is complete on the time scale of a single vibration. In an indirect process, a crossing takes place between two surfaces which, at the level of the Born-Oppenheimer approximation, can be classified as bound and unbound. Because the process depends on the coupling, the dissociation might take considerably longer than a vibrational period. Panels (A) and (B) show direct dissociation processes caused either by the excitation of a region of the excited potential energy surface above the dissociation limit or by excitation of a purely repulsive surface, respectively. Panel (C) shows an example of an indirect process. In this case, excitation is to a level that would be bound if it were not for an avoided crossing with a dissociative state. Along the adiabatic curves that result from the interaction, the molecule dissociates. This indirect process is also known as *predissociation*.

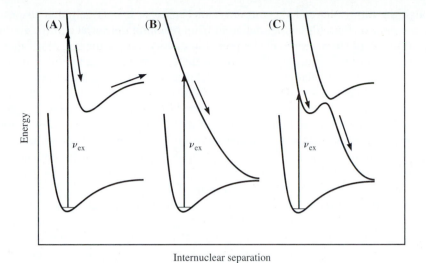

Figure 7.12

Different types of photodissociation: (A) and (B) direct dissociations and (C) predissociation.

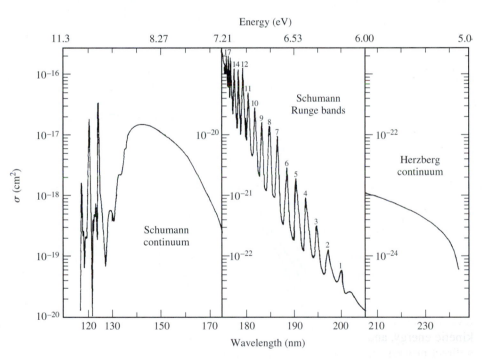

Figure 7.13

The ultraviolet absorption cross
section (in cm²) of O_2.

Figure from "Photochemical Processes and Elementary Reactions" in
CHEMISTRY OF THE NATURAL
ATMOSPHERE by Peter Warneck,
copyright © 1988 by Academic Press,
reproduced by permission of the
publisher. All rights or reproduction
in any form reserved.

As we will see in the next sections, photodissociation has profound practical
consequences. The energy for nearly all chemical reactions in the Earth's atmosphere and on its surface comes from sunlight, so that an understanding of the light-
induced processes, particularly dissociation, is extremely important.

A simple example of the direct dissociation process of panel (A) and the indirect dissociation process of panel (C) in **Figure 7.12** is provided by the O_2, whose
absorption spectrum is shown in **Figure 7.13.** There is virtually no absorption in the
visible region of the spectrum, but the cross section for absorption increases strongly

through the ultraviolet and vacuum ultraviolet regions. It is largely this absorption that prevents vacuum ultraviolet radiation from reaching the surface of the Earth.

Figure 7.14 shows some of the potential energy curves for O_2 (recall that the ground state of O_2 is a triplet: $^3\Sigma_g^-$.) While there is an enormous number of states for even such a simple diatomic molecule, let us concentrate our interest on the lowest five curves. The absorption regions of **Figure 7.13** labeled "Schumann continuum" and "Schumann-Runge bands" correspond to absorption from the $^3\Sigma_g^-$ ground state to the continuum and bound regions of the $^3\Sigma_u^-$ potential curve. Absorption to the continuum region leads to dissociation of O_2 into $O(^3P) + O(^1D)$, which is the process illustrated in panel (A) of **Figure 7.12.** Absorption to the bound region leads to predissociation via the $^3\Pi_u$ repulsive curve and produces 2 $O(^3P)$ atoms, illustrating the process in panel (C) of **Figure 7.12.**

Note from **Figure 7.13** that the absorption by O_2 at wavelengths longer than 200 nm is quite weak, so that O_2 does not very effectively block sunlight in the region, say, near 250 nm, where the principal component of life, DNA, absorbs strongly. However, as we will examine in detail in the next section, the oxygen atoms produced via absorption on the Schumann continuum and on the Schumann-Runge bands can recombine with O_2 to form O_3. Ozone does absorb light in the 200- to 300-nm region, and it is this absorption that protects life on Earth from ultraviolet radiation–induced mutations, as discussed in detail in the next section.

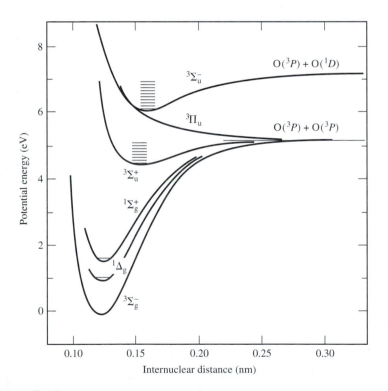

■ Figure 7.14

Calculated potential curves for the triplet valence states of O_2. Note the crossing of the $^3\Sigma_u^-$ state by the repulsive $^3\Pi_u$ state.

Based on curves published by F. R. Gilmore, *J. Quant. Spectrosc. Radiat. Transfer* **5,** 369 (1965). Reprinted from *Journal of Quantitative Spectroscopy and Radiative Transfer,* copyright 1965, with permission from Elsevier Science.

7.4   ATMOSPHERIC CHEMISTRY

The composition of the atmosphere provides an interesting and important application of photochemistry. We start by categorizing the different regions of the atmosphere, using **Figure 7.15** as a reference. As a function of the height above sea level, the solid line in this figure shows the average temperature and the bars give its range of variation. Note that the temperature decreases as the height increases to about 10 km, and then the temperature increases to a height of roughly 50 km. The extremes in temperature are used to define the different regions of the atmosphere, as identified in the figure.

The *troposphere* is the region nearest the surface of the Earth, and the photochemistry in this region is extremely complex because of the large variety of atmospheric components. The light reaching this region is limited primarily to the visible and infrared parts of the spectrum. It is in the troposphere that human activity has had the most effect on the atmosphere. Acid rain, photochemical smog, and global warming, for example, are largely issues of the troposphere.

The *stratosphere* is the region in which the temperature increases with increasing height. The photochemistry of this region is dominated by the dissociation of ozone and, in the upper regions, oxygen. It is here that the issues of ozone depletion by nitric oxide and chlorofluorocarbons are most important. The light reaching this area of the atmosphere is primarily at wavelengths above about 200 nm; light below this wavelength is absorbed by oxygen.

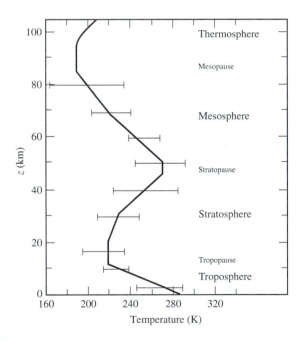

■ **Figure 7.15**

The temperature of the atmosphere as a function of height above sea level.

The *mesosphere* is a region of the upper atmosphere dominated by the photochemistry of small diatomic molecules and the reactions of atoms and ions. The *thermosphere* is the outer limit of the atmosphere, above about 90 km.

The stratosphere provides a good example of how our understanding of photochemistry can lead to an improved understanding of the atmosphere. The major chemical constituents of this region of the atmosphere are N_2 (with a mole fraction of 0.79), O_2 (of mole fraction 0.21), and O_3 (of mole fraction 1.3×10^{-5}). Since we have already remarked that it is the ozone that prevents near ultraviolet light from reaching the surface of Earth, it is of importance to understand the photochemical processes that maintain these steady-state concentrations. A first approximation to a chemical mechanism for the stratosphere was given by Chapman in 1930.[d] In this simple model, the nitrogen is chemically inert, whereas the oxygen and ozone are coupled by two photochemical and two chemical reactions. The first of the photochemical processes is the dissociation of oxygen, already discussed in the preceding section:

$$O_2 + h\nu\,(\lambda < 200\text{ nm}) \rightarrow 2\,O. \tag{R1}$$

The second photochemical process is the dissociation of ozone:

$$O_3 + h\nu\,(200\text{ nm} < \lambda < 310\text{ nm}) \rightarrow O_2 + O. \tag{R2}$$

The absorption spectrum of ozone is shown in **Figure 7.16.** Because the mole fraction of ozone is small relative to that of oxygen and because the oxygen absorbs strongly below 200 nm, the primary region of ozone photochemistry is the so-called Hartley band between 200 and 310 nm.

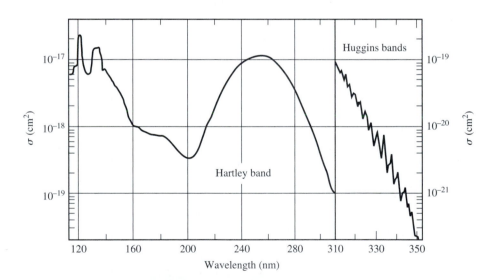

■ **Figure 7.16**

The absorption cross section (in cm^2) of O_3.

[d]S. Chapman, *J. R. Meteorol. Soc.* **3,** 103 (1930); *Philos. Mag.* **10,** 345 (1930).

Two chemical reactions complete the Chapman mechanism, the recombination of O with O_2,

$$O + O_2(+M) \rightarrow O_3(+M), \qquad \textbf{(R3)}$$

and the reaction between O and O_3,

$$O + O_3 \rightarrow 2\,O_2, \qquad \textbf{(R4)}$$

where M is either O_2 or N_2 and serves to take away enough energy to stabilize the O_3. The process **reaction R4** is of minor importance compared to that in **reaction R3.**

Although the Chapman model is a reasonable first approximation, the steady-state concentration of O_3 that it predicts (Problem 7.7) is three times higher than that observed. It should be noted that, given the flux of solar radiation in the stratosphere and the steady-state concentrations, ozone is photodissociated and regenerated by **reaction R2** and **reaction R3** about 30 times for every O_2 molecule dissociated in **reaction R1.** Thus, even minor components of the atmosphere that interfere with the cycle of **reactions R2–R3** can have a large effect on the steady-state concentration of O_3.

Two such minor components of the natural atmosphere are N_2O ($\sim$100 parts per billion by volume, ppbv) and H_2O ($\sim$5 ppmv). N_2O undergoes photodissociation in the stratosphere, while H_2O reacts with oxygen atoms. The resulting NO and HO radicals participate in catalytic reactions that lower the steady-state concentration of ozone. The reaction mechanisms are

$$
\begin{aligned}
NO + O_3 &\rightarrow NO_2 + O_2 \\
NO_2 + O &\rightarrow NO + O_2
\end{aligned}
$$

Net:
$$O_3 + O \rightarrow 2\,O_2,$$

and

$$
\begin{aligned}
HO + O_3 &\rightarrow HOO + O_2 \\
HOO + O_3 &\rightarrow HO + 2\,O_2
\end{aligned}
$$

Net:
$$2\,O_3 \rightarrow 3\,O_2.$$

Notice that the net result of each of these cycles is to destroy ozone and that in each case the initial radical is regenerated so that it continues the cycle.

In the 1970s, the airline industries proposed to develop a fleet of supersonic aircraft that would fly at a height of 20–40 km. Fortunately, enough was known about the chemistry of the stratosphere at that time so that chemists were able to prevent a potential disaster. The nitric oxides injected into the ozone layer by the exhaust from such a fleet of planes might have seriously depleted the steady-state concentration ozone by participating in the first of the catalytic cycles listed above.[e]

Unfortunately, chemists did not have an early enough understanding to head off another anthropogenic cause of ozone depletion. In 1974 Rowland and Molina proposed another possible catalytic cycle involving chlorine atoms[f]

$$
\begin{aligned}
Cl + O_3 &\rightarrow ClO + O_2 \\
ClO + O &\rightarrow Cl + O_2
\end{aligned}
$$

Net:
$$O_3 + O \rightarrow 2\,O_2.$$

---

[e]For an interesting account, see H. S. Johnston, *Ann. Rev. Phys. Chem.* **43,** 1 (1992).

[f]M. J. Molina and F. S. Rowland, *Nature* **249,** 810 (1974). Rowland and Molina shared the 1995 Nobel Prize in Chemistry (with P. Crutzen) for their work in this area.

They warned that chlorofluorocarbons used for industrial cleaning and refrigerants might produce Cl atoms that could destroy ozone. Ironically, for reasons involving human safety, the chlorofluorocarbons used for these purposes were designed to be relatively inert. They turned out to be too inert. Whereas most chlorine-containing compounds dissolve in rain and are returned to the Earth's surface, these substances rise all the way to the upper stratosphere, where they are dissociated to produce chlorine atoms that then catalytically destroy ozone.

Over the past 50 years, stratospheric chlorine concentrations have increased from a background level of 0.5 ppbv to a level today of 3.5 ppbv. The ozone concentration is beginning to show the effects, as demonstrated by satellite measurements. Whereas the average range of ozone concentration during the period 1979–1990 was nearly 300 Dobson units (the equivalent ozone column height in units of 10^{-5} m at standard temperature and pressure), the 1992 average was about 7 units lower and the 1993 average (taken until the *Nimbus-7* satellite ceased to function) was lower by another 7 units.

The effect has been even more dramatic in the region of the polar caps, particularly above Antarctica, where in 1985 an "ozone hole" has been detected just following the return of the solar radiation in the spring of that and each subsequent year.[g] A stream of air known as the polar vortex isolates the region of the stratosphere above the Antarctic continent. During the winter, ice and nitric acid trihydrate condense to form polar stratospheric clouds, and the crystals in these clouds provide reactive surfaces that store chlorine as $ClONO_2$. When the sun reappears in the early spring, the crystals melt and release the both Cl atoms and ClO. The atoms destroy ozone in the catalytic cycle predicted by Molina and Rowland, while the chlorine monoxide participates in its own catalytic cycle:

$$ClO + ClO \rightarrow Cl_2O_2$$

$$Cl_2O_2 + h\nu \rightarrow Cl + ClOO$$

$$ClOO + M \rightarrow Cl + O_3 + M$$

$$2\,(Cl + O_3 \rightarrow ClO + O_2)$$

Net: $\qquad 2\,O_3 + h\nu \rightarrow 3\,O_2.$

Experiments monitoring ClO and ozone concentrations in the region of the Antarctic vortex have further established the connection between chlorine and ozone concentrations. **Figure 7.17** demonstrates the close inverse correlation between the ClO concentration and the ozone concentration just following the spring thaw. As a high-altitude airplane flew its instruments through layers formed by the vortex, it recorded a dip in the ozone concentration at every location where the ClO concentration increased. Within the vortex, the ozone concentration was found to be less than half the concentration outside the vortex.

Fortunately, the end of this story may be in sight. Under the 1990 revisions of the Montreal protocol, agreed to by most nations producing chlorofluorocarbons, the rate of human injection of chlorine into the stratosphere should be cut back enough so that the chlorine concentration should peak around the year 2000 and fall after that. Models predict, however, that the Antarctic ozone hole will not disappear until 2050.

---

[g]J. C. Farman, B. G. Gardiner, and J. D. Shankin, *Nature* **315**, 207 (1985).

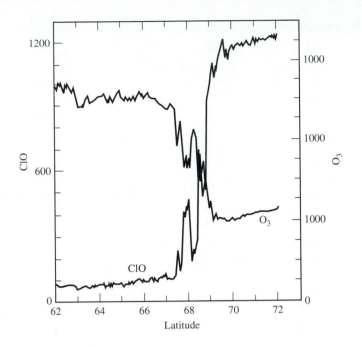

Figure 7.17

Mixing ratio in parts per trillion by volume for ClO and O_3 as a function of latitude in degrees south. Note the close inverse correlation between the two concentrations.

From J. G. Anderson, D. W. Toohey, and W. H. Brune, *Science* **251,** 39 (1991). Reprinted with permission from *Science.* Copyright 1991 American Association for the Advancement of Science.

While the stratospheric ozone concentration illustrates both the importance of photochemistry to the composition of the atmosphere and the importance of understanding the atmospheric implications of anthropomorphically generated chemicals, it is but one of many problems of current global concern. Global warming due to the emission of gases that absorb infrared light, acid rain due to sulfuric acid emissions, and photochemical smog generated by the interaction of sunlight and automobile exhaust are but a few of the problem areas where a knowledge of photochemistry is of practical relevance. These and other problems are discussed in detail in a few of the books listed in the Suggested Readings section.

7.5   PHOTODISSOCIATION DYNAMICS

The goal of a subfield of photochemistry known as photodissociation dynamics is to understand at the molecular level the process by which light induces a dissociation. What causes the absorption? What are the forces on the fragments as they fly apart? What are the final states of the products, and how do they depend on which initial state was created by the photon? To physical chemists, the field of photodissociation dynamics emerged as a science with the development by Norrish and Porter of the flash photolysis or pump-probe technique.[h]

---

[h]G. Porter, *Proc. R. Soc.* **A200,** 284 (1950). R. G. W. Norrish and G. Porter shared the 1967 Nobel Prize in Chemistry (with M. Eigen) for their work in this area.

7.5.1   The Pump-Probe Technique

In the earliest versions of the pump-probe technique, photolysis of a starting material was achieved by one flash lamp while the production of photochemical fragments was monitored by their absorption of continuum light from another flash lamp. In observing spectroscopically the internal energy states of the photofragments, Norrish and Porter demonstrated the first general technique that provided information about the *dynamics* of a photochemical event. The pace of investigation increased rapidly in the 1960s due to two improvements. Tunable lasers with powers high enough to dissociate a large fraction of the parent molecule and with spectral resolution high enough to excite individual internal energy levels gradually replaced the photolysis flash lamp. In addition, fragment detection by absorption from a second flashlamp was replaced by one of three more sensitive laser detection techniques: time-resolved laser absorption, laser-induced fluorescence, or multiphoton ionization. We now briefly explore with an example how results obtained using these experimental techniques compare with theory.

A major goal of the field is to determine the forces acting on the fragments as they separate. This force is simply the negative of the gradient of the potential energy surface. For example, **Figure 7.18** shows the sections of the excited potential energy surface for the water molecule leading to dissociation into OH + H. Each curve gives the potential energy as a function of (HO)–H distance for a fixed value of the OH bond length. The different curves are for different angles of departure; the curve labeled V_0 gives the angle averaged potential. Note that the energy

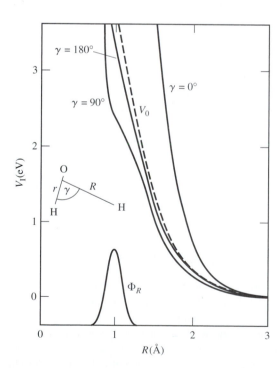

Figure 7.18

Potential energy curves for H_2O as a function of R in the coordinate system shown for different values of γ and for $r = 0.971$ Å.

From R. Schinke, V. Engel, and V. Staemmler, *J. Chem. Phys.* **83**, 4522 (1985). Reprinted with permission from the *Journal of Chemical Physics*. Copyright, American Institute of Physics, 1985.

decreases rapidly with increasing (HO)–H distance, so that there is a large repulsive force between the recoiling fragments. The potential was calculated using *ab initio* techniques. How well does it do in predicting the experimental results?

An experiment performed in 1987 prepared individual rotational levels of vibrationally excited water using one laser, dissociated them by excitation to the repulsive curves of **Figure 7.18** with a second laser, and then probed the OH fragment rotational distribution using laser-induced fluorescence (a technique to be described below) with a third laser. **Figure 7.19** shows the close agreement between the distributions calculated using the *ab initio* potential energy surface of **Figure 7.18** and that measured by the experiment. The combination of experiment and theory gives confidence that the dissociation dynamics of water are understood at the molecular level.

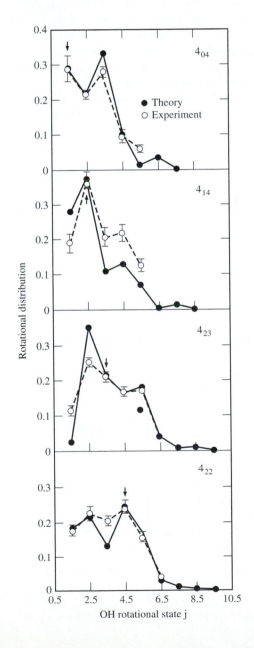

Figure 7.19

Experimental and theoretical OH rotational distributions. The panels give the results for dissociation of different J_{K_a,K_c} levels.

From D. Häusler, P. Andresen, and R. Schinke, *J. Chem. Phys.* **87,** 3949 (1987). Reprinted with permission from the *Journal of Chemical Physics.* Copyright, American Institute of Physics, 1987.

Before proceeding further, it is helpful to understand the two detection techniques of laser-induced fluorescence and multiphoton ionization.

7.5.2   Laser-Induced Fluorescence

One of the most sensitive and commonly used probe techniques is called laser-induced fluorescence (LIF). **Figure 7.20** shows the basic principle of the technique, which was first recognized as a tool for studying dynamics by Schultz, Cruse, and Zare.[i] The physical processes are similar to those discussed in Section 7.3.1. For simplicity, the figure shows schematic-level diagrams for the vibrational and rotational levels of the lower and upper states but omits the electronic potential energy curves. A tunable laser of frequency ν_L excites molecules from vibrational-rotational levels of the ground electronic state to an upper electronic state that subsequently fluoresces. If the total fluorescence intensity is recorded as a function of the frequency or wavelength of the exciting laser, a line spectrum will be observed. At every frequency for which the laser is in resonance with an allowed transition from the lower to upper state, molecules will be excited and will subsequently fluoresce, whereas no excitation or fluorescence will be observed if the laser is not in resonance with an allowed transition.

For example, **Figure 7.21** shows a laser-induced fluorescence spectrum of CO produced in the 193-nm photodissociation of acetone to yield $CO + 2\ CH_3$. The top panel probes a region of the CO absorption band corresponding to excitation from rotational levels of CO in $v = 0$ of the ground electronic state to rotational levels in $v = 2$ of the upper electronic state, whereas the bottom panel probes a region taking CO from $v = 1$ in the lower state to $v = 0$ in the upper state. Q-branch transitions (those with $\Delta J = 0$) are identified in each scan, although P- and R-branch transitions ($\Delta J = -1$ and $\Delta J = +1$) are also present. It is thus qualitatively clear from the spectrum that CO is produced in both $v = 0$ and $v = 1$ and in rotational levels ranging up to at least $J = 50$.

A quantitative analysis of the spectrum can yield accurate relative populations for the CO levels produced. Assuming (as is nearly valid in this case) that the fluorescence intensity is unaffected by nonradiative processes, the signal is proportional to the population in the lower level (because of the high excitation energy, we assume no initial population in the upper level). The quantities involved in the proportionality constant include the Franck-Condon factor for excitation between the two

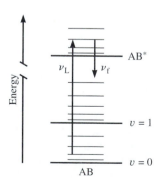

Figure 7.20

The laser-induced fluorescence technique.

[i]A. Schultz, H. W. Cruse, and R. N. Zare, *J. Chem. Phys.* **57**, 1354 (1972); P. J. Dagdigian and R. N. Zare, *Science* **185**, 739 (1974).

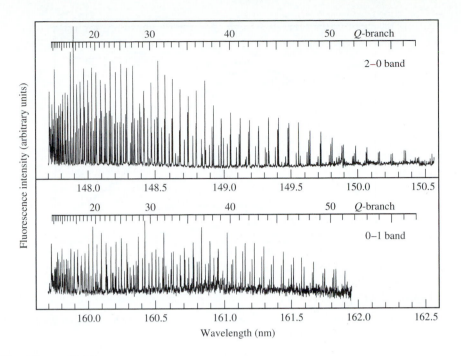

Laser-induced fluorescence spectrum of CO produced in the 193-nm photodissociation of acetone. From K. A. Trentelman, S. H. Kable, D. B. Moss, and P. L. Houston, *J. Chem. Phys.* **91,** 7498 (1989). Reprinted with permission from the *Journal of Chemical Physics.* Copyright, American Institute of Physics, 1989.

vibrational levels and the rotational line strengths, called Hönl-London factors. For CO, these quantities have been accurately measured. Problem 7.12 provides an example of how these are used.

Figure 7.22 shows the populations derived from the spectrum of **Figure 7.21** using the known Franck-Condon and Hönl-London factors for the CO $\tilde{A} \leftarrow \tilde{X}$ band (see Problem 7.12 for details). The figure also includes results based on additional data probing CO in $v = 2$. At room temperature, the most populated level of CO is near $J = 10$. Note, therefore, that very high rotational levels are produced in all vibrational levels, suggesting that the force on the CO fragment due to the departure of the two methyl fragments is unequal. A likely explanation is that the dissociation proceeds in a sequential fashion with an acetyl intermediate, as shown in **Figure 7.23.**

7.5.3   Multiphoton Ionization

A second technique commonly used to probe the products of photodissociation is called multiphoton ionization (MPI). Although the technique was first used to detect atoms, the initial application to molecules was in the detection of Cs_2 by Collins et al.[j] Extension to more chemically interesting species, such as benzene and NO, was subsequently made by Johnson and his coworkers.[k]

[j]C. B. Collins, B. W. Johnson, and M. Y. Mirza, *Phys. Rev. A* **10,** 813 (1974).
[k]P. M. Johnson, M. R. Berman, and D. Zakheim, *J. Chem. Phys.* **62,** 2500 (1975).

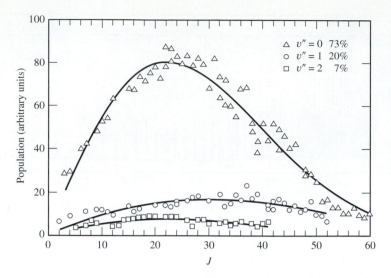

△ $v''=0$ 73%
○ $v''=1$ 20%
□ $v''=2$ 7%

Figure 7.22

Relative populations of CO energy states produced in the photodissociation of acetone determined from spectrum of **Figure 7.21.**

From K. A. Trentelman, S. H. Kable, D. B. Moss, and P. L. Houston, *J. Chem. Phys.* **91,** 7498 (1989). Reprinted with permission from the *Journal of Chemical Physics.* Copyright, American Institute of Physics, 1989.

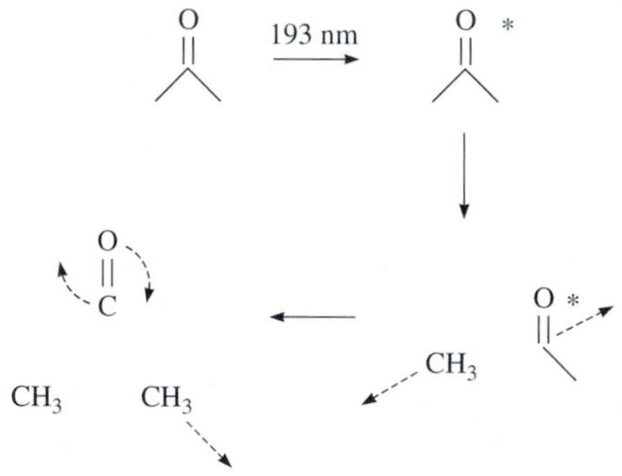

Figure 7.23

Probable dissociation mechanism for acetone at 193 nm.

Figure 7.24 shows the principle of the technique, which is very similar to that used in laser-induced fluorescence. A pulsed, tunable laser at frequency ν_L excites molecules from a specific vibrational-rotational level of the ground electronic state of AB to a level of an excited electronic state, AB^*. The number of AB^* molecules thus produced is again given by the same equation that governs the laser-induced fluorescence signal (see Problem 7.12). If the intensity of the laser is strong enough, a second photon from the same laser pulse can cause further excitation of AB^* to the ionization continuum, producing $AB^+ + e^-$. The ions are accelerated into a detector and counted. If, as is typical, the ionization step has a high cross section, nearly every AB^* will be ionized. Thus the number of ions is proportional to the population in the initially selected vibrational-rotational level of the ground state. Because the detection of ions is often more sensitive than the detection of

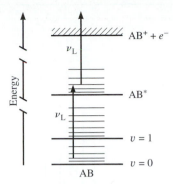

■ **Figure 7.24**

Multiphoton ionization through the resonant intermediate state AB^*. The example is a $1 + 1$ ionization.

photons, the technique has been widely used to study the products of chemical reactions. Note also that even if the excited level of AB^* were, for example, dissociative, it might still be possible to observe a signal if the power of the laser is sufficiently high to cause ionization in competition with the dissociation.

The process illustrated in **Figure 7.24** is the most common type of multiphoton ionization, but other types are also useful. The initial excitation to AB^* might, for example, be caused by a two-photon absorption, whereas the $AB^* \rightarrow AB^+ + e^-$ step might be caused by a one-photon transition. The overall process would then be called a resonant $2 + 1$ multiphoton ionization. In general, an "m + n" multiphoton process means that m photons were used to excite the intermediate AB^* state, while n photons were used to ionize it.

Figure 7.25 shows the resonant $2 + 1$ multiphoton ionization spectrum of N_2 produced in the 193-nm photodissociation of N_2O. Using a formula similar to that used in Problem 7.12, the authors determined the N_2 rotational distribution given in **Figure 7.26**. On average, 57% of the energy available to the products appears in N_2 rotation. Since the ground potential energy surface of N_2O has a minimum in the linear configuration, the experimental results strongly suggest that the upper electronic state induces substantial bending in the N_2O as the N_2–O bond breaks.

7.5.4   Unimolecular Dissociation

An interesting experiment has recently been performed using the pump-probe technique with laser-induced fluorescence detection. A 1-ps pulse from a tunable laser was used to excite NO_2 in the energy region just above its dissociation threshold near 25,130 cm^{-1}. It is known from the optical spectroscopy of NO_2 that the levels excited are strong mixtures of ground and excited state electronic character; that is, the internal conversion from the excited state to the ground state (see Section 7.3.3) is virtually instantaneous. Since the energy is above the barrier to dissociation on the ground-state surface, NO_2 then decomposes to $NO + O$.

A second 1-ps pulse from the laser system was used to probe the NO product by laser-induced fluorescence (Section 7.5.2). The probe laser was tuned to monitor NO in a particular state and the signal was recorded as a function of the time delay between the pump and probe pulses. The results are shown for different pump

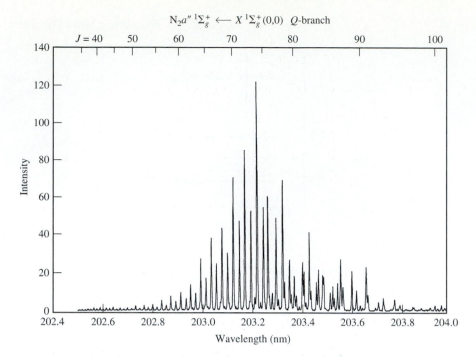

$N_2 a'' \, ^1\Sigma_g^+ \longleftarrow X \, ^1\Sigma_g^+ (0,0) \quad Q\text{-branch}$

Figure 7.25

The 2 + 1 multiphoton ionization spectrum of N_2 produced in the 193-nm photodissociation of N_2O.

From T. F. Hanisco and A. C. Kummel, *J. Phys. Chem.* **97,** 7242 (1993). Reprinted with permission from *The Journal of Physical Chemistry.* Copyright 1993 American Chemical Society.

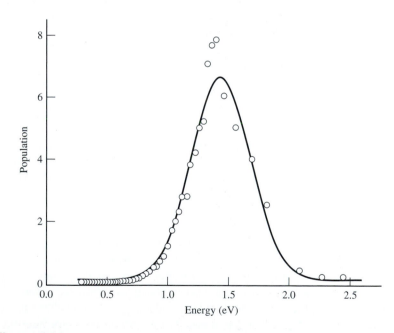

Figure 7.26

The rotational distribution obtained from analysis of the spectrum in **Figure 7.25.** The abscissa gives the energy corresponding to the measured rotational levels. The solid line is a Gaussian fit to the population distribution.

From T. F. Hanisco and A. C. Kummel, *J. Phys. Chem.* **97,** 7242 (1993). Reprinted with permission from *The Journal of Physical Chemistry.* Copyright 1993 American Chemical Society.

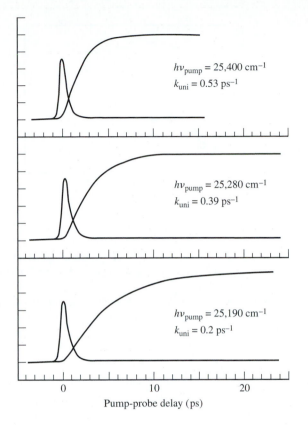

$h\nu_{pump} = 25{,}400 \text{ cm}^{-1}$
$k_{uni} = 0.53 \text{ ps}^{-1}$

$h\nu_{pump} = 25{,}280 \text{ cm}^{-1}$
$k_{uni} = 0.39 \text{ ps}^{-1}$

$h\nu_{pump} = 25{,}190 \text{ cm}^{-1}$
$k_{uni} = 0.2 \text{ ps}^{-1}$

Pump-probe delay (ps)

■   **Figure 7.27**

NO laser-induced fluorescence signal as a function of delay.

From G. A. Brucker, S. I. Ionov, Y. Chen, and C. Wittig, *Chem. Phys. Lett.* **194,** 301 (1992). Reprinted from *Chemical Physical Letters,* copyright 1992, with permission from Elsevier Science.

frequencies in **Figure 7.27.** Two curves are shown in each of the three panels of this figure. The sharp spike in each panel is a measure of the time resolution of the experiment, while the rising curve shows the NO product concentration as a function of time after excitation of the NO_2. Comparison of the three panels shows that as the wave number of the pump laser is tuned above the barrier at 25,130 cm^{-1} the appearance rate of the NO increases from 0.2 ps^{-1} at 25,190 cm^{-1} to 0.53 ps^{-1} at 25,400 cm^{-1}.

The experiment was performed at a variety of pump wave numbers with the result shown in **Figure 7.28,** which plots the rate as a function of energy above the threshold for dissociation. In addition to the general upward trend in the rate with energy, note that the increase actually appears to occur in steps, at least at low energy. What is the explanation for this steplike behavior, and, more generally, how can we account for the rise in the decomposition rate with energy?

We now consider a theoretical treatment of the rate at which an energized molecule decomposes to products. The problem has a long and interesting history, going back to the introduction of the Lindemann mechanism for unimolecular reactions (see Section 2.4.4). A quite successful theoretical treatment of both the collisional activation/deactivation and the unimolecular decomposition of the energized

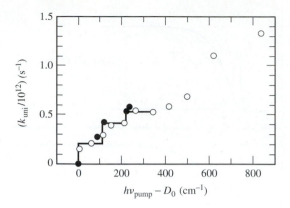

■   **Figure 7.28**

NO_2 unimolecular decay rates as a function of energy above the dissociation threshold. The closed and open circles are for two different laser pulse widths: 1.5 and 0.5–0.9 ps, respectively.

From G. A. Brucker, S. I. Ionov, Y. Chen, and C. Wittig, *Chem. Phys. Lett.* **194,** 301 (1992). Reprinted from *Chemical Physical Letters,* copyright 1992, with permission from Elsevier Science.

molecule is known as the RRKM theory, named after its authors Rice, Ramsperger, Kassel, and Marcus.[l] Our goal here is to focus on the rate, $k_a(E^*)$, at which a molecule with energy E^* decomposes to products.[m]

Figure 7.29 gives the definitions of energies used in the derivation. We consider the decomposition of molecules with internal energies in the range from E^* to $E^* + dE$, where the energy E^* consists of a vibrational contribution, E_v^*, and a rotational contribution, E_r^*, all energies being measured from the zero-point vibrational level of the reactant. In the course of the unimolecular process, the reaction path crosses a critical activated configuration where the energy, as measured between the zero-point vibrational levels of the reactant and the activated complex, is denoted E_0. The remaining energy of the molecule, which we will call E^+, is composed of three parts: a vibrational contribution, E_v^+, a rotational contribution, E_r^+, and a "translational" contribution, x, in the direction along the reaction path.

Following the spirit of Section 3.4 on the activated complex theory, we approximate the rate of the reaction $k_a(E^*)$ by (1) calculating the concentration of activated complexes $[A^+]$ from the equilibrium constant between the reactants and the complexes and (2) multiplying by the rate at which the activated complexes proceed to products. From statistical mechanics, we know that the equilibrium constant is given by the ratio of the partition functions: $K_e = [A^+]/[A^*] = q^+/q^*$. Consequently, $[A^+] = (q^+/q^*)[A^*]$. Assuming that half the activated complexes decompose toward products along the reaction path and half return to reactants, the rate constant for product formation may be written as $\frac{1}{2}k^+$, where k^+ is the total rate

[l]O. K. Rice and H. C. Ramsperger, *J. Am. Chem. Soc.* **49,** 1617 (1927); **50,** 617 (1928); L. S. Kassel, *J. Phys. Chem.* **32,** 225 (1928); **32,** 1065 (1928); R. A. Marcus, *J. Chem. Phys.* **20,** 359 (1952); R. A. Marcus and O. K. Rice, *J. Phys. Colloid Chem.* **55,** 894 (1951).

[m]The discussion here follows the outline and notation of Section 4.5 of P. J. Robinson and K. A. Holbrook, *Unimolecular Reactions* (Wiley-Interscience, London, 1972). Note that k_{uni} of **Figure 7.28** is called k_a here, in accord with Robinson and Holbrook.

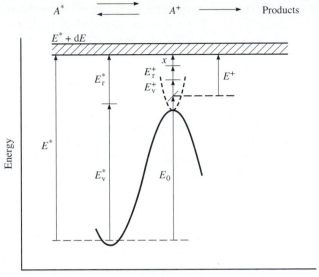

Figure 7.29

Energy definitions for discussion of RRKM theory.

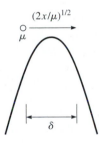

Figure 7.30

Parameters for calculating k^+.

constant for disappearance of A^+ into product and reactant. Thus, the rate of the reaction is $d[\text{products}]/dt = \frac{1}{2}k^+[A^+] = \frac{1}{2}k^+(q^+/q^*)[A^*]$. However, $d[\text{products}]/dt = -d[A^*]/dt = k_a(E^*)[A^*]$, so that

$$k_a(E^*) = \frac{1}{2}k^+\frac{q^+}{q^*}. \tag{7.24}$$

We now evaluate separately the rate constant k^+ and the ratio of partition functions (q^+/q^*) before combining them into an expression for $k_a(E^*)$.

The rate constant k^+ is evaluated by treating the motion along the reaction coordinate as a translation in one dimension (see also Problem 3.9). As shown in **Figure 7.30,** let the reduced mass of the atoms involved in the motion be μ and let δ be a distance along the reaction coordinate that characterizes the activated complex. (Admittedly, neither of these two parameters is particularly well defined, but, as we will see, they cancel in the final result.) The rate constant for decomposition due to motion along the reaction coordinate (in s^{-1}) is given by the velocity (in m/s) divided by the distance (in m). If the translational energy is x (see **Figure 7.29**), then the velocity is $(2x/\mu)^{1/2}$, so that the rate constant is

$$k^+ = \frac{(2x/\mu)^{1/2}}{\delta} = \left(\frac{2x}{\mu\delta^2}\right)^{1/2}. \tag{7.25}$$

The ratio of partition functions is just the ratio between the number of states for the activated complex and the number of states for the reactant accessible to each in the energy range dE near E^+, or equivalently E^*. If the energy range dE is small enough, we can assume that the number of states is given by the density of states, N (the number per unit energy), times the energy range dE, where N is taken to be constant throughout the small range dE. Thus, $q^+ = N^+(E^+)\,dE$ and $q^* = N^*(E^*)\,dE$. The ratio is simply

$$\frac{q^+}{q^*} = \frac{N^+(E^+)}{N^*(E^*)}. \tag{7.26}$$

Since the energy of the reactant, E^*, is typically quite high, it is usually sufficient to use a continuous function to describe $N(E^*)$. Quite good approximations to this function can be obtained using the known spectroscopic constants of the reactant.[n]

The density of states $N^+(E^+)$ is more troublesome. First of all, the energy E^+ is generally small enough so that the discrete nature of the levels of the activated complex must be included explicitly in the calculation. Second, the location of these states is difficult to determine from experiment, so that approximations usually must be made in order to evaluate $N^+(E^+)$. Suppose we know from calculation or from a guess the location of the vibrational-rotational levels of the activated complex. Let $P(E_{vr}^+)$ be the probability of finding a vibrational-rotational level at *exactly* the energy E_{vr}^+, where $E_{vr}^+ = E_v^+ + E_r^+$. Thus, $P(E_{vr}^+)$ will be either unity or zero, depending on whether or not a vibrational-rotational level is present at exactly the energy E_{vr}^+. For any value of E_{vr}^+, the remaining energy needed to make up the total energy E^+ of the activated complex must be in translation along the reaction coordinate: $x = E^+ - E_{vr}^+$. Since the translational energy levels are closely spaced, it will be a good approximation to treat their density as a continuous function. Let $N_{rc}^+(x)\,dE$ be the number of translational energy states along the reaction coordinate in the energy range between x and $x + dE$. Recall that for each value of x there is a corresponding value of E_{vr}^+ given by the equation $x = E^+ - E_{vr}^+$. Thus, for any particular value of E_{vr}^+, the density of states of the activated complex is given by $P(E_{vr}^+)N_{rc}^+(x)$. The total density of states is obtained by integrating $P(E_{vr}^+)N_{rc}^+(x)$ over the possible vibrational-rotational energies. However, since $P(E_{vr}^+)$ is either zero or one, the integral is the same as a sum over those energies for which $P(E_{vr}^+) = 1$:

$$N^+(E^+) = \sum_{E_{vr}^+=0}^{E_{vr}^+=E^+} P(E_{vr}^+)N_{rc}^+(x). \tag{7.27}$$

We next need to determine how the density of translational states, N_{rc}^+, depends on the translational energy x. The energy levels of a particle of mass μ moving in a one-dimensional box of length δ are given by the formula $E_n = n^2h^2/8\mu\delta^2$. Letting

[n]A useful method for calculating $N^*(E^*)$ is the Whitten-Rabinovitch approximation, described in Robinson and Holbrook, among other texts.

the translational energy be x and solving for n gives $n = (8\mu\delta^2 x/h^2)^{1/2}$; this is the number of states with translational energy up to and including the value of x. Now consider the number of states in the region from x to $x + dE$. If dE is large compared to the energy spacing, then the number of states in the desired range can be calculated by multiplying the number of states per unit energy by the width of the energy interval: $dn = (dn/dx)dE$. Since in the preceding paragraph we have called this number of translational states $N_{rc}^+(x)\,dE$, it is clear that $N_{rc}^+(x) = dn/dx$. Performing the derivative gives

$$N_{rc}^+(x) = \left(\frac{2\mu\delta^2}{h^2 x}\right)^{1/2}. \tag{7.28}$$

We finally assemble **equations 7.24** through **7.28** to obtain an equation for the rate at which the reactant with energy E^* decays to products:

$$k_a(E^*) = \frac{1}{2}\left(\frac{2x}{\mu\delta^2}\right)^{1/2} \frac{\displaystyle\sum_{E_{vr}^+=0}^{E_{vr}^+=E^+} P(E_{vr}^+)(2\mu\delta^2/h^2 x)^{1/2}}{N^*(E^*)}, \tag{7.29}$$

or

$$k_a(E^*) = \frac{1}{hN^*(E^*)} \sum_{E_{vr}^+=0}^{E_{vr}^+=E^+} P(E_{vr}^+). \tag{7.30}$$

Equation 7.30, our final result, is actually even simpler than it looks. Recall that $P(E_{vr}^+)$ is simply 1 or 0, depending on whether or not a vibrational-rotational level exists at the energy E_{vr}^+. The summation on the right-hand side is thus simply the number of energy levels of the transition state between the zero-point vibrational level and the energy of reaction E^+. The summation is often abbreviated as $W(E^+)$, so that another way to write **equation 7.30** is

$$k_a(E^*) = \frac{W(E^+)}{hN^*(E^*)}. \tag{7.31}$$

If the positions of the all the vibrational-rotational energy levels of the activated complex were known, then the sum would be evaluated simply by counting the number of levels with energy equal to or less than E^+.

It is now clear why the rate constant for NO_2 decomposition increases in the steplike manner shown in **Figure 7.28.** As the energy of excitation, E^*, increases, new levels of the transition state will become energetically allowed, so that the numerator of **equation 7.30** will increase in a steplike fashion. Of course, the denominator, $N^*(E^*)$, will also increase, but the density of reactant states is so high that the increase of $N^*(E^*)$ is a nearly smooth function of energy. In addition, the fractional increase in $N^*(E^*)$ is much smaller with energy than that of $\Sigma P(E_{vr}^+) = W(E^+)$. While the latter function doubles (increases from 1 to 2) when, for example, the energy becomes just large enough to include the second state of the activated complex, the former quantity, while increasing by a much larger absolute number, exhibits only a very small fractional change. Thus, **equation 7.30** predicts that the rate of the NO_2 decomposition should increase in a steplike fashion near the threshold for dissociation, as observed in **Figure 7.28.**

Because the states of the activated complex for NO_2 have not been calculated, one can only speculate as to the identity of the states that become available. A reasonable assumption is that these are bending states of the ON–O complex, states that eventually transform into rotations of the NO product. A further test of **equation 7.30** is that the size of the first step should be given by $1/[hN^*(E^*)]$. The estimated density of reactant levels is $N^*(E^*) \approx 0.2$ per cm^{-1} at $E^* \approx 25{,}130$ cm^{-1}, so that the increase in k_a at each step should be $\sim 1.5 \times 10^{11}$ s^{-1}, roughly as observed. We conclude that **equation 7.30** should provide an accurate estimate of the rate of unimolecular dissociation.

We complete this section by closing a circle. We started by assuming that the activated complex theory could be applied to the reaction $A^* \rightleftharpoons A^+ \rightarrow$ products, all at a fixed value of the energy. Actually, the activated complex theory was developed for a fixed temperature rather than for a fixed energy. We here make the connection between the RRKM result of **equation 7.30** or **equation 7.31,** which gives the reaction rate $k_a(E^*)$ at a specific energy, and the activated complex theory (Section 3.4), which gives the reaction rate $k_a(T)$ at a specific temperature. As discussed briefly in Chapter 4 and in more detail in Chapter 8, the latter is simply the average of the former over the Boltzmann distribution of energies:

$$k(T) = \int_{E^*=0}^{\infty} k_a(E^*) P(E^*) \, dE^*. \tag{7.32}$$

The probability per unit energy range of having energy E^* is simply the density of states in that energy range times the Boltzmann factor, all divided by the partition function:

$$P(E^*) = \frac{N^*(E^*)\exp(-E^*/kT)}{Q^*}. \tag{7.33}$$

Inserting **equations 7.31** and **7.33** into **equation 7.32,** we obtain

$$k(T) = \frac{1}{hQ^*} \int_{E^*=0}^{\infty} N^*(E^*)\exp\left(-\frac{E^*}{kT}\right)\frac{W(E^+)}{N^*(E^*)} \, dE^*. \tag{7.34}$$

By cancelling $N^*(E^*)$ in the numerator and denominator of the integrand and writing $\exp(-E^*/kT)\, dE^*$ by its equivalent $(-kT)\, d[\exp(-E^*/kT)]$, we can rewrite **equation 7.34** as

$$k(T) = -\frac{kT}{hQ^*} \int_{E^*=0}^{\infty} W(E^+)\, d\left[\exp\left(-\frac{E^*}{kT}\right)\right]. \tag{7.35}$$

Integration by parts gives

$$k(T) = -\frac{kT}{hQ^*} \left\{ W(E^+)\left[\exp\left(-\frac{E^*}{kT}\right)\right]\Big|_{E^*=0}^{\infty} - \int_{E^*=0}^{\infty} \exp\left(-\frac{E^*}{kT}\right) d[W(E^+)] \right\}. \tag{7.36}$$

The first term in the brackets vanishes since $W(E^+)$ is zero when E^* is zero and the exponential term is zero when E^* is infinity. Since the density of states in the activated complex is just the derivative of the number of states with respect to energy, $N^+(E^*) = d[W(E^+)]/dE^*$, we can rewrite **equation 7.36** as

$$k(T) = \frac{kT}{hQ^*} \int_{E^*=0}^{\infty} N^+(E^*)\exp\left(-\frac{E^*}{kT}\right) dE^*. \tag{7.37}$$

However, the integral is simply Q^+, the partition function for the activated complex. Thus, we obtain

$$k(T) = \frac{kT}{h}\frac{Q^+}{Q^*},\qquad(7.38)$$

which is simply the activated complex theory result for the rate constant for the reaction $A^* \rightarrow A^+$, where both Q^* and Q^+ are evaluated from the same zero of energy.[o] As a consequence, we see that the RRKM result is consistent with the activated complex theory. In fact, the RRKM result is simply the microcanonical (E = constant) version of the canonical (T = constant) the activated complex theory rate constant.

7.5.5   Photofragment Angular Distributions

A complete description of the dynamics of a photodissociative event must include not only an understanding of the rate of dissociation and the energy disposal in the products but also an understanding of the photofragment angular distribution. In many dissociations, it is found that the distribution of fragment recoil velocities is not isotropic, but rather is strongly aligned with respect to the polarization vector of the dissociating light.

For example, **Figure 7.31** shows a two-dimensional projection of the speed and angular distribution for methyl radicals produced in the 266-nm photolysis of methyl

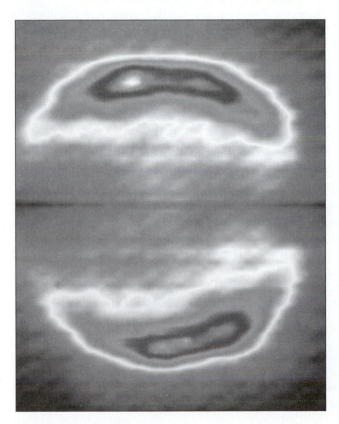

■ **Figure 7.31**

Angle and speed distribution of CH_3 radicals following 266-nm photodissociation of CH_3I.

Reprinted from D. W. Chandler, J. W. Thoman, Jr., M. H. M. Janssen and D. H. Parker, *Chemical Physics Letters,* **156,** 151 (1989). Copyright © 1989 with permission from Elsevier Science.

[o]If Q^+ is evaluated from the zero-point energy of the activated complex while Q^* is evaluated from the zero-point energy of the reactants, then Q^+ in **equation 7.38** must be replaced by $Q^+\exp(-E_0/kT)$, where E_0 is the energy of the zero point of the activated complex relative to the zero point of the reactants.

iodide. The polarization direction of the dissociating light is along the vertical axis of this figure, and the contours of intensity show that the methyl photofragments fly predominantly along this direction. A detailed analysis of the angular distribution shows that the recoiling fragments are distributed approximately in proportion to $\cos^2\theta$, where θ is the angle between the fragment recoil direction and the polarization direction. We now examine the reasons for such an angular distribution.

Consider an experiment in which an isolated molecule is excited with linearly polarized light and subsequently dissociates to yield two fragments recoiling from one another. There are three important frames of reference. The frame most easily understood is the laboratory frame, defined for example by the polarization direction of the light used for molecular excitation. This direction is that of the electric field of the light $\hat{\mathbf{E}}$. A second important frame of reference is that of the parent molecule. It is usually defined by some symmetry element, such as a reflection plane or rotation axis, but for our purposes the important symmetry element will be the direction of the transition dipole moment $\boldsymbol{\mu}$. For an allowed transition from a symmetric ground state, this moment has the same symmetry as the electronic symmetry of the excited molecular orbital. A final frame of reference of importance for our discussion is that of the fragments. This frame is most easily defined as the direction of the recoil velocity, $\mathbf{v}$. For many simple photodissociations, the vector $\mathbf{v}$ is nearly along the direction of the bond that breaks, and since this bond direction is fixed in the molecular frame, there will be a correlation between the direction of $\boldsymbol{\mu}$ and the direction of $\mathbf{v}$.

Zare and Herschbach were the first to point out that there is a relationship between the molecular and laboratory frames of reference.[p] At the start of the photodissociation experiment, the molecular frames are distributed randomly in the laboratory frame. However, immediately after the absorption of the photon, the molecular frames of the *excited* molecules will have a definite laboratory alignment, since the strength of absorption is proportional to the product $|\boldsymbol{\mu} \cdot \hat{\mathbf{E}}|^2$. Thus, the excited molecules will have a distribution of alignments proportional to $\cos^2\theta$, where θ is the angle between the molecular transition dipole and the laboratory electric vector. If these molecules dissociate very rapidly compared to the rotational time scale, then the recoil velocity $\mathbf{v}$ of the fragments, which usually has a fixed relationship to the transition dipole $\boldsymbol{\mu}$, will also be aligned in the laboratory frame. For example, suppose that (1) $\boldsymbol{\mu}$ is along the bond that breaks, (2) the molecule dissociates before it has a chance to rotate, and (3) the fragments get enough recoil energy from the dissociation so that they too fly out along the direction of the breaking bond. Under these assumptions, which are approximately valid in the photodissociation of CH_3I, $\mathbf{v}$ should be parallel to $\boldsymbol{\mu}$. Because $\boldsymbol{\mu}$ is distributed as $\cos^2\theta$ with respect to $\hat{\mathbf{E}}$, $\mathbf{v}$ will likewise be distributed, as observed in **Figure 7.31.** Of course, it is not necessary in general that $\boldsymbol{\mu}$ be aligned in the molecular frame along the bond that breaks. When it is not, the photofragment angular distribution is given as

$$I(\theta) = (4\pi)^{-1}[1 + \beta P_2(\cos\theta)], \qquad (7.39)$$

where $P_2(\cos\theta) = \frac{1}{2}(3\cos^2\theta - 1)$ is the second Legendre polynomial and $\beta = 2P_2(\cos\chi)$ with χ as the angle between $\boldsymbol{\mu}$ and the recoil direction $\mathbf{v}$. The value of β, called the *anisotropy parameter,* lies between -1 and 2. When $\beta = 2$, $\cos\chi = 1$ and $\boldsymbol{\mu}$ is parallel to $\mathbf{v}$, whereas when $\beta = -1$, $\cos\chi = 0$ and $\boldsymbol{\mu}$ is perpendicular to $\mathbf{v}$.

[p]R. N. Zare and D. R. Herschbach, *Proc. IEEE* **51,** 173 (1963); R. N. Zare, Ph.D. Thesis, Harvard University, Cambridge, MA, 1964.

Note that the alignment of the recoil velocities with respect to the polarization direction of the dissociating light would be reduced substantially if the parent molecule rotated before it dissociated. Such rotation would scramble the alignment of μ in the laboratory frame, and since μ and $\mathbf{v}$ are related in the molecular frame, the alignment of $\mathbf{v}$ would be scrambled as well. Thus, the degree of alignment can provide information about the lifetime of the excited parent molecule prior to dissociation, where the "clock" by which the dissociation time is measured is the rotational period of the parent (see Problem 7.9).

A quantitative method for measuring the angular distribution of photofragments was first developed in 1969.[q] In this method, a linearly polarized laser is used to dissociate a molecule, while a mass spectrometer is used to detect the arrival time and angular distribution of the photofragment. A drawing of a typical modern apparatus is shown in **Figure 7.32.** In this figure, the photolysis light propagates into the plane of the drawing with its polarization vector at a specified but variable angle to the flight path. The advantages of this device are that both the direction and magnitude of the recoil velocity can be determined. For example, from the angular distribution of the NO fragment observed in the 347-nm photodissociation of NO_2, Busch and Wilson determined that the lifetime of the NO_2 was roughly 200 fs, corresponding to a dissociation rate of $5 \times 10^{12}\ s^{-1}$.[r] Note that the excitation energy is roughly 3680 cm^{-1} above the dissociation limit, so that this rate is in rough agreement with an extrapolation of the data in **Figure 7.28,** which was measured by a later, more direct technique. Other modern methods are also available for measuring the angular distribution. High-resolution lasers can determine the Doppler profile of a photofragment in a pump-probe experiment. Because the Doppler shift depends on the projection of the velocity along the direction of light propagation,

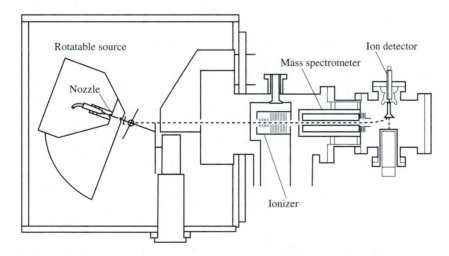

Figure 7.32

Apparatus for measuring angular and time-of-flight distributions for photodissociations.
Reprinted with permission of Dr. Laurie J. Butler.

[q]G. E. Busch, R. T. Mahoney, R. I. Morse, and K. R. Wilson, *J. Chem. Phys.* **51,** 449, 837 (1969); R. W. Diesen, J. C. Wahr, and S. E. Adler, *J. Chem. Phys.* **50,** 3635 (1969).
[r]G. E. Busch and K. R. Wilson, *J. Chem. Phys.* **56,** 3626 (1972); *J. Chem. Phys.* **56,** 3638 (1972).

the shape of the Doppler profile provides information about the angular and speed distribution; in fact, the Doppler profile is simply a one-dimensional projection of the three-dimensional velocity distribution. Alternatively, if multiphoton ionization is used, the three-dimensional velocity distribution can be projected onto a two-dimensional detector, providing an image like that shown for methyl iodide dissociation in **Figure 7.31.**

It is worth digressing for a moment to comment that the apparatus of **Figure 7.32** can measure a second important property, the speed distribution of the photofragments. If the resolution is high enough, conservation of energy and momentum can be used to infer details of the dissociation dynamics. For example, **Figure 7.33** shows the speed distribution of O_2 in the 266-nm photodissociation of ozone. From the structure in the time-of-flight distribution, we see that the dissociation is somewhat more complicated than suggested by **reaction R2** on page 222. The fast peak in the time-of-flight spectrum (near 150 μs) corresponds to dissociation of O_3 into ground state products, $O_2(X\,^3\Sigma) + O(^3P)$, but the largest quantum yield is for dissociation into excited products, $O_2(a\,^1\Delta_g) + O(^1D)$. The products from this latter dissociation channel have less energy available for recoil, so that the O_2 arrives later, at about 250 μs for $O_2(a\,^1\Delta_g)$ in $v = 0$. Subsequent peaks in the time-of-flight correspond to $O_2(a\,^1\Delta_g)$ in $v = 1, 2$, and 3. The quantum yield for production of the $O_2(a\,^1\Delta_g) + O(^1D)$ channel is about 90% at this dissociation wavelength.

The NO_2 and O_3 examples discussed above show the important general features of photodissociation with mass spectrometric detection of the fragments: both the speed and the angular distribution of the photofragments can be measured.

The angular distribution is but one example of a "vector correlation" in molecular dynamics. Here the correlation, called the *anisotropy,* is an angular relationship between the vector describing the polarization direction of the dissociating light and a vector describing the recoil direction of the photofragments. Other vector correlations can also provide information. For example, the correlation between the

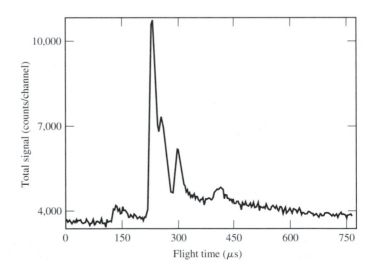

Figure 7.33

Time-of-flight distribution for the O_2 fragment from ozone photodissociation at 266 nm.

From R. K. Sparks, L. R. Carlson, K. Shobatake, M. L. Kowalczyk, and Y. T. Lee, *J. Chem. Phys.* **72,** 1401 (1980). Reprinted with permission from the *Journal of Chemical Physics.* Copyright, American Institute of Physics, 1980.

polarization vector and a vector describing the rotation of one of the photofragments describes the *alignment* of photofragment in the laboratory frame. The correlation between the rotation vector of a photofragment and its velocity vector, sometimes called the *v-J correlation,* provides information about the forces acting on the fragment at the transition state for dissociation. Although we will not consider these correlations in detail, descriptions of how they are measured and examples of what information they provide are given in the review articles listed at the end of the chapter.

example 7.2

Time-of-Flight Distributions

Objective   For the photodissociation of O_3 at 266 nm, calculate the difference in arrival times between O_2 formed in the two channels given the following data. The threshold energy for the $O_2(X\,^3\Sigma) + O(^3P)$ channel is 8064 cm^{-1}, while that the $O_2(a\,^1\Delta_g) + O(^1D)$ channel is 32,258 cm^{-1}. Suppose the detector is 34 cm from the photolysis point and detects fragments recoiling in the direction of the molecular beam, for which the velocity is 800 m/s.

Method   Conservation of energy gives the center-of-mass recoil energies for the two channels, while conservation of linear momentum gives the O_2 and O velocities. The center-of-mass O_2 velocities can thus be calculated. For the geometry indicated, the difference in center-of-mass velocities is also the difference in laboratory velocities. The distance and the laboratory velocity difference can be used to calculate the arrival time difference.

Solution   The energy of a 266-nm photon in wave numbers is 37,594 cm^{-1}. Thus $E_{av} = 37,594 - 8064 = 29,530$ cm^{-1} is available to the triplet channel, while $E_{av} = 37,594 - 32,258 = 5336$ cm^{-1} is available to the singlet channel. If O_2 is formed in $v = 0$ and $J = 0$ in each channel, then all of the available energy goes into the relative recoil velocity v. Consequently, for either channel $\frac{1}{2}\mu v^2 = E_{av}$, where $\mu = m_O m_{O_2}/m_{O_3} = 1.77 \times 10^{-26}$ kg. From conservation of linear momentum $m_O v_O = m_{O_2} v_{O_2}$, so that $v_O = 2v_{O_2}$. The sum of the fragment velocities must be equal to the recoil velocity: $v = v_O + v_{O_2} = 3v_{O_2}$. Consequently, $\frac{1}{2}\mu v^2 = \frac{1}{2}\mu(3v_{O_2})^2$, or $v_{O_2} = [(2E_{av})/(9\mu)]^{1/2}$. The center-of-mass O_2 velocities are thus 2714 m/s for the triplet channel and 1154 m/s for the singlet channel. The velocity for each channel is the speed in the laboratory frame divided into the flight distance, 0.34 m. Thus the difference is $[(0.34\text{ m})/(800 + 1154\text{ m/s})] - [(0.34\text{ m})/(800 + 2714\text{ m/s})] = 77.2\ \mu$s.

Comment   The arrival time difference in **Figure 7.33** is a bit larger because the fragments were detected at an angle 30° to the beam direction, so that a more complicated center-of-mass to laboratory correction is required (see Problem 7.10).

7.5.6 Photochemistry on Short Time Scales

A very recent advance in the study of photodissociation dynamics has been the capability to perform experiments on very short time scales using laser pulses of about 100 fs duration (1 fs = 10^{-15} s).[s] Since this time scale is roughly equal to even the fastest vibrational period, it is clear that any chemistry that can possibly occur must do so on a slower time scale. In favorable circumstances, one can use such short laser pulses to "watch" molecular motion.

An example is the photodissociation of NaI, for which the relevant potential energy curves are shown in **Figure 7.34.** The ground electronic state of NaI is ionic in character, but the potential energy curve for this state is crossed at about 7 Å by

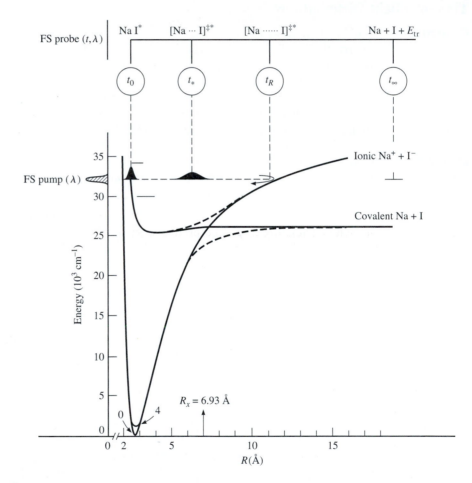

■ **Figure 7.34**

Potential energy curves relevant for an experiment probing NaI photodissociation on a 100-fs time scale.

From T. S. Rose, M. J. Rosker, and A. H. Zewail, *J. Chem. Phys.* **88,** 6672 (1988). Reprinted with permission from the *Journal of Chemical Physics.* Copyright, American Institute of Physics, 1988.

---

[s]Ahmed Zewail won the 1999 Nobel Prize in Chemistry for his work in developing "femtochemistry."

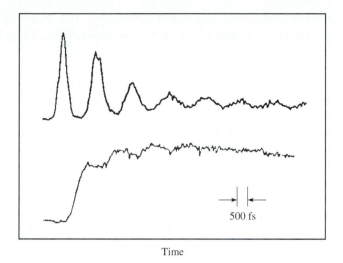

Time

■ **Figure 7.35**

Laser-induced fluorescence on a 50-fs time scale under the two conditions described in the text. From T. S. Rose, M. J. Rosker, and A. H. Zewail, *J. Chem. Phys.* **88,** 6672 (1988). Reprinted with permission from the *Journal of Chemical Physics.* Copyright, American Institute of Physics, 1988.

a valence curve leading to ground-state Na + I products. Strong mixing between these two states leads to an upper electronic state correlating to ionic products and a ground electronic state correlating to atomic products, as shown by the dashed lines in the figure. Excitation from the ground state at time t_0 with a 50-fs laser pulse creates a superposition state that can be described by a wave packet moving back and forth on the upper electronic state. Although most of the amplitude of the wave packet oscillates with the vibrational frequency of NaI in the excited state, each time the wavepacket enters the crossing region there is a finite probability that some amplitude will leak out the dissociation channel to Na + I.

Suppose now that a second 50-fs laser pulse is used to perform laser-induced fluorescence. If the probe laser is tuned to excite from a particular configuration of Na–I, say that depicted at time t_R, pulses will be observed as the wave packet moves in and out of the configuration probed, as shown in the upper trace of **Figure 7.35.** The amplitude decreases in time since after about 10 vibrational periods the NaI has dissociated. If the Na product is probed, then the pulses of probability amplitude are integrated into the step-like rise shown in the bottom trace of **Figure 7.35.**

Although detailed theoretical interpretations for results such as those shown above have since been performed, it is clear even on a qualitative level that short pulsed lasers will greatly extend our understanding of photodissociation dynamics.

7.6   SUMMARY

We began this chapter by discussing the interaction of light with matter. The intensity of light passing through a sample of density ρ and length ℓ is given by

$$\frac{I}{I_0} = \exp[-\sigma(\nu)\ell\rho], \tag{7.3}$$

where $\sigma(\nu)$ is the absorption cross section at frequency ν. This absorption cross section is related to the Einstein coefficients, which are also related among themselves:

$$B_{12} = B_{21}\frac{g_2}{g_1},$$

$$A_{21} = \frac{8\pi h\nu^3}{c^3}B_{21},$$

(7.6)

$$B_{12} = \frac{c\sigma(\nu)\,d\nu}{h\nu}.$$

(7.12)

They are also related to the transition dipole moment, μ_{12}, through the equation

$$A_{21} = \frac{64\pi^2\nu^3}{3hc^3}|\mu_{21}|^2.$$

(7.13)

Under the Born-Oppenheimer approximation, which assumes that the electron motion is much more rapid than the nuclear motion, the transition moment can be written as

$$\mu_{21} = -e\int \psi_{e_2}(\mathbf{r})\mathbf{r}\psi_{e_1}(\mathbf{r})\,d\tau_{el}\int \psi_{v_2}\psi_{v_1}\,d\tau_{nuc}.$$

(7.15)

The second of the integrals in this last equation is known as the Franck-Condon factor. Because it is a multiplicative factor in the transition moment, the maximum absorption or emission will occur when the vibrational wave functions of the upper and lower states have the maximum overlap.

Absorption lines are not infinitely narrow, due both to the Doppler effect and the finite lifetime of the upper level. The Doppler width is given by

$$\Delta\nu = 2\frac{\nu_0}{c}\sqrt{\frac{2kT\ln 2}{m}},$$

(7.18)

whereas the width due to the finite lifetime is given by $\Delta\nu = 1/(\pi\tau)$.

A major section of the chapter dealt with photophysical processes. Perhaps the simplest process is re-emission of energy as fluorescence, but under the presence of collisions the emission can be quenched. The kinetics of quenching is described by the Stern-Volmer equation:

$$\frac{1}{I_f} = \frac{1}{I_a\sigma M_0}\left(\frac{k_q}{k_f}[Q] + 1\right).$$

(7.21)

Intramolecular vibrational energy redistribution was also shown to be an important process, particularly in larger molecules where the density of vibrational states, ρ_v, is high enough that several states fall within the energy of the matrix element V coupling their motion. When $V\rho_v \approx 1$, then the fluorescence spectrum from the isolated molecule is normally structured, and the fluorescence decay is an exponential decay, perhaps modulated with quantum beats. When $V\rho_v \gg 1$, energy redistribution is rapid, so that the fluorescence spectrum is more continuous and the decay is rapid and often nonexponential.

Internal conversion and intersystem crossing can be understood as processes that are similar to internal vibrational energy redistribution, except that the energy transfer is between two different electronic states. If the two states have the same multiplicity, the process is called internal conversion, whereas if they have different multiplicities the process is called intersystem crossing. Internal conversion and intersystem crossing can take place in isolated molecules as a result of coupling terms normally omitted from the Hamiltonian or as collisional processes as a result of the perturbation introduced by the collision partner.

A final process we considered was photodissociation, which can take place in a direct fashion, when a repulsive part of the potential is reached by excitation, or in an indirect fashion, when some energy redistribution must take place before dissociation.

Atmospheric chemistry, particularly the chemistry of stratospheric ozone, provides an application of photochemistry. The photochemical steady state can be used to calculate the concentration of ozone on the basis of the simple Chapman mechanism, but the steady-state concentration so calculated is higher than that observed. The reason is that there are several catalytic cycles involving minor species, such as NO, OH, Cl, and ClO, that destroy ozone without consuming the minor component. The Antarctic "ozone hole" is likely to be a result of increased concentrations of stratospheric chlorine due to the photodissociation of chlorofluorocarbons injected into the atmosphere by human activity.

A major section of this chapter concerned photodissociation dynamics. The pump-probe technique for probing these dynamics was introduced, and two sensitive probe methods were described. In laser-induced fluorescence, specific vibrational-rotational levels of the photochemical product are excited to an upper electronic state by a probe laser, and the resulting fluorescence is detected as the probe frequency is scanned. In multiphoton ionization, the excitation scheme is similar except that the probe laser is sufficiently intense to cause absorption of more than a single photon, resulting in ionization. The ion yield is then recorded as a function of probe frequency. When the ionization step is saturated, as is typically the case, then the ion intensity is related to the populations by the same equation used for laser-induced fluorescence (see Problem 7.12).

After describing these techniques and some examples of the results they produce, we turned our attention to the rate of unimolecular reactions. By extending activated complex theory we showed that the rate of a unimolecular reaction depends on the number of states accessible in the transition state to dissociation:

$$k_a(E^*) = \frac{1}{hN^*(E^*)} \sum_{E_{vr}^+ = 0}^{E_{vr}^+ = E^+} P(E_{vr}^+). \tag{7.30}$$

However, the rate of a photochemical reaction and the energy disposal in the photofragments are not the complete story. Information about the photodissociation can also be obtained from the photofragment angular distribution, described by the anisotropy parameter β in the equation

$$I(\theta) = (4\pi)^{-1}[1 + \beta P_2(\cos\theta)]. \tag{7.39}$$

A photofragment spectrometer was described that enables not only measurement of the angular distribution but also determination of the photofragment speed distribution.

We closed the photodissociation dynamics section by considering photochemistry on short timescales. Ultrafast lasers are now providing measurements of chemical events as they happen.

suggested readings

M. N. R. Ashfold and J. E. Baggott, eds., *Molecular Photodissociation Dynamics* (Royal Society of Chemistry, London, 1987).

T. Baer and W. L. Hase, *Unimolecular Reaction Dynamics: Theory and Experiments* (Oxford University Press, New York, 1996).

L. J. Butler and D. M. Neumark, "Photodissociation Dynamics," *J. Phys. Chem.* **100,** 12801 (1996).

J. G. Calvert and J. N. Pitts, *Photochemistry* (Wiley, New York, 1966).

W. Demtröder, *Laser Spectroscopy—Basic Concepts and Instrumentation* (Springer-Verlag, Berlin, 1981).

B. J. Finlayson-Pitts and J. N. Pitts, Jr., *Atmospheric Chemistry: Fundamentals and Experimental Techniques* (Wiley-Interscience, New York, 1986).

W. Forst, *Theory of Unimolecular Reactions* (Academic, New York, 1973).

R. G. Gilbert and S. C. Smith, *Theory of Unimolecular and Recombination Reactions* (Blackwell Scientific, Boston, 1990).

G. Herzberg, *Molecular Spectra and Molecular Structure III. Electronic Spectra and Electronic Structure of Polyatomic Molecules* (Van Nostrand, Princeton, 1966).

P. L. Houston, "Snapshots of Chemistry: Product Imaging of Molecular Reactions," *Acc. Chem. Res.* **28,** 453 (1995).

P. L. Houston, "New Laser-Based and Imaging Methods for Studying the Dynamics of Molecular Collisions," *J. Phys. Chem.* **100,** 12757 (1996).

J. L. Kinsey, "Laser-Induced Fluorescence," *Ann. Rev. Phys. Chem.* **28,** 349 (1977).

B. A. Lengyel, *Lasers* (Wiley, New York, 1971).

S. R. Leone, "Photofragmentation Dynamics," *Ann. Rev. Phys. Chem.* **50,** 255 (1982).

A. Mooradian, T. Jaeger, and P. Stockseth, eds., *Tunable Lasers and Applications* (Springer-Verlag, Berlin, 1976).

C. B. Moore, ed., *Chemical and Biochemical Applications of Lasers, Vols. 1–5* (Academic, New York, 1974–1980).

D. J. Nesbitt and R. W. Field, "Vibrational Energy Flow in Highly Excited Molecules: Role of Intramolecular Vibrational Redistribution," *J. Phys. Chem.* **100,** 12735 (1996).

R. W. G. Norrish and G. Porter, "The Application of Flash Techniques to the Study of Fast Reactions," *Disc. Farad. Soc.* **17,** 40 (1955).

W. A. Noyes, Jr., and P. A. Leighton, *The Photochemistry of Gases* (Dover, New York, 1966).

H. Okabe, *Photochemistry of Small Molecules* (Wiley-Interscience, New York, 1966).

P. J. Robinson and K. A. Holbrook, *Unimolecular Reactions* (Wiley-Interscience, London, 1972).

H. Sato, *Photodissociation of Simple Molecules in the Gas Phase* (Bunshin, Tokyo, 1992).

R. Schinke, *Photodissociation Dynamics* (Cambridge University Press, 1993).

Y. R. Shen, *The Principles of Nonlinear Optics* (Wiley, New York, 1984).

J. P. Simons, "Photodissociation, A Critical Survey," *J. Phys. Chem.* **88,** 1287 (1984).

J. I. Steinfeld, ed., *Laser and Coherence Spectroscopy* (Plenum, New York, 1981).

J. I. Steinfeld, *Molecules and Radiation* (MIT Press, Cambridge, MA, 1985).

N. J. Turro, *Modern Molecular Photochemistry* (Benjamin-Cummings, Menlo Park, CA, 1978).

P. Warneck, *Chemistry of the Natural Atmosphere* (Academic Press, New York, 1988).

A. Yariv, *Quantum Electronics* (Wiley, New York, 1975).

R. N. Zare and P. J. Dagdigian, "Tunable Laser Fluorescence Method for Product State Analysis," *Science* **185,** 739 (1974).

A. H. Zewail, "Femtochemistry: Recent Progress in Studies of Dynamics and Control of Reactions and Their Transition States," *J. Phys. Chem.* **100,** 12701 (1996).

Photofragment angular distributions and other vector correlations

R. N. Dixon, "The Determination of the Vector Correlation between Photofragment Rotational and Translational Motions Form the Analysis of Doppler-Broadened Spectral Profiles," *J. Chem. Phys.* **85,** 1866 (1986).

G. E. Hall and P. L. Houston, "Vector Correlations in Photodissociation Dynamics," *Ann. Rev. Phys. Chem.* **40,** 375 (1989).

P. L. Houston, "Vector Correlations in Photodissociation Dynamics," *J. Phys. Chem.* **91,** (1987).

P. L. Houston, "Correlated Photochemistry: The Legacy of Johann Christian Doppler," *Accounts of Chem. Res.* **22,** 309 (1989).

problems

7.1 The fractional absorption of a light beam traveling through a uniform sample depends on which of the following: (a) the length of the sample, (b) the cross-sectional area of the beam, (c) the concentration of the sample, (d) the frequency of the light, or (e) the molecular velocity?

7.2 Is it true or false that laser action requires the number of molecules in the upper of two radiatively coupled levels be larger than the number of molecules in the lower one?

7.3 What is the physical basis for the assertion that optical transitions occur "vertically," i.e., without a change in geometry?

7.4 Suppose you measured the intensity of an absorption feature for a transition between rovibrational levels in two different electronic states. Name three quantities that you would need to know to relate the absorption intensity to the population difference between the two levels.

7.5 In addition to depending on the strength of a coupling interaction involving terms normally omitted from the Hamiltonian, on what other feature of the molecule do the rates of processes such as intramolecular vibrational redistribution, internal conversion, and intersystem crossing depend?

7.6 Use the steady-state approximation to show that the kinetic scheme of **equation 7.20** leads to **equation 7.21.**

7.7 The Chapman mechanism (Section 7.4) for ozone production and loss can be written as

a. $O_2 + h\nu \xrightarrow{j_a} O + O.$

b. $O + O_2 + M \xrightarrow{k_b} O_3 + M.$

c. $O_3 + h\nu \xrightarrow{j_c} O_2 + O.$

d. $O + O_3 \xrightarrow{k_d} O_2 + O_2.$

In these equations j_a and j_c are the effective rates of dissociation of O_2 and O_3, respectively, given the spectrum of the solar flux in the stratosphere and the cross sections for absorption. Letting n_1, n_2, and n_3 be the concentrations for O, O_2, and O_3, respectively, write the differential equations for these three species. By adding the differential equations for O and O_3, show that

$$\frac{dn_1}{dt} + \frac{dn_3}{dt} = 2j_a n_2 - 2k_d n_1 n_3 \approx 0.$$

Under stratospheric conditions, processes b and c dominate the destruction and production of O atoms, so that the remaining terms in dn_1/dt can be neglected in calculating the steady-state concentration of O, which is found to be

$$n_1 = \frac{j_c n_3}{k_b n_2 n_M}.$$

Show that the steady-state concentration of O_3 is then given by

$$n_3 = n_2 \sqrt{\frac{k_b n_M j_a}{k_d j_c}}.$$

7.8   Another manifestation of **equation 7.30** is shown in **Figure 7.36,** which gives the yield, measured spectroscopically, of CH_2 in its lowest vibrational-rotational energy level as a function of photolysis energy in the photodissociation of ketene, CH_2CO. The threshold for dissociation is roughly 30125 cm^{-1}, as marked on the ordinate of the figure. Note that the yield of CH_2 increases in a step-like fashion with increasing ketene photolysis energy. The largest steps take place at locations where new rotational states of the sibling product CO become energetically allowed. The experiment was performed by probing the CH_2 at a delay time short compared to the decomposition rate. Write the kinetic equations for the production of this level of ketene and show why in such an experiment the signal is roughly proportional to the rate constant for decomposition. Given that proportionality, what states of the activated complex are likely to be responsible for the large steps labeled $J_{CO} = 0, 1, 2, 3$? [Note that the smaller steps, labeled by $J_K(CH_2CO)$ are due to excitation of ketene from different initial states in the molecular beam used for the experiment; that is, they are an experimental artifact that does not need to be explained by **equation 7.30.**]

7.9   As mentioned in Section 7.5.5, the anisotropy parameter β is decreased if the parent molecule rotates before dissociating. Jonah [*J. Chem. Phys.* **55,** 1915 (1971)] has shown that in the presence of parent rotation, the formula for the anisotropy parameter can be written as

$$\beta = 2P_2(\cos \chi)\left(\frac{1 + \omega^2\tau^2}{1 + 4\omega^2\tau^2}\right),$$

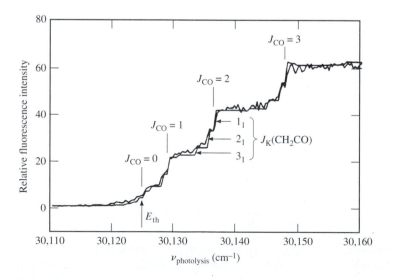

Figure 7.36

Photofragment yield of CH_2 as a function of ketene photolysis energy.

From I.-C. Chen, W. H. Green, Jr., and C. B. Moore, *J. Chem. Phys.* **89,** 314 (1988). Reprinted with permission from the *Journal of Chemical Physics.* Copyright, American Institute of Physics, 1988.

where χ is the angle between the parent transition dipole moment and the recoil velocity vector of the fragment, ω is the frequency of the parent rotation, and τ is the lifetime of the parent. If the above equation is averaged over a Boltzmann distribution of initial parent rotational states, the result is[t]

$$\beta_{\text{eff}} = \frac{1}{2}P_2(\cos \chi)\left(1 + 3\gamma e^{\gamma}\int_{\gamma}^{\infty}\frac{e^{-t}}{t}\,dt\right),$$

where

$$\gamma = \frac{I}{8kT\tau^2},$$

with the moment of inertia, I, the Boltzmann factor, k, and the temperature, T. Using as I the average of the two largest moments of inertia for NO_2, plot β_{eff} as a function of τ from 1 to 10^5 fs for $T = 370$ K and $T = 15$ K. The former temperature was used by Busch and Wilson, who found $\beta \approx 0.7$, while the latter temperature was used by Hradil et al., who found $\beta \approx 1.4$. Show that the two results are actually consistent with one another. [*Hint:* The solution is shown in V. P. Hradil, T. Suzuki, S. A. Hewitt, P. L. Houston, and B. J. Whitaker, *J. Chem. Phys.* **99**, 4455 (1993).]

7.10   As mentioned in the comment at the end of **Example 7.2,** the time-of-flight in **Figure 7.33** is slightly longer than that calculated in the example because the products are detected at 30° from the beam direction. A laboratory to center-of-mass conversion is thus necessary. Suppose that the beam velocity is $v_B = 800$ m/s. What is the difference in arrival times for O_2 from the two different channels? The velocity vector diagram of **Figure 7.37** will be useful in your analysis. You are given two values of u_{O_2}, $\alpha = 30°$, and $v_B = 800$ m/s. You need to calculate the two corresponding values of v_{O_2} and then the time difference, given that the distance to the detector is 34 cm.

7.11   Based on the data shown in **Figure 7.35,** (a) what is the vibrational frequency of NaI and (b) what is the predissociation rate for NaI $\rightarrow$ Na + I?

7.12   The complete formula for conversion of relative intensities for laser-induced fluorescence lines into relative populations is

$$N(v'',J'') = \frac{I(v', J' - v'', J'')g(J'')}{S(J',J'')F(v',v'')},$$

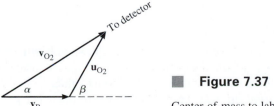

Figure 7.37

Center-of-mass to laboratory conversion.

[t]S. Yang and R. Bersohn, *J. Chem. Phys.* **61,** 4400 (1974).

where v'', J'' and v', J' denote the vibrational and rotational quantum numbers of the lower and upper states, respectively; $I(v', J' - v'', J'')$ is the measured intensity of the laser-induced fluorescence line (where the quantum numbers indicate the states involved in the absorption); $g(J'')$ is the rotational degeneracy of the lower state (usually $2J'' + 1$); $S(J', J'')$ is the rotational line strength (Hönl-London factor), and $F(v', v'')$ is the Franck-Condon factor. We assume here either that all fluorescence is detected or that the fraction detected is independent of the upper level. Other correction factors are needed if this assumption is not valid.

The following table gives the R-branch J'' values and corrected integrated intensities I for CO transitions starting in $v = 0$ and exciting to $v = 2$. The intensities were obtained from the spectrum in **Figure 7.21** after correction for some minor mixing between the fluorescing state and a nearby triplet state. The Hönl-London factor for this transition is equal to $J'' + 2$. Calculate and plot the relative populations $N(v'', J'')$ and compare to the curve in **Figure 7.22.**

J''	I	J''	I	J''	I
8	36.04	27	49.64	39	24.53
10	42.04	28	43.56	40	33.50
12	42.08	28	43.56	41	24.83
13	39.07	29	52.04	42	22.1
14	41.08	30	52.12	43	22.5
15	50.10	31	45.45	44	25
16	47.09	32	42.32	45	19
17	51.15	33	41.75	46	11
20	53.26	34	47.81	47	18
21	53.42	35	38.50	48	22
22	52.79	36	35.38	49	17
23	59.49	36	36.40	50	14
24	40.96	37	34.80	51	8
25	42.71	37	33.77	53	9
26	49.12	38	22.68	54	8
27	48.63	39	21.47	56	12

7.13   It was asserted in the text that combination of **equation 7.5** with the Boltzmann distribution and the blackbody radiation density $\rho(\nu)$ lead to **equation 7.6.** Show that, assuming A_{21}, B_{21}, and B_{12} not to depend on temperature, this must be the case. The Boltzmann distribution gives $N_2/N_1 = (g_2/g_1) \times \exp(-h\nu/kT)$ and the blackbody radiation density is given by $\rho(\nu) = (8\pi h\nu^3/c^3)/[(\exp(h\nu/kT) - 1]$.

7.14   Consider the quenching of fluorescence in I_2 shown in **Figure 7.5.** The fluorescence intensity, I_f, from the excited level ($v = 25, J = 34$) is measured as a function of helium concentration relative to its value, I_f^0, in the absence of helium. From **equation 7.21** we note that at zero quencher pressure $1/I_f \equiv 1/I_f^0 = 1/(I_a\sigma M_0)$, so that $(I_f^0/I_f) - 1 = (k_q/k_f)[Q]$, where what we have called k_q corresponds to the sum of all processes leading to the depopulation of the initially excited level. A table of results is given below. Determine k_q given that k_f is 1.4×10^6 s^{-1}.

[He] $10^{-5}\,M$	$(I_f^0/I_f) - 1$	[He] $10^{-5}\,M$	$(I_f^0/I_f) - 1$
0.1	0.20	1.0	1.25
0.2	0.25	1.1	1.15
0.2	0.30	1.5	1.95
0.22	0.40	1.5	2.05
0.6	0.75	2.2	3.15
0.6	0.80	2.2	3.50
0.8	1.05	3.2	4.25

7.15 In the dissociation of a molecule to two fragments, the energy available to the products can be calculated from the bond dissociation energy and the photon energy. If the internal energy and center-of-mass recoil velocity of one fragment is known, show how conservation of momentum and energy can be used to determine the internal energy of the second fragment.

7.16 The photodissociation of acetylene by 49,291 cm^{-1} photons produces hydrogen atoms that have a maximum velocity of 8970 m/s. What is the bond dissociation energy of acetylene, D_0^0(HCC–H)?

7.17 Another convincing experimental confirmation of intramolecular vibrational redistribution has been observed in the molecule p-difluorobenzene. **Figure 7.38** displays the emission spectrum of p-difluorobenzene as a function of the fluorescence lifetime of the upper state. The excitation is to

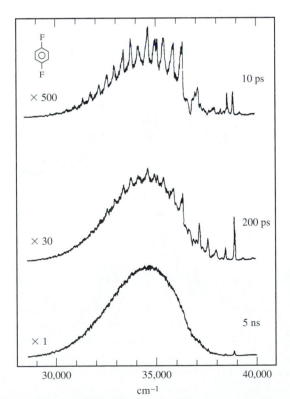

Figure 7.38

Emission spectrum of p-difluorobenzene as a function of the fluorescence lifetime of the upper state.

From R. A. Coveleskie, D. A. Dolson, and C. S. Parmenter, *J. Phys. Chem.* **89,** 645 (1985). Reprinted with permission from *The Journal of Physical Chemistry.* Copyright 1985 American Chemical Society.

a specific normal vibrational mode 3310 cm^{-1} above the ground level of S$_1$. Collisions with O$_2$ quench the fluorescence (by relaxing the excited singlet state) but are ineffective in changing the vibrational level in this system. In the top panel, enough O$_2$ has been added so that the *p*-difluorobenzene can radiate for only 10 ps before being quenched, whereas in the middle panel it can radiate for 200 ps and in the lower panel for 5 ns. Explain the changes in the spectrum.

7.18   **Figure 7.39** shows the B-A fluorescence emission spectrum of XeBr created in various upper vibrational levels. Explain the major features in the spectrum. (*Hint:* The lower state of this molecule is dissociative.)

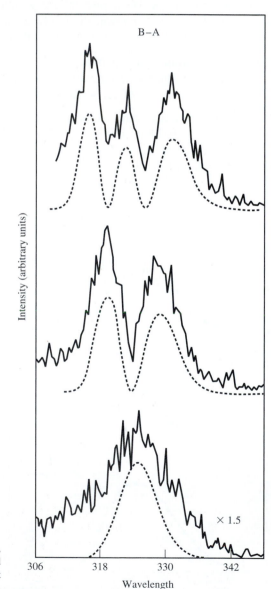

Figure 7.39

The B $\rightarrow$ A emission spectrum of XeBr following creation of this molecule in $v = 2$ (*top*), $v = 1$ (*middle*), and $v = 0$ (*bottom*).

From J. K. Ku, D. W. Setser, and D. Oba, *Chem. Phys. Lett.* **109,** 429 (1984). Reprinted from *Chemical Physics Letters,* copyright 1984, with permission from Elsevier Science.

7.19   RRKMWSU is a program to calculate RRKM rates written by L. Zhu and
       W. L. Hase. It is available from the Quantum Chemistry Program Exchange
       at Indiana University (#644). The following problem assumes that this pro-
       gram and its accompanying guide are available.
           Consider the unimolecular dissociation of CF_3Br. The molecule has no
       internal rotors (so that lines 12 and 13 as well as lines 19 and 20 should be
       omitted from the input file). The molecular parameters for the molecule are
       available in the following references: W. F. Edgell and C. E. May, *J. Chem.
       Phys.* **22,** 1808 (1954); A. H. Sharbaugh, B. D. Pritchard, and T. C. Madi-
       son, *Phys. Rev.* **77,** 302 (1950); J. Sheridan and W. Gordy, *J. Chem. Phys.*
       20, 591 (1952).

   a. Perform a "standard" calculation for the dissociation of this molecule
       assuming that the transition state occurs at a location which gives a
       moment of inertia (perpendicular to the figure axis) of twice the origi-
       nal one and assuming that the bending frequency (the "rock") falls to 75
       cm^{-1} at the transition state. Calculate the dissociation rate for 10 ener-
       gies ranging from the dissociation threshold of 69.4 kcal/mole to 71.3
       kcal/mole. Calculate the result for five values of initial rotational quan-
       tum number J from 10 to 14. Plot the log of the rate constant as a func-
       tion of collision energy (starting at 71.4 kcal/mole).
   b. Investigate the effect of different assumptions for the transition state.
       What is the effect of having a lower vibrational rocking frequency if the
       position of the transition state is held constant? What is the effect of
       starting with more rotational excitation on the rate constant?
   c. Choose another molecule of interest to you, find the correct constants
       for the molecule, make some reasonable assumption about the transition
       state, and calculate the rate constant as a function of energy near the
       threshold.

7.20   When bromine and methane are irradiated with visible light at 500 K, the
       rate of formation of CH_3Br is proportional to the square root of the light
       intensity, and to the first power of the methane pressure, for a fixed pres-
       sure of bromine vapor. Devise a kinetic scheme to account for these obser-
       vations.

7.21   Calculate the concentration (in molecules cm^{-3}) of iodine atoms when an
       excimer laser pulse of 10 mJ at 248 nm is incident on a cubical cell of vol-
       ume 1 cm^3 containing 3×10^{15} molecules cm^{-3} of ethyl iodide: $C_2H_5I \rightarrow$
       $C_2H_5 + I$. The absorption coefficient of ethyl iodide at 248 nm is $\epsilon = 63$
       L/mole/cm. (What assumption do you need to make about the quantum
       yield for this photodissociation?)

7.22   **Figure 7.40** is based on the work of A. C. Terentis and S. H. Kable [*Chem.
       Phys. Lett.* **258,** 626 (1996)] and shows the rotational levels observed and
       not observed for HCO produced in the photopredissociation of H_2CO via
       various rotational levels of a particular vibrational band of the H_2CO upper
       singlet state. The total H_2CO energies are given on the abscissa, while the
       energies of the rotational states of HCO observed are shown on the ordinate.
       The full circles represent states observed, while the open ones represent
       states that were not observed. Estimate the dissociation energy of H_2CO to
       give H + HCO. (The H atom is in its electronic ground state.)

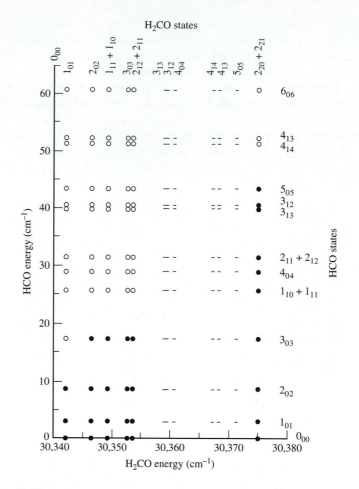

■ **Figure 7.40**

Rotational levels for HCO produced in the photopredissociation of H₂CO.

From A. C. Terentis and S. H. Kable, *Chem. Phys. Lett.* **258,** 626 (1996). Reprinted from *Chemical Physics Letters,* copyright 1996, with permission from Elsevier Science.

Chapter Eight

8

Molecular Reaction Dynamics

Chapter Outline

8.1 INTRODUCTION

The underlying goal of this book has been to understand kinetic processes at the molecular level. We started in Chapter 1 to look at the details of molecular collisions. In Chapter 2 we broke down overall reactions into their component elementary steps, and in Chapter 3 we examined how our understanding of collisions could be extended to predict and understand rate constants for reactions. In this chapter we will go one step further by breaking down elementary steps into *state-to-state* reaction rates and by seeing how these are related to the potential energy surface that controls the reaction. This field, called *reaction dynamics,* seeks to understand the dynamics of chemical systems at a molecular level.

In addition to the intellectual challenge of learning how a rate constant depends on both the initial state of the reactant and the final state of the product, there are ample practical reasons to explore the field of reaction dynamics. Many reactions take place under nonequilibrium conditions, so that if the reaction rate depends on the initial state of the reactant, the rate constant for the process might be quite different from that for reactants in Boltzmann equilibrium. Perhaps by understanding the dynamics we can direct the reaction to produce either desired, unconventional products or products with a particular final state distribution. Furthermore, if a reaction, even one starting from thermal equilibrium, selectively produces a nonequilibrium distribution of products, we might be able to use that distribution in some practical way, for example, to convert chemical energy into another useful form.

In this chapter we start by considering a simple reaction system that has had both historical and practical importance in the of field reaction dynamics, $F + D_2 \rightarrow DF + D$. After learning how the state-resolved reaction rate constants are related to the overall thermal rate constant in Section 8.3, we look in a more detailed way at molecular scattering in Section 8.4. In Section 8.5 we examine how the angular distribution of scattering and the state-resolved rate constants are related to the potential energy surface. Not all of the processes of interest to chemists involve reactions, so we spend some effort in Section 8.6 investigating processes in which the reactants and products have the same chemical identity but in which energy has been exchanged between various degrees of freedom. We end the chapter by exploring several examples to convey some of the excitement of the field and to point to directions for future effort.

8.2   A MOLECULAR DYNAMICS EXAMPLE

The goals of this chapter are illustrated with a simple example. In 1961, John Polanyi suggested that it might be possible to use a chemical reaction to produce a population inversion that could be used to power a new type of "chemical" laser.[a] The idea was to find a reaction that, for example, produced more product in a high vibrational level than in a lower one, so that stimulated emission might occur between the two levels to produce a laser in the infrared region of the spectrum. It was clear that one wanted preferentially to consider exothermic reactions, which could provide enough energy to populate several different internal states of the products. Several examples were found,[b] but we will focus our attention on the $F + H_2$ and $F + D_2$ systems. For some practical reasons, $F + D_2$ has been more carefully studied.

Figure 8.1 shows, on the left, the energetics of the reaction; the DF product can be produced in any of five vibrational levels from $v = 0$ to $v = 4$. The right-hand side of the figure gives the distribution of population among the vibrational levels as measured by three techniques: observing the chemiluminescence from the reaction, observing the behavior of a DF laser based on the reaction, and observing the velocity distribution of the DF product following a crossed molecular beam reaction. This latter technique, whose results are summarized in **Table 8.1,** will be discussed in more detail in Section 8.4. The important point to note at this stage is that the reaction produces a higher population of DF($v = 3$) than DF($v = 2$). Because the $v = 3 \rightarrow v = 2$ transition is optically allowed, this population "inversion" can form the basis for a chemical laser. Indeed, one reason that this reaction was studied in so much detail is that there had been interest at one time on the part of the military in using chemical lasers as weapons.[c] **Figure 8.2** shows a TRW-built high-energy laser that successfully targeted and shot down an operational short-range rocket in a 1996 flight test—on the first firing.

---

[a]J. C. Polanyi, *J. Chem. Phys.* **34,** 347 (1961). Polanyi shared the 1986 Nobel Prize in Chemistry (with D. R. Herschbach and Y. T. Lee) for his work in this area.

[b]The first was discovered by J. V. V. Kasper and G. C. Pimentel, *Phys. Rev. Lett.* **14,** 352 (1965).

[c]American Physical Society Study Group (N. Bloembergen and C. K. N. Patel, cochairmen), "Report to the American Physical Society of the Study Group on Science and Technology of Directed Energy Weapons," *Rev. Mod. Phys.* **59**(3), Part II (July 1987); P. W. Boffey, W. J. Broad, L. H. Gelb, C. Mohr, and H. B. Noble, *Claiming the Heavens* (Time Books, New York, 1988).

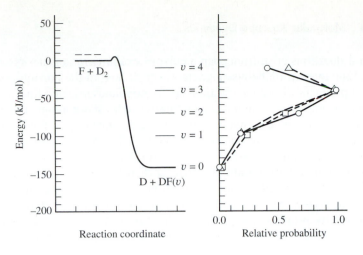

(*left*) Energetics for the F + D$_2$ reaction showing possible final vibrational states. (*right*) Population of final vibrational levels relative to $v = 3$.

TABLE 8.1	Population of Final Vibrational Levels v in the Reaction F + D$_2$ → DF(v) + D	
	v	$P(v)^*$
	0	0
	1	0.19
	2	0.67
	3	1.00
	4	0.41

*Data normalized to $P(v = 3) = 1.00$ and taken from D. M. Neumark, A. M. Wodtke, G. N. Robinson, C. C. Hayden, K. Shobotake, R. K. Sparks, T. P. Schafer, and Y. T. Lee, *J. Chem. Phys.* **82**, 3067 (1985).

■ **Figure 8.2**

Chemical laser based on hydrogen/fluorine reaction claimed by TRW to have shot down a short-range rocket.

Courtesy of TRW, Inc.

Several questions arise from even a cursory examination of this system. What features of the interaction between reactants might lead to the nonequilibrium distribution of products? Since the reaction clearly populates selected final states, does the reaction depend as well on the reactant initial state? If we assign a rate constant for the reaction between a particular state i of the reactants and a final state f of the products, how are these rate constants related to the overall rate constant, $k(T)$, measured at thermal equilibrium? Finally, what happens if we look at this reaction in even more detail by examining the angular distribution of the products or their rotational energy levels? These are the questions that motivate the field of reaction dynamics. The goal is to learn about the molecular mechanism of elementary chemical processes by probing rate processes directly at molecular level; ideally, we would like to "watch" the atoms and molecules move as they undergo reaction.

8.3   MOLECULAR COLLISIONS—A DETAILED LOOK

If we wish to understand a reaction such as $F + D_2 \rightarrow DF + D$ at a fundamental level, we must start by recognizing that there are many internal energy levels of both the reactants and products, and that the rate constant for reaction might depend both on the initial state, i, of the reactant and on the final state, f, of the products, where i and f here each symbolize a set of quantum numbers representing, for example, vibration, rotation, and electronic degrees of freedom. Moreover, this "state-to-state" reaction might itself depend on the relative velocity of the reactants, or equivalently on the collision energy. In Section 3.3 we saw that the rate constant at a particular energy ϵ_r could be expressed as the product of the relative velocity, v_r, and the energy-dependent cross section $\sigma(\epsilon_r)$: $k(\epsilon_r) = v_r \sigma(\epsilon_r)$. Since even at a given collision energy, the rate constants from a particular initial state to a selected final state might differ, we now need to be more specific. Let $\sigma(\epsilon_r, i, f)$ be the reaction cross section at energy ϵ_r for going from state i of the reactants to state f of the products. The rate constant for this process is then $k(\epsilon_r, i, f) = v_r \sigma(\epsilon_r, i, f)$.

We now wish to address two questions. The first is how the overall rate constant at ϵ_r, $k(\epsilon_r)$, is related to $k(\epsilon_r, i, f)$. The second is how the final energy distribution in the DF product is related to these "state-to-state" rate constants.

We approach the first question in two steps. First we use our knowledge of parallel reactions, Section 2.4.2, to show that the total rate for reaction from a particular initial state, $k(\epsilon_r, i)$ is just the sum of the state-to-state reaction rates over the final states, f. Suppose for simplicity that there are only two final states of the products, $f = 1$ and $f = 2$. Then for a given initial state, we have the parallel processes

$$[i] \xrightarrow{k(\epsilon_r, i, f = 1)} [f = 1]$$

and

$$[i] \xrightarrow{k(\epsilon_r, i, f = 2)} [f = 2].$$

The situation is exactly analogous to the $A \rightarrow B$ and $A \rightarrow C$ parallel reactions of Section 2.4.2, so that the total rate of disappearance of i should be given by the sum of the two rate constants; that is, $k(\epsilon_r, i) = k(\epsilon_r, i, f = 1) + k(\epsilon_r, i, f = 2)$. If there are many final states, we can generalize this result by summing over them all:

$$k(\epsilon_r, i) = \sum_{f=1}^{f=f_{max}} k(\epsilon_r, i, f), \tag{8.1}$$

where f_{max} is the highest energy level of the products consistent with the relative collision energy ϵ_r.

The second step in the answer to how $k(\epsilon_r)$ is related to the state-to-state rate constants is taken by recognizing that $k(\epsilon_r)$ is an average over the rate constants starting in different initial states. Of course, when taking the average we need to weight the different initial states[d] according to their populations, $P(i)$. Thus,

$$k(\epsilon_r) = \sum_{i=1}^{\infty} P(i)k(\epsilon_r,i)$$

$$= \sum_{i=1}^{\infty} \sum_{f=1}^{f_{max}} P(i)k(\epsilon_r,i,f). \tag{8.2}$$

It should be noted that $P(i)$ will generally depend on a temperature. For example, if the reactants are at equilibrium with a heat bath at temperature T, then $P(i)$ will simply be the Boltzmann distribution:

$$P(i) = \frac{g_i \exp(-\epsilon_i/kT)}{Q}, \tag{8.3}$$

where g_i and ϵ_i are the degeneracy and energy of the ith level and Q is the partition function at temperature T:

$$Q = \sum_{i=1}^{\infty} g_i \exp\left(-\frac{\epsilon_i}{kT}\right). \tag{8.4}$$

Equation 8.2 is then the answer to the first of our two questions:

The rate constant at a particular energy is the average of the state-selected rate constants over initial states and the sum over final states.

What of the second question? How are the rate constants related to the distribution of population in the products? Our understanding of parallel reactions can again come to our aid. Recall from Section 2.4.2 that the branching ratio to a particular product, or equivalently the probability of forming that product, is given by the rate constant for the desired channel divided by the sum of the rate constants for all of the possible channels. Let $P(\epsilon_r,f)$ be the probability of forming the final state f of the product(s). Then,

$$P(\epsilon_r,f) = \frac{k(\epsilon_r,f)}{\sum_{f=1}^{f_{max}} k(\epsilon_r,f)}, \tag{8.5}$$

where each of the $k(\epsilon_r,f)$ is an average over the possible initial states for reaction at relative energy ϵ_r:

$$k(\epsilon_r,f) = \sum_{i=1}^{\infty} P(i)k(\epsilon_r,i,f). \tag{8.6}$$

We close this section by recalling from Section 3.3 that the rate constant at a given temperature, $k(T)$, is obtained from $k(\epsilon_r)$ by averaging over the thermal distribution of energies:

$$k(T) = \int_0^{\infty} G(\epsilon_r)k(\epsilon_r)\,d\epsilon_r, \tag{8.7}$$

[d]A similar weighting was used in **equation 1.8** to determine the average grade on an examination.

where $G(\epsilon_r)$ is the Boltzmann distribution of energy at temperature T (see **equation 1.37**). Substitution of **equation 8.2** and the definition $k(\epsilon_r,i,f) = v_r\sigma(\epsilon_r,i,f)$ leads to

$$k(T) = \int_{\epsilon=0}^{\infty} \sum_{i=1}^{\infty} \sum_{f=1}^{f_{max}} P(i)v_rG(\epsilon_r)\sigma(\epsilon_r,i,f)\,d\epsilon_r. \qquad (8.8)$$

This last equation relates the macroscopic rate constant for an elementary process to the state-to-state energy-dependent cross sections. The field of reaction dynamics seeks to measure these cross sections and to interpret them at the molecular level. The interpretation is most compactly expressed by the potential energy surface for the system, whose relationship to $\sigma(\epsilon_r,i,f)$ is discussed in Section 8.5. Before exploring this relationship, however, we examine the cross section at a deeper level to learn about its dependence on recoil angles, as measured by scattering experiments.

example 8.1

Calculating $k(T)$ from $\sigma(\epsilon,i,f)$

Objective   A reaction has two initial states. State $i1$ is the ground state and state $i2$ lies 100 cm^{-1} higher. The products have four final states: $f1, f2, f3,$ and $f4$. The cross sections as a function of energy are given by $T(i,f) = \pi d^2[1 - \epsilon^*(i,f)/\epsilon_r]$, where $T(i,f)$ and $\epsilon^*(i,f)$ are given in the table, and $d = 0.10$ nm (1.0 Å). If the reduced mass for the collision is 5 amu, calculate the overall rate constant for the reaction at a temperature of 300 K; i.e., $k(T)$ for $T = 300$.

i,f	$T(i,f)$	$\epsilon^*(i,f)$ (cm^{-1})
$i1,f1$	0.05	1000
$i1,f2$	0.20	1000
$i1,f3$	0.10	1000
$i1,f4$	0.05	1000
$i2,f1$	0.1	500
$i2,f2$	0.3	500
$i2,f3$	0.1	500
$i2,f4$	0.1	500

Method   Use **equation 8.8**, recalling that we have already performed the integration over the thermal energy distribution for a cross section of this functional form in the equations preceding **equation 3.7**.

Solution   Since each of the cross sections has the same functional form with respect to ϵ_r, we can perform the integral in **equation 8.8** before doing the sum. From the equations leading to **equation 3.7** we see that $\int \pi d^2[1 - \epsilon^*(i,f)/\epsilon_r]G(\epsilon_r)d\epsilon_r = \pi d^2 v_r\exp[-\epsilon^*(i,f)/kT]$. The sum over final states and average over initial states is then expressed as:

$$\Sigma_i\Sigma_f P(i)T(i,f)\pi d^2 v_r\exp\left[-\frac{\epsilon^*(i,f)}{kT}\right].$$

We next calculate $P(i)$, noting that kT at 300 K is 207 cm^{-1}: the number of reactants in the ground state is proportional to $\exp(-0/kT) = 1.0$. The number in state $i2$ is proportional to $\exp(-100\text{ cm}^{-1}/kT) = \exp[-100/207] = 0.617$. Thus $P(i1) = 1.0/(1.0 + 0.617) = 0.618$ and $P(i2) = 0.617/(1.0 + 0.617) = 0.382$.

Next, note that $\pi d^2 = 3.14 \times 10^{-2}$ nm^2 (molecule^{-1}) and that at room temperature $v_r = [8kT/\pi\mu]^{1/2} = [8(1.38 \times 10^{-24}$ J molec^{-1} K^{-1})(6.02 $\times$ 10^{23} molec/mol)(300 K)/π(5.0 g/mol) (1 kg/1000 g)]$^{1/2}$ = 1.13 km/s; thus $\pi d^2 v_r = (3.14 \times 10^{-20}$ m^2 molec^{-1}) (1.13 $\times$ 10^3 m/s) = 3.55 $\times$ 10^{-17} m^3 molec^{-1} s^{-1} = 2.14 $\times$ 10^{10} L mol^{-1} s^{-1}. We also evaluate $\exp(-1000/207) = 7.98 \times 10^{-3}$ and $\exp(-500/207) = 8.93 \times 10^{-2}$.

Finally we perform the sum and average to obtain $k(T = 300$ K) = (2.14 $\times$ 10^{10} L mol^{-1} s^{-1}) $\times$ {[(7.98 $\times$ 10^{-3})(0.618)(0.05 + 0.20 + 0.10 + 0.05)] + [(8.93 $\times$ 10^{-2})(0.382)(0.1 + 0.3 + 0.1 + 0.1)]} = 4.80 $\times$ 10^8 L mol^{-1} s^{-1}.

8.4 MOLECULAR SCATTERING[e]

Ever since Rutherford obtained information about the structure of the atom from measurement of the angular distribution of deflected alpha particles, scattering experiments have been used by both physicists and chemists to explore fundamental processes in their respective fields. The objective of this section is to investigate what can be learned about chemical interactions by examination of the angular distribution of products following reaction. How do we measure the angular distribution? As shown in **Figure 8.3,** the angle of scattering in Rutherford's case was measured relative to the velocity of the incoming alpha particles that struck a stationary metal foil. Chemists face a somewhat more complicated situation because neither of the reacting molecules is stationary in the laboratory frame. In

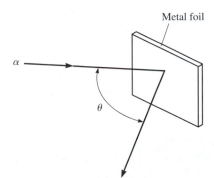

Metal foil

α

θ

Figure 8.3

The Rutherford scattering experiment in which alpha particles (He^{2+} with large kinetic energy) were scattered from a foil of metal atoms.

---

[e]D. R. Herschbach and Y. T. Lee shared the 1986 Nobel Prize in Chemistry (with J. C. Polanyi) for their work using molecular beams to study the dynamics of elementary chemical processes.

the ideal experiment, the observer of a chemical reaction would like to be located in advance at the position where two reactants would collide. From this center-of-mass observation point, shown in **Figure 8.4,** one could then watch both the approach of the reactants along their relative velocity vector and the departure of the products along a new line making an angle θ with the relative velocity vector. Unfortunately, the center-of-mass point for two reacting species is always itself moving in the laboratory frame, so we first need to investigate how the laboratory velocities are related to the velocities in the center-of-mass frame.

8.4.1   The Center-of-Mass Frame—Newton Diagrams

We assume that the initial velocities of two reactants of masses m_1 and m_2 are defined in the laboratory frame by the use of molecular beam techniques. In a typical case, the reactants might each be coexpanded in a dilute mixture with a light rare gas from a nozzle source and collimated by skimmers, as shown in **Figure 8.5.** In such expansions, the enthalpy of the gas behind the nozzle is converted to bulk translational energy, and the distribution is shifted to higher energies. A simple way

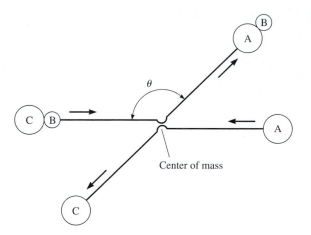

Figure 8.4

A + BC → AB + C reaction as seen by an observer from the center-of-mass point.

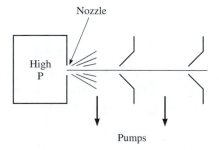

Figure 8.5

In a supersonic expansion source the random thermal motions on the high pressure side of the nozzle are converted into directed motion along the beam axis.

of thinking about what happens is that the very fast molecules are slowed by collisions with those in front of them, while the slow molecules are sped up by collisions with faster molecules behind them. The result is a beam whose velocity distribution is peaked sharply compared to a Maxwell-Boltzmann distribution, as illustrated in **Figure 8.6.** One can estimate the stream velocity, v_0, of the beam for a monatomic carrier gas by setting its enthalpy, $5kT/2$, equal to the kinetic energy of the molecules, $\frac{1}{2}mv_0^2$, so that one obtains $v_0 = (5kT/m)^{1/2}$. For helium, or for heavier gases seeded in helium, and for expansion from a room-temperature reservoir, this velocity is $v_0 = 1.76 \times 10^3$ m/s.

Now let two such beams, one containing each reactant, intersect as in the apparatus of **Figure 8.7.** We demonstrated in Appendix 1.4 using a diagram like that in **Figure 8.8** (often called a "Newton" diagram) that the total energy,

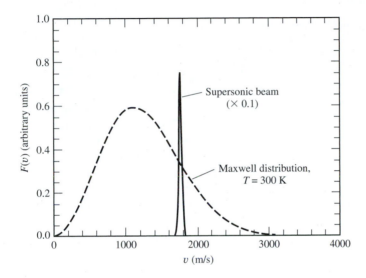

Figure 8.6

Comparison between velocity distributions for a Maxwell-Boltzmann distribution and the distribution obtained from a supersonic expansion. The example shown is for He at 300 K.

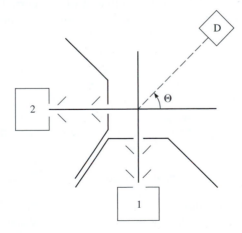

Figure 8.7

Crossed molecular beam apparatus. The products of the reaction are detected (D) as a function of the variable laboratory angle Θ.

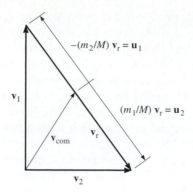

Figure 8.8

Newton diagram for intersecting beams.

$E = \frac{1}{2}m_1 v_1^2 + \frac{1}{2}m_2 v_2^2$, could also be written as $E = \frac{1}{2}\mu v_r^2 + \frac{1}{2}Mv_{com}^2$, where $M = m_1 + m_2$, $\mu = m_1 m2/M$, and $\mathbf{v}_r$ and $\mathbf{v}_{com}$ are shown in the figure. The speed and direction of $\mathbf{v}_{com}$, recall, do not change during the collisions provided that the only forces act between the two particles. Specifically, $\mathbf{v}_r = \mathbf{v}_2 - \mathbf{v}_1$, and $\mathbf{v}_{com}$ is a vector from the origin to the center-of-mass point on $\mathbf{v}_r$ located a distance $(m_2/M)\mathbf{v}_r$ from the intersection of $\mathbf{v}_1$ and $\mathbf{v}_r$. Thus, if the two molecular beams of velocities $\mathbf{v}_1$ and $\mathbf{v}_2$ define directions in the laboratory frame, then their angle of intersection, their magnitudes, and the masses of the reactants define the direction both of the center of mass and of their relative motion.

Figure 8.8 shows that an observer riding along the center-of-mass point would then see reactant 1 approaching from the lower right in the diagram with velocity $\mathbf{u}_1$ and reactant 2 approaching from the upper left with velocity $\mathbf{u}_2$. Note that in this frame of reference, the momentum of particle 1, $m_1 \mathbf{u}_1 = -m_1(m_2/M)\mathbf{v}_r$, is equal in magnitude and opposite in direction to the momentum of particle 2, $m_2 \mathbf{u}_2 = m_2(m_1/M)\mathbf{v}_r$. Because the center of mass of the system will keep moving in the same direction at the same speed whatever the interaction between the reactants, the amount of energy available to the collision of the particles is not their total energy, but rather just that measured in the moving center-of-mass frame: $\epsilon_r = \frac{1}{2}m_1 \mathbf{u}_1^2 + \frac{1}{2}m_2 \mathbf{u}_2^2 = \frac{1}{2}m_1(m_2 \mathbf{v}_r/M)^2 + \frac{1}{2}m_2(m_1 \mathbf{v}_r/M)^2 = \frac{1}{2}\mu v_r^2$.

After the reaction takes place, the product molecules will move away from the center of mass along a new direction, as already illustrated in **Figure 8.4.** What will their new relative velocity be, and how will it be partitioned between the new fragments? The second question is answered by applying conservation of momentum to the problem. In the center-of-mass frame, the momenta of the particles must sum to zero. If we label the product masses by m_3 and m_4 and their velocities with respect to the center of mass as $\mathbf{u}_3$ and $\mathbf{u}_4$, then

$$m_3 \mathbf{u}_3 + m_4 \mathbf{u}_4 = 0 \qquad (8.9)$$

by conservation of momentum.

Conservation of energy can help us to answer the first question. The new relative velocity will depend on the amount of energy available for translational motion. Suppose that the energy release for the specific reaction from state i of the reactants to state f of the products is $\epsilon_{ex}(i,f)$; then the total energy available for translation is that available before the reaction, ϵ_r, plus the exoergicity, $\epsilon_{ex}(i,f)$. In the center-of-mass frame, the products will then have an energy $\epsilon_r + \epsilon_{ex}(i,f)$. Then, by conservation of energy we have

$$\frac{1}{2}m_3 \mathbf{u}_3^2 + \frac{1}{2}m_4 \mathbf{u}_4^2 = \epsilon_r + \epsilon_{ex}(i,f) = \frac{1}{2}\mu' \mathbf{v}_r'^2. \qquad (8.10)$$

If $\mathbf{v}_r'$ is the relative velocity between m_3 and m_4 after the collision, $\mathbf{v}_r' = \mathbf{u}_4 - \mathbf{u}_3$, then the solution for $\mathbf{u}_3$ and $\mathbf{u}_4$ that satisfies **equations 8.9** and **8.10** is

$$\mathbf{u}_3 = -\left(\frac{m_4}{M}\right)\mathbf{v}_r',$$

$$\mathbf{u}_4 = \left(\frac{m_3}{M}\right)\mathbf{v}_r'. \qquad (8.11)$$

By conservation of mass, $M = m_3 + m_4$ is the same as $M = m_1 + m_2$.

Figure 8.9 illustrates the relationships just described. For the example drawn, $m_3 > m_4$.

The task remaining is to find out where the product molecules will appear in the laboratory frame of reference. Let us first concentrate on the product with mass m_4, recognizing that generalization to the other product would follow similar arguments. The conservation laws of energy and momentum limit the length of $\mathbf{u}_4$ but not its direction, so that, as shown in **Figure 8.10,** m_4 could in principle be found anywhere on a sphere of radius $\mathbf{u}_4$ centered on the center-of-mass position. Of course, the product will not necessarily be isotropically distributed on this sphere; its distribution is what we would like to learn. As shown in the figure, if we set the detector at a laboratory angle of Θ and in the plane of the two molecular beams, it will detect products that have been scattered by an angle θ in the center-of-mass frame. By scanning Θ and transforming from the laboratory to the center-of-mass frame, it is then possible to map out the center-of-mass angular distribution of products. This distribution is called the *differential reaction cross section,* because it tells us how the cross section varies with solid angle. Specifically, we will symbolize the differential cross section by $d^3\sigma(v_r,\theta)/d^2\omega \, dv_r'$. It provides the cross section for product molecules that are scattered per unit time into a solid angle $d^2\omega$ with

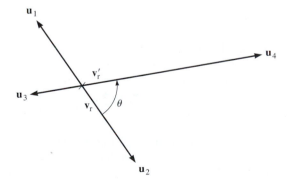

Figure 8.9

Products separate along their new relative velocity vector $\mathbf{v}_r'$ making an angle θ with respect to the reactant relative velocity vector $\mathbf{v}_r$.

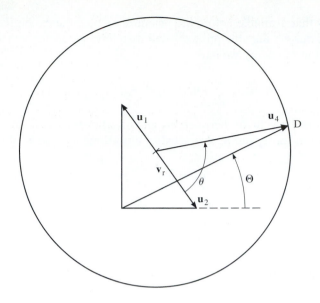

■   **Figure 8.10**

The reaction product m_4 will be found on a sphere of radius $\mathbf{u}_4$. Products scattered by θ in the center-of-mass frame are scattered by Θ in the laboratory frame.

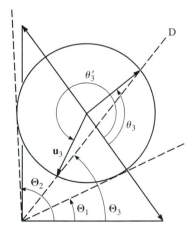

■   **Figure 8.11**

Newton diagram for the heavier product, m_3.

final relative velocities in the range from $\mathbf{v}_r'$ to $\mathbf{v}_r' + d\mathbf{v}_r'$. We use the notation $d^2\omega$ for the solid angle to mean $\sin\theta \, d\theta \, d\phi$, where θ and ϕ are the spherical coordinates in the center-of-mass frame.

It is interesting to note from **Figure 8.10** that to measure the complete differential cross section for m_4, the detector would have to be moved in a complete circle around the scattering center. What would the Newton diagram look like for the other product, m_3? **Figure 8.11** shows that this product is located on a much smaller sphere, corresponding to its smaller velocity. The entire scattering angle distribution from $\theta = 0$ to $\theta = 2\pi$ in the center-of-mass frame can be measured by rotating the detector between $\Theta = \Theta_1$ and $\Theta = \Theta_2$ in the laboratory frame. Note, however, that the scattering at a particular laboratory angle, say Θ_3, measures reaction products usually corresponding to two center-of-mass angles. If the detector is equipped to measure the laboratory velocity of the products by using, for example, a velocity selector, then it would detect fast fragments due to scattering at center-of-mass angle θ_3, and slow fragments due to scattering at center-of-mass angle θ_3'.

Before proceeding to an example of what can be learned from measurement of the differential cross section, we note that much of the scattering to the sphere centered on the center-of-mass point is directed toward points that are outside of the plane formed by the crossed molecular beams and in which the detector is typically located. For a given center-of-mass scattering angle θ and for randomly oriented reactants, we can deduce that the scattering should be cylindrically symmetric about the relative velocity vector $\mathbf{v}_r$. The reason is shown in **Figure 8.12.** The angle of scattering θ should depend on the impact parameter b, defined in Section 1.7, and on the orientation of the two reactants as they collide. If the orientation is random, then θ depends just on b. However, for a given magnitude of b between b and $b + db$, there are an equal number of trajectories passing through any part of an annulus centered on $\mathbf{v}_r$. Because the annulus has circular symmetry about $\mathbf{v}_r$, the scattering should also be symmetric about $\mathbf{v}_r$; that is, the intensity may depend on θ but it does not depend on ϕ, the azimuthal angle about $\mathbf{v}_r$.

With this cylindrical symmetry in mind, let us note that the differential reaction cross section, $d^2\sigma(v_r,\theta)/d^2\omega\,dv_r'$, is defined in such a way that its integral over final velocities and angles gives the total reactive cross section at a particular relative energy:

$$\sigma_R(\epsilon_R) = 2\pi \int_{\theta=0}^{\pi} \int_{v_r'=0}^{\infty} \frac{d^3\sigma_R(v_r',\theta)}{d^2\omega\,dv_r'} \sin\theta\,d\theta. \tag{8.12}$$

A word of caution is necessary here. Molecular beam reaction experiments sometimes report $\sigma_R(v_r,\theta)$, but they more often report the "product contour diagram." The two are related. The product contour diagram plots in polar coordinates the derivative of the cross section with respect to both angles and product velocity; that is, they plot $d^2\sigma(v_r,\theta)/d^2\omega\,dv_r'$, where $d^2\omega$ is shorthand for $\sin\theta\,d\theta\,d\phi$ and v_r' is the product relative velocity.

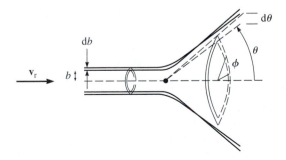

Figure 8.12

Diagram showing that scattering should be cylindrically symmetric about $\mathbf{v}_r$ because of the cylindrical symmetry of b about $\mathbf{v}_r$.

8.4.2   Reactive Scattering: Differential Cross Section for F + D$_2$

We now consider measurements of the differential cross section for the F + D$_2$ reaction. A product contour diagram, already converted to the center-of-mass frame, is shown in **Figure 8.13.** What can be learned from such a diagram? First, note the dashed circles superimposed on the diagram. These correspond to the maximum velocities consistent with the conservation laws for production of DF in various vibrational levels. For example, the circle marked "$v = 1$" shows the velocity expected for DF($v = 1$, $J = 0$). Higher rotational levels of this vibrational state would have less energy available for translation and so would lie within this circle. There are clear peaks in the contours near the energetic limits for the various vibrational levels, and these must then correspond to products formed in the indicated levels with varying amounts of rotational excitation. The contour peak for the $v = 3$ level is the highest, so most of the DF product must be formed in this state. A careful analysis produces the relative populations already presented in **Table 8.1.**

Note also that the DF products are "backward" scattered, i.e., that with $\theta = 0°$ defined as the direction in which the attacking F atom moves, the DF product is scattered primarily to angles near $\theta = 180°$. The picture of the reaction which emerges is that the F atom must hit the D$_2$ or H$_2$ nearly head-on, and then bounce backward taking one of the hydrogen atoms with it. We might call this a "rebound" reaction.

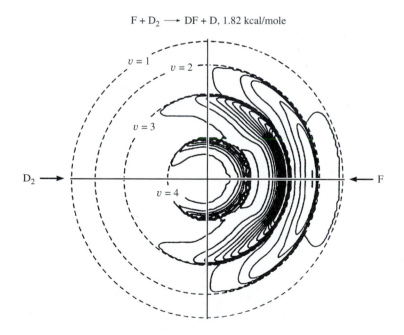

■   **Figure 8.13**

Product contour map for the F + D$_2$ reaction at a collision energy of 1.86 kcal mol^{-1}. The contours show the DF velocity distribution.

From D. M. Neumark, A. M. Wodtke, G. N. Robinson, C. C. Hayden, K. Shobotake, R. K. Sparks, T. P. Schafer, and Y. T. Lee, *J. Chem Phys.* **82,** 3067 (1985). Reprinted with permission from the *Journal of Chemical Physics.* Copyright, American Institute of Physics, 1985.

Reactions of other sorts are also possible. Product contour maps, or the related differential cross section, are valuable precisely because they tell us what sort of reaction we are observing. For example, in reactions such as $K + I_2 \rightarrow KI + I$, the newly formed product, KI, is found predominantly in the *forward* direction, as shown in **Figure 8.14.** Reaction occurs at large impact parameters where an electron is transferred from the K atom (whose ionization potential is low) to the iodine molecule (whose electron affinity is high). The K^+ and I_2^- are then drawn toward one another by the electrostatic force, and the more energetically stable $KI + I$ products are formed with the KI moving off in the direction of the original K atom. This type of reaction mechanism is known as the *harpoon mechanism,* because, in effect, the potassium has used its electron as a harpoon to pull in an iodine atom along the line of electrostatic force. The theory of such reactions will be discussed in Section 8.6.4. Harpoon reactions are examples of a more general class of reaction, called *stripping reactions,* in which the attacking atom or radical carries off part of the attacked molecule in the forward direction.

Figure 8.15 shows the product contour plot for a reaction involving a *collision complex.* Note that there is backward-forward symmetry in the differential cross section. The interpretation is that the O and Br_2 have "stuck" together for a time long compared to their mutual rotation period. To see why such complex-forming reactions might lead to forward-backward symmetry in the product contour plots, we consider the collision of two structureless particles that form a complex. **Figure 8.16** shows the two particles coming together at a given impact parameter and orbiting one another in a plane. If the collision complex lives longer than a rotational period, then the reaction products are likely to be flung out radially like water spinning off a wet frisbee. Thus, if $dN(\theta)/dt$ is the number of products created per unit

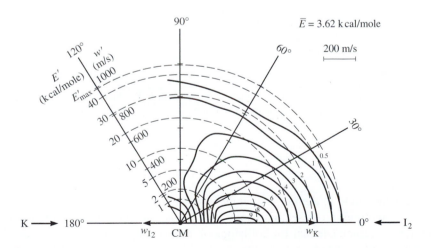

Figure 8.14

KI contour map for the reaction $K + I_2 \rightarrow KI + I$. Note that the KI product is scattered in the forward direction with respect to the K velocity.

From K. T. Gillen, A. M. Rulis, and R. B. Bernstein, *J. Chem. Phys.* **54,** 2831 (1971). Reprinted with permission from the *Journal of Chemical Physics.* Copyright, American Institute of Physics, 1971.

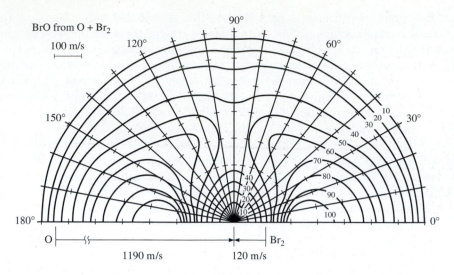

Figure 8.15

BrO contour map for $O + Br_2 \rightarrow BrO + Br$ reaction. Note that the scattering distribution is symmetric in the forward and backward directions.

From D. D. Parish and D. R. Herschbach, *J. Am. Chem. Soc.* **93,** 6133 (1973). Reprinted with permission from *The Journal of the American Chemical Society.* Copyright 1973 American Chemical Society.

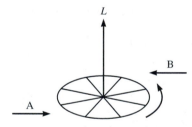

Figure 8.16

Two molecules forming a complex with angular momentum L. Products are "sprayed" out with a uniform distribution of angles in the plane of rotation.

time near the angle θ, then $d^2N(\theta)/dt\,d\theta$ = constant. Although it might thus seem that the differential cross section should have equal intensity in all directions, it is important to remember that for collisions of structureless particles, and indeed for the average collisions of structured particles if they are not oriented on average with respect to one another, the scattering is cylindrically symmetric about the relative velocity vector because of the cylindrical symmetry of ϕ in **Figure 8.12.** In the frisbee analogy, we need to think of rotating the already spinning frisbee about an axis in the plane of the frisbee: the water will be more dense along this relative velocity axis because many values of ϕ for θ near 0° or 180° contribute to flinging water in the same direction.

In summary, the differential cross section, obtained by analysis of molecular scattering, is useful because it gives us a detailed picture of how the reaction proceeds. Common examples are rebound reactions, where the scattering is predominantly in the backward direction, stripping reactions, where the scattering is predominantly in the forward direction, and collision complex reactions, where the scattering exhibits forward-backward symmetry corresponding to peaking in the forward and backward directions.

Our focus in this chapter has been on reactive collisions, but scattering experiments also provide important information about other kinds of collisions. We digress briefly here to discuss elastic and inelastic collisions.

8.4.3   Elastic Collisions

Even for collisions that do not result in reaction, the differential cross section, obtained from scattering experiments, provides us with the most detailed information about the collision mechanism. As we will see, the distribution of scattering angles for nonreactive collisions can be used rather directly to calculate the potential of interaction between the colliding species. We consider in this section collisions for which there is neither reaction nor energy transfer; that is, collisions for which the energy of the outgoing particles is exactly the same as that of the incoming particles, only the direction of their motion has changed. Such collisions are called *elastic collisions*; collisions between billiard balls are a familiar approximation. It was essentially collisions of this type which we considered in Chapters 1 and 4 when we introduced the concept of mean free path and then used it to examine transport properties.

Example 8.2 shows how a Newton diagram can be used to relate the laboratory and center-of-mass scattering angles for elastic processes.

example 8.2

Newton Diagrams for Elastic Collisions

Objective   A beam of argon (MW = 40) intersects a beam of krypton (MW = 84) at right angles. Calculate the laboratory angle between the krypton direction and the direction of a detector placed so as to monitor krypton scattered elastically in the backward center-of-mass direction. Assume that both collision species are traveling at the same speed.

Method   Draw a Newton diagram for the collision, noting that for elastic scattering the krypton product velocity will be located on the surface of a sphere whose radius is the original relative velocity of the krypton.

Solution   **Figure 8.17** shows the Newton diagram.
Because the original species are traveling at the same velocity, the relative velocity vector lies at an angle of 45° to each of the primary beams. The center-of-mass velocity vector is located between the origin and a point 40/(40 + 84) of the distance from the end of the Kr velocity and the end of the Ar velocity. The elastically scattered Kr will be on a sphere of radius $(40/124)v_r$, where

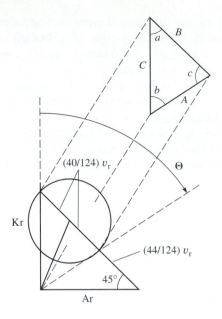

Figure 8.17

Newton digram for elastic scattering of krypton and argon.

v_r is the length of the relative velocity vector. Kr that is exactly backward scattered in the center-of-mass frame has its velocity vector in the plane of the two beams and lying between the center-of-mass point and the point where the sphere intersects v_r; i.e., a point on v_r located 80/124 of the distance from the tip of the original Kr velocity vector. Direct measurement of the diagram gives $\Theta \approx 60°$. Alternatively, one can use the following relationships for a triangle whose sides are A, B, and C, and whose opposite angles are a, b, and c, respectively: $a + b + c = 180°$; $A/\sin a = B/\sin b = C/\sin c$. Letting b be the angle we want, we find from the last two relationships that $\sin b = \sin(180 - a - b)B/C$, where $B = (80/124)v_r$, $C = v_{Kr}$, and $a = 45°$. Realizing that $v_r = \sqrt{2}v_{Kr}$, we obtain $\sin b = \sin(135 - b) \times 0.9124$; a few iterations on a calculator gives $b = \Theta = 61°$.

The bottom portion of **Figure 8.18** shows on a log scale the angular distributions for scattering between argon and krypton and argon and xenon obtained from crossed beam experiments. The scattering distributions exhibit many features common to all elastic scattering. First, the scattering is strongly peaked in the forward direction. Second, for each collision partner there is a series of oscillations in the intensity as the angle increases. These lead to a broad relative maximum centered on the so-called *rainbow angle*. Finally, the intensity falls off strongly at angles approaching 180°. Such features can be understood both qualitatively and quantitatively as the consequence of the attractive and repulsive forces between the collision partners. Let us examine the qualitative aspects first.

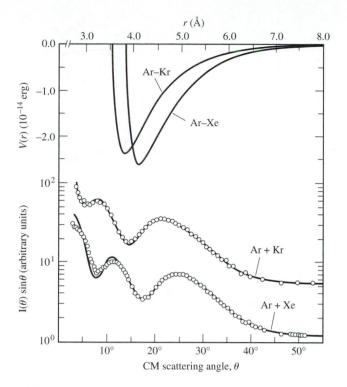

(*bottom*) Scattering distributions for Ar + Kr and Ar + Xe at $E_{col} = 1 \times 10^{-20}$ J. (*top*) Potential energy functions.

From J. M. Parson, T. P. Schafer, P. E. Siska, F. P. Tully, Y. C. Wong, and Y. T. Lee, *J. Chem. Phys.* **53**, 3755 (1970). Reprinted with permission from the *Journal of Chemical Physics*. Copyright, American Institute of Physics, 1970.

The top portion of **Figure 8.18** shows a good approximation to the inter-atomic potential between the two closed-shell species, either argon and krypton or argon and xenon in this example. Note that the potential energy is negative (attractive) for large distances and repulsive (this part of the potential is not drawn) for small distances. It is these forces that ultimately cause the breakdown of the ideal gas approximation. By knowing these forces, it is possible to calculate the coefficients in the virial expansion for a real gas.

The consequences of the potential functions shown in **Figure 8.18** are depicted in **Figure 8.19,** where the deflection angle χ is plotted for various reduced impact parameters, $b^* = b/R_e$, where R_e is the distance at which the potential energy is a minimum. For collisions involving large impact parameters, there is a slight attraction between the two species which leads to slightly negative value of χ. As b^* decreases, this negative deflection increases until reaching a maximum when b is equal to b_r, the impact parameter where there is the maximum negative deflection, χ_r. This impact parameter is called the *rainbow impact parameter* for reasons that will be described below. Further reduction in b^* leads to increasing (less negative) deflection. At a particular impact parameter b_g called the *glory impact parameter,* the deflection is zero. For smaller impact parameters the deflection increases rapidly until for $b^* = 0$ there is backward scattering, $\chi = 180°$.

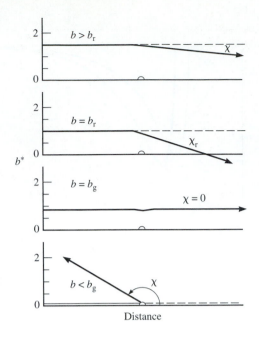

Figure 8.19

Trajectories giving the deflection angle for different reduced impact parameters.

A qualitative plot of the deflection angle χ as a function of impact parameter is given in **Figure 8.20.** Of course, a scattering experiment cannot tell whether a given center-of-mass scattering angle θ resulted from a positive or negative deflection χ; what is measured is $\theta = |\chi|$, as shown in the dashed line.

There are several interesting features of this plot. First, note that a very large number of (large) impact parameters lead to small deflection. Qualitatively, this means that the scattering should be strongly peaked in the forward (zero deflection) direction, as observed in **Figure 8.18.** Quantitatively, note that $d\theta/db \rightarrow 0$ as $b \rightarrow \infty$.

Second, note that at the so-called *rainbow angle,* $\theta = \theta_r$, a range of impact parameters lead to scattering at the same laboratory angle. Qualitatively, we might expect a relative maximum at this angle, as, indeed, is observed in **Figure 8.18.** Quantitatively, we note that near the rainbow angle we again have $d\theta/db = 0$. The same phenomenon causes rainbows of the celestial sort, where light of a given color entering raindrops at different locations is scattered to the same angle.

Third, note that there are three impact parameters that contribute to the scattering at every angle θ smaller than the rainbow angle. From quantum mechanics we know that having more than one trajectory leading to the same final state always results in *interference,* so we might expect the differential cross section to exhibit oscillations for $\theta < \theta_r$, as indeed are observed. Finally, we see that only the very smallest impact parameters lead to large deflections near 180°, so that we expect there will not be very many scattering events that produce this deflection. This is the reason that the differential cross section in **Figure 8.18** falls dramatically as $\theta \rightarrow 180°$.

A more quantitative analysis can be performed with the aid of **Figure 8.12.** Note that the contribution to scattering between angles θ and $\theta + d\theta$ comes from collisions occurring within an annulus between impact parameters b and $b + db$. Thus, $d\sigma = I(\theta)\sin \theta \, d\theta \, d\phi = 2\pi b \, db$, or, after integrating over $d\phi$,

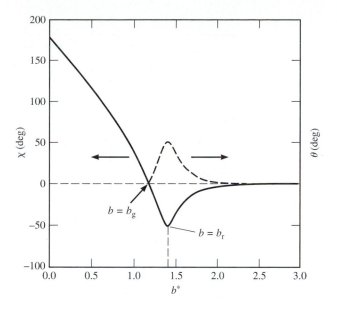

Figure 8.20

Deflection angle as a function of reduced impact parameter. Note that $\theta = |\chi|$ is the angle observed.

$$I(\theta) = \frac{b}{\sin \theta \, |d\theta/db|}, \tag{8.13}$$

where the absolute value has been introduced in recognition that the intensity is always positive. **Equation 8.13** shows that, at least according to classical mechanics, the scattering intensity $I(\theta)$ should be infinite when $d\theta/db$ goes to zero. **Figure 8.21** shows the expected results. The top panel reproduces the curve of θ as a function of b^* (**Figure 8.20**) but turned on its side so that θ is the abscissa. The bottom panel plots the logarithm of the derivative $d\theta/db$. We have already noted that $d\theta/db$ is zero at $\theta = 0$ and at $\theta = \theta_r$. Experiments are necessarily performed with finite resolution, and quantum mechanics makes it impossible to specify θ closely enough for $I(\theta) \rightarrow \infty$. These considerations avoid the infinities, but there are still peaks in $I(\theta)$ at these locations. In particular, the rainbow peak in the scattering distribution is clearly evident at the angle marked by the dashed line. Note also the oscillations in $I(\theta)$ for $\theta < \theta_r$ corresponding to multiple impact parameter collisions contributing intensity at the same angle.

Because the scattering function $I(\theta)$ depends on the deflection function $\chi(b)$, and because $\chi(b)$ depends on the potential $V(r)$, it should be clear that a measurement of $I(\theta)$ can be used to determine $V(r)$. In practice, however, the "inversion" from $I(\theta)$ to $V(r)$ is far from straightforward. While new techniques are still under development, the potential is usually determined iteratively by a "forward convolution" technique. A functional form is assumed, classical or quantum mechanical scattering calculations are then performed, and the results are convoluted with experimental parameters and compared to the experimental results. The functional form of the potential is then adjusted, and the cycle is repeated. For example, the solid line fits to the $I(\theta)$ curves in **Figure 8.18** come from forward convolution of

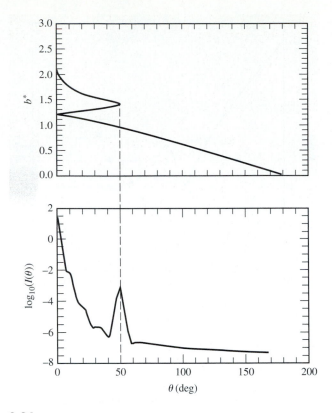

Figure 8.21

(*top*) Relationship between scattering angle and impact parameter. (*bottom*) $I(\theta)$ from **equation 8.13.**

the potentials shown at the top of the figure. Much of our knowledge concerning the potential of interaction between closed shell species comes from scattering experiments such as those just described.

8.4.4   Inelastic Collisions

Although we will consider inelastic collisions in more detail in Section 8.6, it is important to note that scattering techniques provide important information about energy transfer. The Newton diagram for inelastic scattering is simpler than that for reactive scattering because the masses of the products are the same as the masses of the reactants. On the other hand, unlike the diagram for elastic scattering, the final relative velocity will be different from the initial one, because, in general, the inelastic collision will transfer energy between translational and internal degrees of freedom.

Consider, for example, the rotational excitation of NO by collision with argon. **Figure 8.22** shows a Newton diagram for the collision. In this diagram, the argon beam is traveling from top to bottom and the NO beam from right to left. Ionization of components in the two beams and projection of the ions onto a screen gives a picture of the beams and a measurement of the velocities. The circle shows the range of NO velocities expected if the scattering were elastic. For processes such as $Ar + NO(J_i) \rightarrow Ar + NO(J_f)$ with the final rotational quantum number larger than the initial one, $J_f > J_i$, energy will be transferred from the translational to rotational

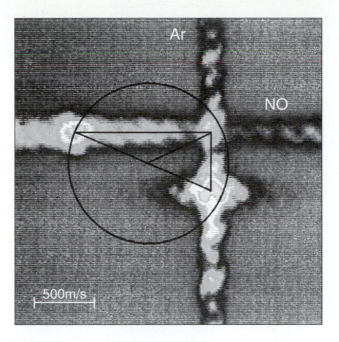

Figure 8.22

Image of molecular beams and the superimposed Newton diagram for Ar + NO.

Reprinted with permission from L. S. Bontuyan, A. G. Suits, P. L. Houston, and B. J. Whitaker, *Journal of Physical Chemistry,* **97,** 6342, 1993. Copyright © American Chemical Society.

degree of freedom, so that the final relative velocity should be less than the initial one. We would thus expect that final NO velocity vectors should lie inside of the indicated circle.

Figure 8.23 provides images of the final velocity vector distribution obtained by a technique in which the product NO molecules are state-selectively ionized and projected onto a screen. The image is such that the observer is looking down on the Newton sphere of **Figure 8.22.** The circles in the three panels are the same diameter, and represent the size of the scattering sphere that would be expected if all collisions were elastic (see **Example 8.3**). In these experiments the initial state is $NO(J_i = 0.5)$, while the final state is indicated in the caption.[f] There are three important points to note. First, as J_f increases, the size of the scattering decreases; in the top panel the scattering is nearly at the elastic limit, whereas in the bottom panel it is substantially more confined. This observation is a simple consequence of conservation of energy (see **Example 8.3**). Second, as J_f increases the distribution of scattering moves from the forward part of the scattering sphere to the backward part. A little thought shows why this should be so. Collisions at large impact parameter produce very little change in either the direction of the NO or its rotational state. Thus, low rotational levels with J_f just a little above J_i should be forward scattered. To appreciably change the rotation of the NO, the argon must hit nearly head-on, after which it will rebound nearly along its initial direction. Thus, high J_f products are

---

[f]Here J denotes the total angular momentum, which is half-integral because it is composed of both the angular momentum due to nuclear spin and the angular momentum of the unpaired electron.

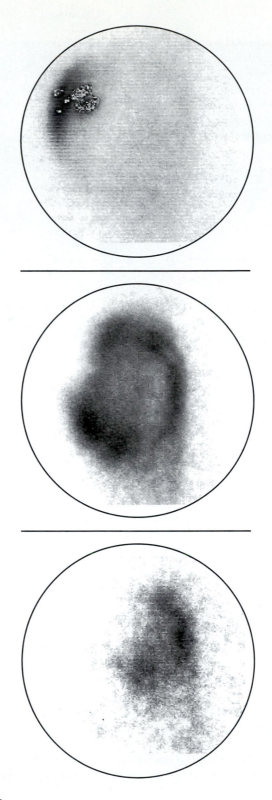

■ **Figure 8.23**

Differential cross section for Ar + NO. (*top*) $J_f = 7.5$; (*middle*) $J_f = 18.5$; (*bottom*) $J_f = 24.5$.

Reprinted with permission from L. S. Bontuyan, A. G. Suits, P. L. Houston, and B. J. Whitaker, *Journal of Physical Chemistry,* **97,** 6342, 1993. Copyright © American Chemical Society.

associated with backward scattering. Third, the scattering is concentrated at specific locations, called *rotational rainbows*. Their origin is similar to the origin of the rainbows in elastic scattering. At the rainbow angles, many trajectories contribute to production of NO in the specified state with the same final scattering angle.

The way in which the rainbow angle moves from the forward to the backward part of the scattering sphere as the final rotational state increases can be used to obtain detailed information about the shape of the NO molecule. Specifically, at the high collision energies of these experiments, the collision samples the repulsive part of the potential between NO and Ar, and this potential looks nearly like an ellipse centered on the NO and having a difference between the major and minor axes of about 0.031 nm.

example 8.3

Newton Spheres for Inelastic Collisions

Objective   Calculate the velocity corresponding to the radius of the Newton sphere for NO($v = 0$, $J = 0.5$) in an elastic collision with argon if the center-of-mass collision energy is 0.25 eV. What would be the radius if an inelastic collision produced NO($v = 0$, $J = 10.5$) given that the NO rotational constant is 1.7 cm^{-1}?

Method   For the elastic case, we can calculate the center-of-mass relative velocity from the energy and the reduced mass. The NO velocity will then be given by conservation of momentum. For the inelastic case, the final energy will be just the collision energy minus the energy needed to rotationally excite the NO. The NO velocity is then calculated in the same manner as for the elastic collision.

Solution $\frac{1}{2}\mu_{NO-Ar}v_r^2 = (0.25$ eV$)(8066$ cm^{-1} eV$^{-1})(hc$ J cm$) = 0.40 \times 10^{-19}$ J. Thus, $v_r = [2(0.401 \times 10^{-19}$ J$)(6.02 \times 10^{23})(1000$ kg/g$)/(30 * 40/70$ g$)]^{1/2} = 1678$ m/s. The NO speed is thus $(40/70) * 1678$ m/s $= 959$ m/s. For the inelastic case, the final available energy is $[(0.25$ eV$)(8066$ cm^{-1} eV$^{-1}) - (10)(11)(1.7$ cm$^{-1})] * (hc$ J cm$) = 0.362 \times 10^{-19}$ J. Thus, the new NO speed is $[(0.362)/(0.401)](959$ m/s$) = 866$ m/s.

8.5   POTENTIAL ENERGY SURFACES

Having explored how molecular scattering can reveal the details of reactive, elastic, and inelastic collisions, we now return to a main theme, the connection between the potential energy surface and the measured product state distributions. We have already seen in **Figure 8.18** how the potential energy function for elastic collisions is related to the angular distribution of products. Of course, because more atoms are involved, the potential function is more complicated for reactive collisions, but it still controls both the angular and product state distributions. The potential energy surface for a system of atoms can often be determined spectroscopically for some regions of coordinate space (see, for example, Section 8.7.3), but most often these measurements have to be supplemented with calculations. In this section we consider briefly how a potential energy surface is constructed and how to use it to calculate the outcome of collisional events.

Potential energy functions are calculated from the Schrödinger equation using the Born-Oppenheimer approximation. Because the motion of the electrons is much faster than that of the nuclei, it is a reasonable approximation to assume that the electrons adjust rapidly to any change in nuclear configuration. Thus, the electronic energy will be nearly independent of the nuclear motion and can be calculated for each desired geometry of the atoms. The electronic calculation typically employs a self-consistent field approach in which the energy of each electronic orbital is optimized assuming an average field of the remaining electrons. As the electrons are each considered in turn, the average field is improved. Such calculations of the electronic energy are performed for several geometries of the atoms involved in the reaction, and the resulting points on the energy surface are interpolated by a smooth fitting function.

Figure 8.24 gives some early results obtained for the $F + H_2$ reaction in the collinear configuration of lowest energy. The valley at the lower right corresponds to the reactants $F + H_2$, while that at the upper left corresponds to the products $H + HF$. The zero of energy has been taken as the energy of $F + H_2$, and the contours are labeled in units of kcal/mole. A typical reaction would proceed from the valley on the lower right to the one on the upper left over a slight barrier. Note that the barrier to the reaction is located near the entrance channel to the reaction, so that much of the exothermicity of the reaction is released before the H–H bond has stretched very far and before the HF bond has fully formed. The effect of this energy release is to accelerate the products toward short HF distances before the H–H bond has stretched, resulting in vibrational excitation of the HF product. We defer a more detailed explanation of this effect until the next section.

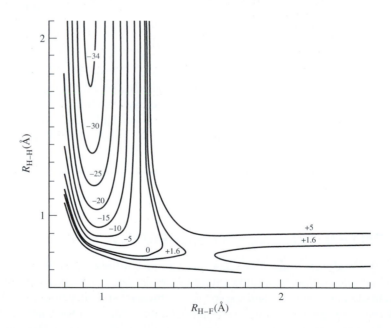

Figure 8.24

Potential energy surface for the $F + H_2$ reaction.

From C. F. Bender, S. V. O'Neil, P. K. Pearson, and H. F. Schaefer III, *Science* **176**, 1412 (1972). Reprinted with permission from *Science*. Copyright 1972 American Association for the Advancement of Science.

In the Born-Oppenheimer approximation the electrons sense only the charges of the nuclei and not their masses. Thus, the potential energy surface for F + D_2 is exactly the same as that for F + H_2; it is only the zero-point energies and vibrational level spacings that change with the differing reduced masses. Thus, we would also expect the product acceleration to excite DF vibrationally. It is precisely this feature of the potential energy surface that leads to the vibrational inversion in the DF product that we have already observed in **Figure 8.1** and **Figure 8.13.**

Before generalizing this result, we see how a semiclassical approach to the dynamics can provide us with greater insight into the dynamics.

8.5.1   Trajectory Calculations by Classical Mechanics

If particles behaved classically according to Newton's laws of motion, then the relationship between the potential energy surface and the dynamics of a reaction would be perfectly straightforward. One would simply specify the initial conditions for the reaction, for example the collision energy, the impact parameter, the angle of approach and so on, and then integrate Newton's laws to find the "trajectory" for the collision, i.e., the coordinates as a function of time. We know from the development of quantum mechanics that such an approach is deficient. Particles penetrate into regions of the potential energy that are classically forbidden, and their vibrational and rotational energies can have only quantized values rather than the continuous ones allowed in classical mechanics. Nonetheless, classical mechanics can give us a conceptual picture of the dynamics, and it often indicates the correct trends, for example, how the product state distribution might depend on the relative reactant velocity. We thus briefly explore here how this classical approach might help us to understand molecular dynamics.

The equations that we need to describe the system are basically simple, but the notation is somewhat cumbersome. Let $k = 1, 2, \ldots, N$ index the atoms in the system, and let q_i^k with $i = 1, 2, 3$ index the three Cartesian coordinates x, y, and z for atom k. The potential energy is a function of the $3N$ coordinates: $V = V(q_i^k; i = 1–3, k = 1–N)$. Consider starting at some location on the potential energy surface. The force along any coordinate is simply the negative of the change in potential along that coordinate: $F_i^k = -(\partial V/\partial q_i^k)$. However, by Newton's law $F_i^k = m_k a_i^k = d(p_i^k)/dt$. Thus,

$$\frac{d}{dt}(p_i^k) = -\frac{\partial V}{\partial q_i^k}. \tag{8.14}$$

The kinetic energy of the system is simply given by the sum of the kinetic energies in all of the coordinates:

$$T = \sum_{i=1}^{3} \sum_{k=1}^{N} \frac{1}{2m_k}(p_i^k)^2. \tag{8.15}$$

By taking the partial derivative of both sides of this equation with respect to p_i^k, we find that

$$\frac{\partial T}{\partial p_i^k} = \frac{1}{m_k}p_i^k = \frac{d}{dt}(q_i^k), \tag{8.16}$$

or

$$\frac{d}{dt}(q_i^k) = \frac{\partial T}{\partial p_i^k} = \frac{p_i^k}{m_k}. \tag{8.17}$$

Suppose we start the system in a particular configuration of positions and velocities corresponding to a particular kinetic and potential energy; the initial values of V, T, p_i^k, and q_i^k are thus known. **Equations 8.14** and **8.17** then enable us to calculate the change in momentum and position along each coordinate. By taking small time steps and integrating these equations numerically, we can then develop the trajectory of the system. A check on the numerical accuracy of the method can be made by comparing the total energy, $E = (T + V)$, at any time with the initial total energy. Although this method is numerically straightforward, there are several features that must be considered.

The first feature of the classical mechanics approach to notice is that it requires us to specify some parameters over which a typical experiment might have no control. For example, even collision experiments that measure cross sections from a given state of the reactants to a specified state of the products still provide no control over such parameters as the initial impact parameter, the orientation of the reactants, the phase of the reactant vibration(s) with respect to the time of collision, and so on. If classical calculations are to provide guidance in understanding our experiments they must average appropriately over these uncontrolled parameters of the collision. A typical method for such averaging is the *Monte Carlo* technique. Initial conditions are chosen at random from the appropriate distribution. For example, the initial phase of reactant vibration might be chosen by picking a number at random between 0 and 2π.

A second feature to note is that, although we will be interested in such quantum mechanical features as the vibrational or rotational product state distribution, classical mechanics does not recognize that the products have quantum states. A given integration of Newton's equations will likely produce a product with vibrational energy different from that allowed in quantum mechanics. Of course, if many states of the products are produced we might expect the overall classical distribution of energy in a particular degree of freedom to be similar to the quantum distribution, but in order to obtain a correspondence we must use a *binning* process to assign quantum states to a particular classical outcome. For example, if the allowed quantum mechanical vibrational energies are given by $(v + \frac{1}{2})h\nu$, we might assign energies $0 \leq \epsilon < h\nu$ to $v = 0$, energies $h\nu \leq \epsilon$ $2h\nu$ to $v = 1$, and so on.

Once we recognize the need for averaging over initial conditions and for binning the final results, it is quite simple to perform a number of trajectories to see how particular features of the potential energy surface influence the dynamics of a reaction. The force a mass feels along any direction r is simply the negative derivative of the potential, $F = -dV/dr$, and the acceleration that the mass experiences in this direction is simply given by Newton's law, $F = ma$. Given the initial values for the positions and velocities, final values can be obtained by calculating the changes in positions and velocities during each of a large number of small time increments.

Figure 8.25 shows typical results for distance as a function of time in the abstraction of an H atom from $H_2(v = 0, J = 1)$ by $O(^1D)$. In this figure the curves labeled R_{OA} and R_{OB} give the distances between the oxygen atom and the two hydrogen atoms, denoted by A and B, while the curve labeled R_{AB} gives the H–H distance. The rapid oscillations on all three curves correspond to the H_2 zero-point vibrational motion, while the slower oscillations on R_{OA} and R_{OB} correspond to H_2 rotational motion. At $t \approx 30 \times 10^{-14}$ s, the collision results in a reaction between the oxygen and the hydrogen labeled A to produce OH, so that R_{OA} now oscillates with the $v = 2$ vibrational motion of OH, while R_{OB} and R_{AB} increase as hydrogen B moves away from the OH product.

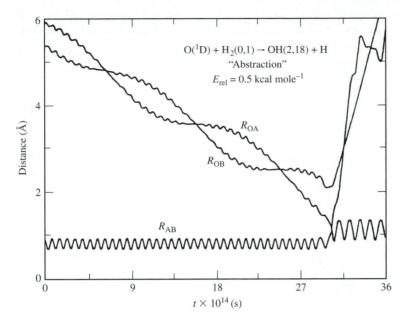

$O(^1D) + H_2(0,1) \rightarrow OH(2,18) + H$
"Abstraction"
$E_{rel} = 0.5$ kcal mole^{-1}

R_{OA}

R_{OB}

R_{AB}

Distance (Å)

$t \times 10^{14}$ (s)

■ **Figure 8.25**

A trajectory for the $O(^1D) + H_2$ abstraction reaction.

From P. A. Whitlock, J. T. Muckerman, and E. R. Fisher, *J. Chem. Phys.* **76,** 4468 (1982). Reprinted with permission from the *Journal of Chemical Physics.* Copyright, American Institute of Physics, 1982.

For a reaction involving $N = 3$ atoms there are $3N - 6 = 3$ coordinates necessary to specify all the relative atomic positions. For example, in the $O(^1D) + H_2$ abstraction of **Figure 8.25** the three chosen coordinates were R_{OA}, R_{OB}, and R_{AB}. Although all the coordinates can be described in a plot like that of **Figure 8.25,** it is often more instructive to fix all but two of the $3N - 6$ coordinates and to plot the trajectory as a point moving along the potential energy surface expressed as a function of the two chosen coordinates. For example, if we constrain the atoms to be in a collinear geometry, then we could plot the trajectory for the $F + H_2$ reaction as a point moving along the potential energy surface of **Figure 8.24.** Such a trajectory might look like that in the upper left-hand panel of **Figure 8.26,** which offers some insight into energy consumption and deposition in chemical reactions.

Note in panels (A) and (B) of **Figure 8.26** that the barrier, like that for $F + H_2$, is close to the reactant valley (A + BC, lower right), a so-called *early* barrier. The top left panel shows a typical trajectory for the case when the reactants A + BC approach one another with translational energy. The translational energy propels the reactants toward the barrier, and the energy released as the trajectory leaves the barrier compresses the AB bond and results in vibrational energy of the products. By contrast, if an equivalent amount of energy were placed in the vibration of the BC reactant rather than in translation, the trajectory might be similar to that shown in panel (B), where not enough energy is provided along the A–B coordinate to attain the top of the barrier. We thus conclude that *early* barriers favor production of vibrationally excited products and require that the reactant energy be in translation rather than vibration.

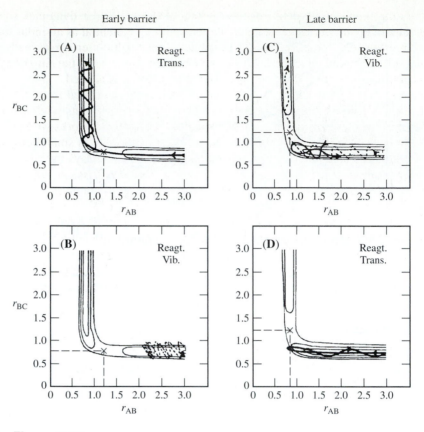

Figure 8.26

The position of the barrier controls what type of energy is deposited into product degrees of freedom as well as what type of reactant energy is required.

From J. C. Polanyi, *Acct. Chem. Res.* **5**, 161 (1972). Reprinted with permission from the *Accounts of Chemical Research.* Copyright 1972 American Chemical Society.

The opposite is true for the *late* barrier depicted in panels (C) and (D) of the figure. In panel (C) we see that, to turn the corner and attain the barrier, the trajectory needs not only to have its energy in BC vibration but also to have the correct phase (dotted trajectory rather than solid one). Translational energy, shown in panel (D), results in deflection back toward the reactant valley. Those trajectories that do react, such as the dotted one in (C), that do react produce primarily translational energy in the products, because much of the reaction exothermicity is released along the B–C direction. We thus conclude that *late* barriers favor production of translationally excited products and require that the reactant energy be in vibration rather than translation.

8.5.2   Semiclassical Calculations

The energy consumption and deposition propensities described in the previous section provide an example where classical calculations give a qualitative and often even quantitative connection between the potential energy surface and the dynamics.

On the other hand, more detailed descriptions of the molecular dynamics sometimes require quantum mechanical calculations with the potential as an input function. Unfortunately, the effort needed to achieve the results often far exceeds the amount of understanding gained. An alternative route that couples the simplicity of classical trajectories with the rigor of quantum mechanics is needed.

One such *semiclassical* approach uses the trajectory calculated from classical mechanics to describe the relative motion of reactants and products while treating the internal energy levels quantum mechanically. This approach works well as long as the amount of energy exchanged in the collision or reaction is small compared to the original translational energy; unfortunately this is not always the case. Even so, it is often instructive to examine the results of semiclassical approaches, because they provide a great deal of insight into what controls energy exchange processes.

For simplicity, we consider an inelastic collision $A + BC(i) \rightarrow A + BC(f)$ taking a molecule initially in state i to a final state f. Assume that the trajectory $R(t)$ is known from classical mechanics, where R here represents the distance of A from the center-of-mass of BC. Then the potential $V(R)$ can be treated as a time-dependent interaction, $V(R(t))$, and the wave function for the system will then satisfy the time-dependent Schrödinger equation

$$i\hbar \frac{\partial \psi(r,t)}{\partial t} = H(r,t)\psi(r,t). \tag{8.18}$$

Of course, at $t \rightarrow \infty$ and $t \rightarrow -\infty$, the potential goes to zero, and $H(r,t)$ becomes the time-independent Hamiltonian for the system, $H_0(r)$, with solutions

$$H_0(r)\phi_j(r) = E_j\phi_j(r), \tag{8.19}$$

where r represents the internal coordinate of BC.

Let us expand the general solution $\psi(r,t)$ in terms of the solutions to the time-independent problem with time-dependent amplitudes $a_k(t)$:

$$\psi(r,t) = \sum_k \phi_k(r)a_k(t)\exp\left(-\frac{iE_k t}{\hbar}\right). \tag{8.20}$$

Note that at $t = -\infty$ all the coefficients a_k are equal to zero except for the one multiplying the initial state of the system ϕ_i, which has a value $a_i = 1$. At some later time t the wave function will be a superposition of the basis states, with the probability of finding the system in state f given by the projection of the superposition state, $\psi(r,t)$, onto the wave function for state f:

$$\left| \int \phi_f^* \psi(r,t)\,dr \right|^2 = |a_f(t)|^2, \tag{8.21}$$

where all other terms in the sum for the expansion **equation 8.19** vanish because of the orthogonality of the solutions to the time-independent problem:

$$\int \phi_j^* \phi_k \, dr = \delta_{jk}. \tag{8.22}$$

The Kronecker delta function, δ_{jk}, is equal to unity if $j = k$ and is zero otherwise.

The probability of finding the system at $t = \infty$ in state f is simply $|a_f(t = \infty)|^2$, where the coefficient is determined by the condition that the wave function of **equation 8.20** be a solution to **equation 8.18.**

To calculate $|a_f(t = \infty)|^2$, let us substitute the expansion of **equation 8.20** into the time-dependent Schrödinger **equation 8.18** using the fact that the Hamiltonian can be written as $H = H_0 + V(t)$:

$$i\hbar \frac{\partial}{\partial t}\left\{\sum_k \phi_k(r)a_k(t)e^{-iE_kt/\hbar}\right\} = [H_0 + V(t)]\left\{\sum_k \phi_k(r)a_k(t)e^{-iE_kt/\hbar}\right\}. \quad (8.23)$$

Taking the derivative on the left-hand side and operating with H_0 on the right-hand side, we obtain

$$i\hbar\left\{\sum_k \phi_k(r)\frac{da_k}{dt}e^{-iE_kt/\hbar} + \sum_k \phi_k(r)a_k(t)\left(-\frac{iE_k}{\hbar}\right)e^{-iE_kt/\hbar}\right\}$$
$$= \left\{\sum_k E_k\phi_k(r)a_k(t)e^{-iE_kt/\hbar} + \sum_k V(t)\phi_k(r)a_k(t)e^{-iE_kt/\hbar}\right\}. \quad (8.24)$$

Note that the second summation on the left-hand side when multiplied by $i\hbar$ is exactly equal to the first summation on the right-hand side, so that

$$i\hbar\sum_k \phi_k(r)\frac{da_k}{dt}e^{-iE_kt/\hbar} = \sum_k V(t)\phi_k(r)a_k(t)e^{-iE_kt/\hbar}. \quad (8.25)$$

We now multiply both sides of **equation 8.25** by ϕ_f^* and integrate over the internal coordinate r:

$$i\hbar\sum_k\left\{\int \phi_f^*(r)\phi_k(r)\,dr\right\}\frac{da_k}{dt}e^{-iE_kt/\hbar}$$
$$= \sum_k\left\{\int \phi_f^*(r)V(t)\phi_k(r)\,dr\right\}a_k(t)e^{-iE_kt/\hbar}. \quad (8.26)$$

Because of the orthogonality of the ϕ_k given in **equation 8.22,** the sum on the left-hand side reduces to the single term for which $k = f$. On the right-hand side, we now make the approximation that, because the probability for excitation is small and because the initial state of the system was characterized by $a_i = 1$, the only coefficient in the sum likely to be important is a_i; all the rest will remain close to zero. Thus,

$$i\hbar\frac{da_f}{dt}e^{-iE_ft/\hbar} = \left\{\int \phi_f^*(r)V(t)\phi_i(r)\,dr\right\}e^{-iE_it/\hbar},$$
$$\frac{da_f}{dt} = \frac{1}{i\hbar}V_{fi}(t)e^{-i(E_i-E)_ft/\hbar}, \quad (8.27)$$

where

$$V_{fi}(t) \equiv \int \phi_f^*(r)V(t)\phi_i(r)\,dr. \quad (8.28)$$

Letting $\hbar\omega = E_f - E_i$, integrating both sides of **equation 8.27** from $t = -\infty$ to $t = \infty$, and squaring both sides, we obtain the time-dependent perturbation theory result known as the Born approximation:

$$|a_f(\infty)|^2 = \left|\frac{1}{i\hbar}\int_{-\infty}^{\infty} V_{fi}(t)e^{-i\omega t}\,dt\right|^2. \quad (8.29)$$

We will find that this equation is quite important in helping to understand energy transfer.

8.6 MOLECULAR ENERGY TRANSFER

At thermal equilibrium the population of each molecular energy level is determined by the Boltzmann distribution and depends on the energy of that level and on the number of states that have that energy:

$$N_i = \frac{g_i\exp(-\epsilon_i/kT)}{Q},\qquad (8.30)$$

where Q is the partition function, g_i is the degeneracy, and ϵ_i is the energy. Suppose, however, that by a reaction, radiative excitation, or some other mechanism, a system is perturbed from equilibrium. At what rate will it return to the distribution given by **equation 8.29,** and how does the rate depend on which degree of freedom is altered from its equilibrium value? In the process of relaxation, motion in a particular degree of freedom might be transferred during a collision to motion in a different degree of freedom. If we distinguish broadly between electronic (E), vibrational (V), rotational (R), and translational (T) degrees of freedom, then **Table 8.2** gives a matrix of possible types of energy transfer, where the column on the left lists the initial type of energy in one collision partner and the row across the top indicates what type of energy is excited in the other. That is, every collision is effective in exchanging translational energy, but up to 10^6 collisions may be needed to remove a quantum of vibration and convert that energy to translation. The numbers for transfer involving electronic energy vary greatly with the individual system and have therefore been omitted.

Although it is impractical to attempt to cover all of the types of energy transfer indicated in the matrix given above, it is instructive to examine a few key features of translational, vibrational, rotational, and electronic energy transfer.

8.6.1 Translational Energy Transfer

An important case of energy transfer is elastic scattering, in which, while no energy is lost from translational degrees of freedom, momentum is exchanged between the collision partners. **Figure 8.27** shows the speed distribution measured by Doppler spectroscopy for $S(^1D)$, produced in the 222-nm photodissociation of OCS, as a function of the number of collisions with argon at $T = 300$ K.[g] After about ten collisions the distribution is essentially indistinguishable from a Maxwell-Boltzmann

TABLE 8.2	**Energy Transfer between Different Degrees of Freedom**		
	Accepting Degree of Freedom (DOF)		
Initial DOF	**V**	**R**	**T**
V	10^2	10^4	10^6
R	10^4	1	10^2
T	10^6	10^2	1

The numbers in the table indicate roughly how many collisions would be required to effect the indicated energy transfer.

[g]G. Nan and P. L. Houston, *J. Chem. Phys.* **97,** 7865–7872 (1992).

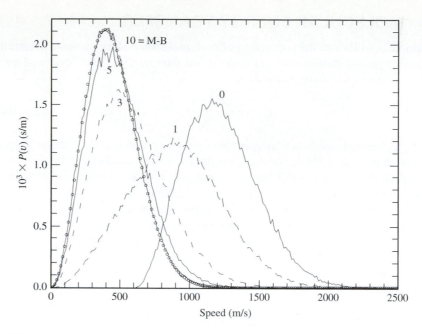

■ **Figure 8.27**

Relaxation of S(^{1}D) by argon. The curves give the speed distribution of sulfur as a function of the number of collisions with an argon gas at $T = 300$ K.

The data are taken from G. Nan and P. L. Houston, *J. Chem. Phys.* **97**, 7865–7872 (1992).

distribution at this temperature, shown as the open circles in the plot. Note that even after one collision, the average velocity of the sulfur has changed substantially toward the Boltzmann average.

A limiting case for elastic collisions is the hard-sphere approximation, where the potential of interaction, $V(r)$, is approximated as being infinite for $r < b_{max}$ and zero otherwise, where b_{max} is the distance of closest approach and is equal to the sum of radii for the collision partners. For such a potential, the scattering angle and momentum transfer are easy to calculate. **Figure 3.5** shows that when two hard spheres collide, the angle between the relative velocity vector and a vector along the distance of closest approach is given by $\sin \alpha = b/b_{max}$. Because no energy is lost, the trajectory for each collision partner must be symmetric about the vector along the distance of closest approach, so that the scattering angle, χ, is given by the equation

$$\chi = \pi - 2\alpha$$

$$= \pi - 2 \sin^{-1} \frac{b}{b_{max}}. \tag{8.31}$$

The momentum exchange between the two particles is governed by the conservation law

$$m_1 \mathbf{v}_1 + m_2 \mathbf{v}_2 = m_1 \mathbf{v}_1' + m_2 \mathbf{v}_2', \tag{8.32}$$

where the primed quantities are values after the collision and the unprimed ones are the initial values. For hard spheres, of course, the cross section for collision is constant as a function of energy and equal to $\pi b_{max}^2 = \pi(r_1 + r_2)^2$.

TABLE 8.3	**Collision Cross Sections for S(^{1}D) + Ar**		
P_{argon} (torr)	$n^\dagger$	$<v>$ (m/s)	σ_{col} (Å^2)
0.305	0.52 ± 0.02	1275	21
0.693	1.2 ± 0.1	980	22
1.151	2.2 ± 0.2	800	27
1.665	3.0 ± 0.3	800	28
2.220	4.0 ± 0.4	705	31
4.020	7.5 ± 1.0	600	38 ± 3
4.645	8.5 ± 1.0	597	39 ± 3
5.280	9.5 ± 2.0	595	39 ± 3

†Calculated number of collisions under experimental conditions, obtained from comparison of theory and experiment.

Table 8.3 compares the experimental results with a calculation based on the hard collision theory. If the theory were correct, the collision cross section would be a constant. However, it is clear that the cross section increases as the average relative velocity for the colliding pair decreases.

A much better fit to the data can be obtained by assuming a more realistic interaction potential, $V(r)$. For an arbitrary potential it can be shown that the scattering angle for elastic collisions is given by a somewhat more complicated formula[h]

$$\chi = \pi - 2b \int_{r_m}^{\infty} \frac{dr}{r^2[1 - V(r)/E_0 - b^2/r^2]^{1/2}}, \tag{8.33}$$

where $E_0 = \frac{1}{2}\mu v_r^2$ is the collision energy, and r_m, the distance of closest approach, depends on both E_0 and b. Given the scattering angle, the change in momentum can be calculated for each particle.

A potential often used to describe such collisions is the Lennard-Jones 6–12 potential

$$V(r) = 4\epsilon \left[\left(\frac{\sigma}{r} \right)^{1/2} - \left(\frac{\sigma}{r} \right)^6 \right], \tag{8.34}$$

where ϵ is the well depth and σ is the distance at which the potential is zero, as shown in **Figure 8.28**. Note from the figure that (for zero impact parameter) the distance of closest approach, r_m, increases as the collision energy decreases, qualitatively accounting for the increase in cross section with decreasing velocity described in **Table 8.3.** Detailed analysis of the data shows that the interaction potential is adequately described by the parameters $\sigma = 0.36 \pm 0.05$ nm and $\epsilon = 2.5 \pm 0.5$ kJ/mol.

---

[h]See, for example, R. E. Weston, Jr., and H. A. Schwarz, *Chemical Kinetics* (Prentice-Hall, Englewood Cliffs, NJ, 1972), pp. 46ff.

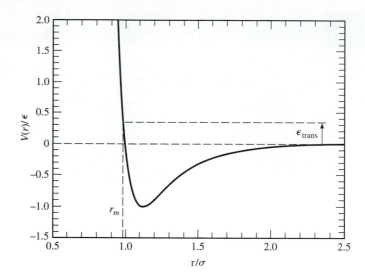

Figure 8.28

The Lennard-Jones 6–12 potential.

8.6.2   Vibrational Energy Transfer

The transfer of energy between translational and vibrational degrees of freedom is a rather inefficient process, usually requiring about 10^4–10^7 collisions (see **Table 8.2**). For example, as shown in **Figure 8.29,** the relaxation of vibrationally excited HCl by helium has a probability of about 3×10^{-7} at room temperature, and the probability increases strongly with increasing temperature. What is the cause of this strong increase in probability for energy transfer with temperature, and why does relaxation of CO_2 by N_2 behave so differently than that of HCl by Ar or He?

The essential physics of the exchange between vibrational and translational energy can be understood by considering the excitation of a piano string under two quite different circumstances, one in which the excitation mechanism is rapid compared to the vibrational period of the string and one in which it is slow. If a hammer strikes the string rapidly it is likely to induce vibration. On the other hand, no vibration results if one pushes and releases the string slowly compared to its vibrational period. The same considerations apply to the excitation of vibration in a molecule BC due to collision with an atom A. If, as is usual, the time scale of the collision is long compared to the vibrational period, then the oscillator has time to adjust to the perturbation, and little excitation is achieved. Such a collision is called *adiabatic*. We now examine the situation more quantitatively, using the formalism developed in Section 8.5.2.

Let the displacement coordinate for the BC oscillator from its equilibrium position be denoted as r, and let the distance between A and B be denoted as R_{AB}. We consider a collinear collision in which the collision of A with B compresses the BC bond so as to cause vibrational excitation. It is reasonable to assume that the potential energy rises steeply as a function of R_{AB}, and a common approximation employed in what is known as the Landau-Teller model states that $V(R_{AB}) = A \exp(-R_{AB}/\rho)$, where A is a constant and ρ is a distance known as the range parameter for the potential. In terms of R_{AB}, **Figure 8.30** shows that the relative separation, R, of A from the center of mass of BC can be written as $R = R_{AB} + \gamma(r + r_{eq})$, where $\gamma = m_C/(m_B + m_C)$ and r_{eq} is the equilibrium separation in the oscillator. Thus,

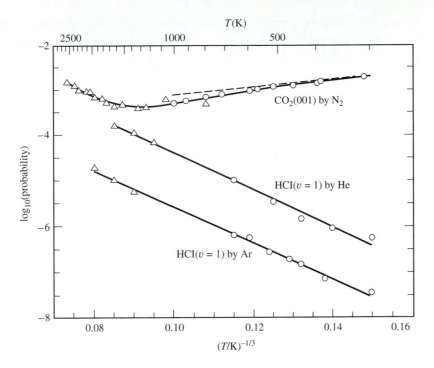

Figure 8.29

Probability of vibrational relaxation as a function of temperature.

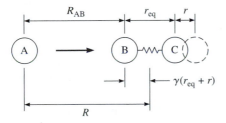

Figure 8.30

Geometry for collinear collision of A + BC showing that $R = R_{AB} + \gamma(r_{eq} + r)$.

$$
\begin{aligned}
V(R_{AB}) &= A \exp\left(-\frac{R_{AB}}{\rho}\right) = A \exp\left[-\frac{R - \gamma(r + r_{eq})}{\rho}\right] \\
&= A \exp(\gamma r_{eq}/\rho)\exp\left(-\frac{R}{\rho}\right)\exp\left(\frac{\gamma r}{\rho}\right) \\
&\approx A' \exp\left(-\frac{R}{\rho}\right)\left(1 + \frac{\gamma r}{\rho} + \cdots\right),
\end{aligned}
\qquad (8.35)
$$

where in the last line we have approximated the exponential by the first two terms in an expansion under the reasonable assumption that $\gamma r/\rho$ is small. The right-hand side of the last line gives the potential acting on the oscillator as a function of the relative separation R.

We now specialize the general result of **equation 8.29** to the situation for vibrational excitation of an oscillator by collision with an atom. Substitution of **equation 8.35** into **equation 8.28** gives

$$V_{fi}(t) = \int \phi_f^* A' e^{-R(t)/\rho} \left(1 + \frac{\gamma r}{\rho} \right) \phi_i \, dr$$

$$= A' e^{-R(t)/\rho} \left(\int \phi_f^* \phi_i \, dr + \frac{\gamma}{\rho} \int \phi_f^* r \phi_i \, dr \right). \tag{8.36}$$

The first integral on this last line is zero because of the orthogonality of the ϕ_k, while the second integral is a nonzero constant. Thus,

$$V_{fi}(t) = C e^{-R(t)/\rho}. \tag{8.37}$$

In the spirit of the semiclassical approximation we approximate the trajectory $R(t)$ as a straight line given by the velocity times time: $R(t) = |vt|$. Substitution of this trajectory into **equation 8.37** and then substitution of the result into **equation 8.29** gives

$$P_{f \leftarrow i} = |a_f(\infty)|^2$$

$$= \left| \frac{C}{i\hbar} \int_{-\infty}^{\infty} e^{-v|t|/\rho} e^{-i\omega t} \, dt \right|^2. \tag{8.38}$$

The integrand on the right-hand side of this equation consists of a rapidly oscillating function, $\exp(-i\omega t)$, times a more slowly varying one, $\exp(-v|t|/\rho)$, so that the integral will be exceedingly small unless the latter function changes appreciably on the time scale of the oscillation. A rule of thumb, then, is that the transfer of energy will be small unless $v/\rho \geq \omega$.

Another way of understanding how the energy transfer depends on the time scale of the collision is to recognize that the integral on the right-hand side of **equation 8.38** is simply proportional to the Fourier transform of the time-dependent potential. The equation tells us that to have appreciable energy transfer, the time-dependent perturbation must have Fourier components at the frequency of the oscillator. This result was anticipated by our consideration of the excitation of piano strings.

In the limit of small probabilities, where **equation 8.38** is valid, the integration yields

$$P_{f \leftarrow i} \propto \exp(-\xi), \tag{8.39}$$

where $\xi \equiv 2\pi\omega\rho/v$ is known as the *adiabaticity parameter*. Because ω is proportional to the reciprocal of the vibrational period, τ_v, and $\rho/v = \tau_c$, the collision time, the adiabaticity parameter is proportional to τ_c/τ_v, the ratio of the two characteristic times. When the adiabaticity parameter is much larger than unity, the collision is slow on the time scale of the oscillation, and the probability for vibration-translation exchange is small. In this limit, the collisional perturbation produces only small amplitude Fourier components at the frequency of the oscillator. When $\xi \approx 1$, the probability is appreciable, and when $\xi < 1$, we approach the *sudden* limit, where the collision is fast on the time scale of the oscillation. In the sudden limit, the perturbation has many high-frequency Fourier components.

To see how the vibration-translation energy transfer depends on temperature, we must average **equation 8.39** over velocity. The resulting *Landau-Teller* equation gives the probability as a function of temperature:[i]

$$P_{f \leftarrow i} = \exp\left[-3 \left(\frac{2\pi^4 \mu v^2}{\rho^2 kT} \right)^{1/3} \right], \tag{8.40}$$

---

[i]L. Landau and E. Teller, *Phys. Z. Sowjetunion* **10**, 34 (1936).

where ν is the vibrational frequency. Note that the logarithm of the probability should be linear if plotted as a function of $T^{-1/3}$, as is indeed observed for HCl($v = 1$) relaxation by helium and argon in **Figure 8.29.**

example 8.4

Mass Dependence of Vibration-Translation Energy Transfer

Objective   **Figure 8.29** shows that the relaxation of HCl($v = 1$) by He is more efficient than by Ar. What is the qualitative reason for this effect? Can we account for it quantitatively?

Method   The qualitative effect can be understood be considering the adiabaticity parameter, while the quantitative effect might be calculated from the Landau-Teller formula, **equation 8.40.**

Solution   The adiabaticity parameter is $\xi \equiv 2\pi\omega\rho/v$, where v is the relative velocity between the colliding partners. This relative velocity will be larger for HCl–He collisions than for HCl–Ar collisions, so the adiabaticity parameter will be smaller for the former than for the latter. The less adiabatic the collision, the higher the probability for vibration-to-translation energy transfer. Another way of looking at the qualitative result is to note that the faster collision with He produces higher amplitude Fourier components at the frequency of the oscillator.

From the Landau-Teller formula, **equation 8.40,** we see that the ratio of probabilities should go as $\exp[-(\mu_{He}/\rho_{He}^2)^{1/3}]/\exp[-(\mu_{Ar}/\rho_{Ar}^2)^{1/3}]$. If we assume that the range of the potential is the same for He and Ar (probably not valid, but close), then the ratio is $\exp[(-\mu_{He} + \mu_{Ar})^{1/3}] = \exp[(-3.58 + 18.51)^{1/3}] = \exp(2.44) = 11.5$. This is about the ratio in the figure.

One might then wonder from **Figure 8.29** what causes the relaxation of the antisymmetric stretching motion in CO_2 by N_2 to behave so differently from the relaxation of HCl by helium and argon. At high temperatures, the probability for relaxation falls with the expected $T^{-1/3}$ dependence, but at lower temperature it actually increases. The answer to this behavior appears to be that vibrational energy is being transferred from the CO_2 to the first vibrational level of N_2 via the process

$$CO_2(001) + N_2(v = 0) \rightarrow CO_2(000) + N_2(v = 1) + \Delta E = 18 \text{ cm}^{-1}.$$

To return to the piano string analogy, this collision is similar to bringing a vibrating piano string close to another nonvibrating string that has nearly the same resonance frequency. In the case of the piano strings, the coupling of the two nearly resonant oscillators by the viscosity of the air between them causes transfer of vibration from one string to another. In the case of CO_2 and N_2, it is the weak attractive force between the molecules that couples the vibrations.

A more quantitative explanation can be obtained by returning to **equation 8.29.** Recall that $\hbar\omega = E_f - E_i = \Delta E$, so that the $\exp(-i\omega t)$ term in this equation varies much more slowly for CO_2/N_2 than for HCl/Ar or HCl/He. We might thus expect that transfer would be effected by collisions with much lower Fourier components in the former case where $\Delta E = 18$ cm^{-1} than in the latter where $\Delta E = 2886$ cm^{-1}. Sharma and Brau evaluated the $V_{fi}(t)$ term in **equation 8.29** and found that the temperature dependence at low temperatures should be given by the dashed line in **Figure 8.29.**[j] Their calculation is seen to be in good agreement with experiment.

To summarize, vibration-to-translation energy transfer is often quite inefficient and typically increases with temperature because the collision becomes less adiabatic. Vibration-to-vibration energy transfer is fairly efficient when the resonance frequencies are similar.

8.6.3 Rotational Energy Transfer

The adiabaticity of the collision is also important in rotational energy transfer, particularly since rotational frequencies span the range from the adiabatic to the nonadiabatic limits for typical collision velocities. It is thus possible to have either a limit in which the molecule is rotating very rapidly during the approach of the collision partner or one in which the collision is so rapid that the rotor barely changes angle during the collision. In some cases, both limits can be observed in the same experiment, as in the $Xe + Na_2^*$ results shown in **Figure 8.31.** The vertical axis gives the detailed, state-to-state rate constant for the process $Xe + Na_2^*(J_i) \rightarrow Xe + Na_2^*(J_f)$, where the asterisk indicates that the process takes place in the excited $A^1\Sigma$

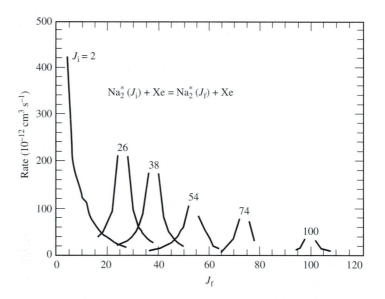

Figure 8.31

Rates of rotational relaxation of Na_2 from specific initial levels to the indicated first levels.

The data are taken from T. A. Brunner, N. Smith, A. W. Karp, and D. E. Pritchard, *J. Chem. Phys.* **74,** 3324 (1981).

[j]R. D. Sharma and C. A. Brau, *J. Chem. Phys.* **50,** 924 (1969).

state of Na_2. At the 350°C temperature of the experiment, collisions of Xe with low rotational levels of Na_2 are essentially in the sudden limit, whereas collisions with high rotational levels are in the adiabatic limit. As is obvious from the figure, the rate constants for relaxation of lower initial levels are much greater than those for relaxation of higher levels.

The other striking feature of the data is the way in which the rate constant falls off with the difference $|J_f - J_i|$. This is an example of a *scaling law* that represents large sets of rate data in terms of a single physical variable and a small number of adjustable parameters. Several scaling laws have been proposed for rotationally inelastic (and other energy transfer) processes,[k] including exponential energy-gap

$$k(J_i \rightarrow J_f) = a \exp\left(-\frac{b|E_f - E_i|}{kT}\right) \qquad (8.41)$$

and statistical power-gap laws

$$k(J_i \rightarrow J_f) = a'(|E_f - E_i|)^{-b'}. \qquad (8.42)$$

The best representation of rotational energy transfer data is often given by an angular-momentum-based scaling law, which is too complex to present here in detail but is discussed in the cited references. All of these scaling laws possess a common feature, namely, the inelastic rates decrease uniformly as either the amount of rotational energy or number of rotational quanta increases. This is another instance of the adiabaticity principle discussed in the preceding section.

8.6.4   Electronic Energy Transfer

Since we have seen that the concept of adiabaticity is important in determining the rates of vibrational and rotational energy transfer, one might wonder if this same concept is applicable to electronic energy transfer. Landau and others showed in 1932 that adiabaticity is indeed important. To see why, we consider the electronic energy transfer process in which $Ca(^1P_1)$ is relaxed to $Ca(^3P_J)$ when colliding with helium, shown schematically in **Figure 8.32.** Here, J refers to the group of three closely spaced fine structure levels $J = 0$, 1, and 2. In contrast to the ground state of calcium, which has its two outermost electrons in $4s$ orbitals, the $Ca(^1P_1)$ level has one electron in an excited $4p$ orbital. If this orbital is pointed toward the incoming helium atom, the electron in the $4p$ orbital will have a predominantly repulsive interaction with the $1s$ electrons on the helium, leading to an increase in the potential energy. On the other hand, if the $4p$ orbital is aligned perpendicular to the trajectory of the incoming helium atom, the ion core of the calcium can attract the He $1s$ electrons, giving rise to a slight decrease in the potential energy. In terms of the diatomic "molecule" HeCa, the Σ state, where the Ca p orbital is parallel to the incoming He trajectory, is repulsive, whereas the Π state, where the Ca orbital is perpendicular, is slightly attractive. Since similar curves arise from the interaction of the $Ca(^3P_J)$ atom with helium, it is likely that the repulsive curve for $Ca(^3P_J)$ + He crosses the attractive curve for $Ca(^1P_1)$ + He. A route for electronic energy exchange is then that marked by the arrows; on the outgoing trajectory a crossing between the attractive He + $Ca(^1P_1)$ curve and the repulsive He + $Ca(^3P_J)$ curve is

[k]T. A. Brunner and D. E. Pritchard, *Adv. Chem. Phys.* **50**, 589 (1982); A. J. McCaffery, M. J. Proctor, and B. J. Whitaker, *Ann. Rev. Phys. Chem.* **37**, 223 (1986); J. I. Steinfeld, P. Rutteberg, G. Millot, G. Fanjoux, and B. Lavorel, *J. Phys. Chem.* **95**, 9638 (1991).

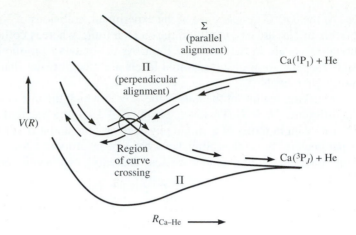

$V(R)$

$R_{\text{Ca–He}}$

▪ Figure 8.32

Schematic diagram of potential energy curves for the deactivation of $Ca(^1P_1)$ to $Ca(^3P_J)$ on collision with He.

From S. R. Leone, *Accts. Chem. Res.* **25,** 71 (1992). Reprinted with permission from the *Accounts of Chemical Research.* Copyright 1992 American Chemical Society.

responsible for the electronic energy relaxation. (Note that the crossing could have also occurred on the incoming trajectory with the same result, provided that only one crossing takes place.) The point of the article from which the figure is reproduced is that the relaxation takes place preferentially if the orbital of the $Ca(^1P_1)$ is aligned perpendicularly to the incoming He. We will return to this point, but for the moment let us concentrate on the curve crossing.

Our objective is to calculate the probability that reactants approaching one another on one potential energy curve might cross to another surface. Furthermore, we would like to understand how this crossing depends on the nature of the curves and on the relative velocity of the reactants.

The curve crossing problem has been considered in great detail by Landau, Zener, and Stueckelberg,[1] but the general features of the answer can be understood by considering the adiabaticity parameter. Recall from **equation 8.39** that the probability for making a transition between two vibrational states is proportional to $\exp(-\xi)$, where ξ is the adiabaticity parameter, $\xi = 2\pi\omega\rho/v$. Here $\omega = \Delta E/\hbar$ is proportional to the energy separation between the two states and ρ/v is the duration of the collision. We now adapt this result for electronic energy transfer.

Figure 8.33 shows in more detail the region of curve crossing (the region enclosed by the circle in **Figure 8.32**). The two crossing curves, which intersect at R_o, are shown as solid lines labeled 1°–1° and 2°–2°. These are the potential energy curves that would be calculated using a zeroth-order Hamiltonian, one that neglects possible interactions between the two surfaces caused, for example, by spin-orbit coupling or some other higher-order term in the Hamiltonian. The dashed curves labeled 1–1 and 2–2 would be the solutions of the exact Hamiltonian in the limit of the Born-Oppenheimer approximation and are called the "adiabatic" potential

[1]L. Landau, *Phys. Z.* **2,** 46 (1932); C. Zener, *Proc. R. Soc.* **A137,** 696 (1932); E. C. C. Stueckelberg, *Helv. phys. Acta* **5,** 369 (1932).

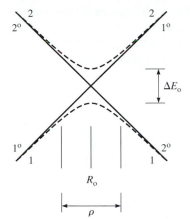

Figure 8.33

The region of the crossing of two potential energy curves 1°–1° and 2°–2°.

energy curves.[m] A trajectory proceeding very slowly through the location R_o would follow one of the adiabatic curves.

Suppose, however, that the trajectory traversed the region near R_o sufficiently rapidly that the electrons could not completely adjust during the time of passage, given roughly by the range ρ divided by the velocity v. There might then be some finite probability that the trajectory would follow, for example, the curve 1°–1° from lower left to upper right rather than the curve 1–1. In this case, a nonadiabatic transition might occur, so that the trajectory would remain on the approximate curves. It is reasonable to expect, in analogy to vibrational energy transfer, that the probability for such a nonadiabatic electronic transition should be dependent on the adiabaticity parameter. Thus, P_{12}, the probability of making a transition between the adiabatic curves 2–2 and 1–1, should be proportional to $\exp(-\xi)$ with $\xi = 2\pi\rho\Delta E/\hbar v$. We now simplify this expression by developing a relationship between ΔE and ρ.

Let the curves 1°–1° and 2°–2° be denoted by $V_1(R)$ and $V_2(R)$, respectively. Near enough to the region of the crossing these potential energy curves may be represented by straight lines, as shown in the figure. The value $|V_1 - V_2|$ is the difference in energy between the two curves. Now consider the curves 2–2 and 1–1. The difference in energy between these curves, ΔE, attains a minimum ΔE_o at R_o and increases slowly with $|R - R_o|$. Let us suppose that ΔE_o can be approximated by the average value of $|V_1 - V_2|$ over the region $R_o \pm \rho/2$. That is

$$
\begin{aligned}
\Delta E_o &= \int_{R_o-\rho/2}^{R_o+\rho/2} d|V_1 - V_2| \\
&= \int_{R_o-\rho/2}^{R_o+\rho/2} \frac{d|V_1 - V_2|}{dR}\, dR \\
&= \int_{R_o-\rho/2}^{R_o+\rho/2} \left| \frac{dV_1}{dR} - \frac{dV_2}{dR} \right| dR.
\end{aligned}
\tag{8.43}
$$

[m]There is substantial disagreement in the literature over the use of the word "adiabatic." Most would agree with K. J. Laidler, *Can. J. Chem.* **72,** 936 (1994) that the adiabatic curve is the one that is followed if the motion is infinitely slow. However, there is disagreement as to whether the adiabatic curve is for the full Hamiltonian (including, for example, spin-orbit coupling) or for some approximate Hamiltonian. Here, we use adiabatic to refer to the curve that would be followed for the full Hamiltonian and for infinitely slow motion.

If, as we have supposed, the potential energy curves V_1 and V_2 can be represented by straight lines, then the absolute value within the integrand is simply the absolute magnitude of the difference in slopes and is constant over the range $R_o \pm \rho/2$. Letting this difference in slopes be denoted ΔF and performing the integration, we obtain

$$\Delta E_o = \Delta F \int_{R_o - \rho/2}^{R_o + \rho/2} dR = \Delta F \rho, \tag{8.44}$$

or

$$\rho = \frac{\Delta E_o}{\Delta F}. \tag{8.45}$$

Substituting this value for ρ into the expression for ξ gives $\xi = 2\pi \Delta E_o^2/(\hbar \Delta F \upsilon)$, or that P_{12} is proportional to $\exp[-4\pi^2 \Delta E_o^2/(h \Delta F \upsilon)]$. The exponent determined from this simple treatment is actually too high by a factor of 4. With the correct exponent as first calculated by Zener, the transition probability is given by the Landau-Zener equation:

$$P_{12} = \exp\left(-\frac{\pi^2 \Delta E_0^2}{h \Delta F \upsilon}\right). \tag{8.46}$$

In collisions between atoms, such as that shown in **Figure 8.32,** the intersection point is usually crossed twice, once on the ingoing trajectory and once on the outgoing one. For deactivation to take place the trajectory must cross the first time and not the second or not cross the first time and cross the second. The total probability is thus

$$P = 2P_{12}(1 - P_{12}). \tag{8.47}$$

In collisions between molecules, the curves of **Figure 8.32** will represent cuts through higher-dimensional surfaces, and the intersection will also be of higher dimensionality than a point. The line (or surface) of intersection might be crossed many times during a collision trajectory, with a probability of P_{12} for transition at each crossing. Methods for treating such complicated problems are still under development. One approach propagates the classical trajectory until a crossing and then has the trajectory "hop" from surface to surface depending on the probability P_{12}. A more advanced method represents the trajectory as a wave packet and propagates part of the probability amplitude on each surface.

Returning to **Figure 8.32,** we note that the deactivation of Ca($^1P_1$) by He to Ca(^3P_J) is predicted to take place more efficiently if the p orbital of the initial calcium atom is aligned perpendicular to the trajectory of the incoming helium. By using molecular beams to define the relative velocity and a polarized laser to align the p orbital, Leone and his colleagues have demonstrated the validity of this prediction. **Figure 8.34** shows the relative deactivation probability as a function of the angle β between the p orbital and the relative velocity vector. Deactivation takes place preferentially with the perpendicular alignment. We consider alignment and orientation in more detail in Section 8.7.1.

The curve crossing ideas we used to develop the Landau-Zener formula are also useful in estimating the magnitude of cross sections for so-called "harpoon" reactions, as discussed in the next example. In addition, as shown in Problem 8.9, the Landau-Zener formula can also be used to predict the velocity dependence of the harpoon mechanism cross section.

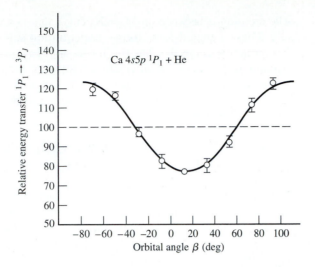

Figure 8.34

Deactivation probability of $Ca(^1P_1)$ by He as a function of the alignment between the calcium p orbital and the relative velocity vector.

From S. R. Leone, *Accts. Chem. Res.* **25,** 71 (1992). Reprinted with permission from the *Accounts of Chemical Research.* Copyright 1992 American Chemical Society.

example 8.5

The Harpoon Mechanism

Objective   Consider the reaction $M + X_2 \rightarrow M^+ + X_2^-$. The ionization potential (IP) of M is relatively small, whereas the electron affinity, EA, of X_2 is high. The relative curves are shown in **Figure 8.35.** At the point where the two potential energy curves cross, the electron can hop from M to X_2. Based on electrostatics, derive an estimate for the cross section for the reaction.

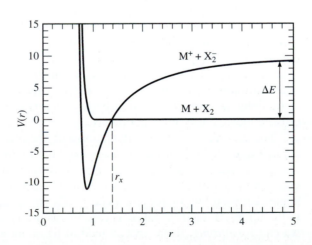

Figure 8.35

Potential energy curves appropriate for a discussion of the harpoon mechanism.

Method   The crossing is between an M + X_2 curve that is nearly flat (actually, it has a Lennard-Jones form—see **equation 8.34**) and an M^+ + X_2^- attractive curve that has the form $-e^2/(4\pi\epsilon_0 r)$ at long range, where e is the charge of an electron. However, the curves are displaced by ΔE, where $\Delta E = IP(M) - EA(X_2)$. We need to determine how r_x depends on these parameters and then estimate the cross section as πr_x^2.

Solution   At the crossing point, $\Delta E - [e^2/(4\pi\epsilon_0 r_x)] \approx 0$, or $r_x = e^2/(4\pi\epsilon_0\Delta E)$. Finally, $\sigma = \pi r_x^2 = e^4/(16\pi\epsilon_0^2\Delta E^2)$.

Comment   This type of curve crossing is very important in the production of light in rare gas-halogen excimer lasers. In these lasers, electrons excite a rare gas, Rg, to a electronic state, Rg^*, in which the outer electron behaves very similarly to that in an alkali metal. When Rg^* collides with a halogen, there is a curve crossing that produces Rg^+X^-, a bound excimer state. Because the ground state of the system, RgX, is repulsive, emission of light causes dissociation of RgX; there is thus always a population inversion.

8.7   MOLECULAR REACTION DYNAMICS— SOME EXAMPLES

The field of reaction dynamics is a rapidly expanding one, and in a short chapter, it is impossible to give much sense of its breadth. Nonetheless, a few examples from areas not previously covered can serve to convey some of the excitement of the field and to point to directions for future effort.

8.7.1   Reactive Collisions: Orientation

When considering collision theory we remarked in Section 3.3.2 that many reaction rate constants depend on the orientation of the reactants, and we introduced a *steric factor, $p < 1$,* to account qualitatively for the reduced reaction probability due to unfavorable orientations. By considering a simple model in which the barrier to reaction depends on the angle of approach, we showed that if the energy requirement along the line of centers varies with approach angle as $\epsilon^* + \epsilon'(1 - \cos\gamma)$, then p can be interpreted as being equal to kT/ϵ'.

Experimental measurement of the reaction rate as a function of approach angle, while difficult, is possible using a variety of techniques including strong electrostatic fields or laser excitation. What we seek is to *orient* or *align* the reactants in crossed molecular beam experiments.

The difference between orientation and alignment can be understood with the aid of **Figure 8.36.** Consider a vector fixed in the frame of a molecular reactant. In the case of methyl iodide, for example, we might consider a vector along the C–I bond and pointing toward the iodine. Now suppose that there is a distribution of angles between this vector and a chosen line in space, for example, a vertical line in the figure. If the distribution is random, as in the left panel of **Figure 8.36,** then we call it *isotropic*. If the vectors tend to lie along a particular line but without a

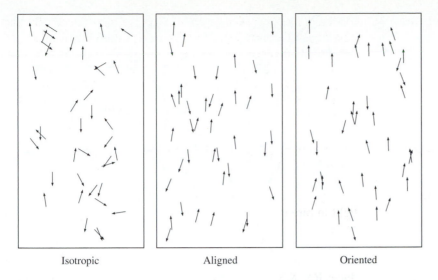

| Isotropic | Aligned | Oriented |

■ **Figure 8.36**

Isotropic, aligned, and oriented distributions of vectors.

bias as to whether they are parallel or antiparallel to the line, as for example, in the center panel of **Figure 8.36,** then we say that the distribution is *aligned.* If in addition to being aligned, the arrows favor a particular direction, such as the up direction in the right panel of **Figure 8.36,** then we say that the distribution is *oriented.*

In the case when the fixed vector in the frame of the molecular reactant is its rotational vector **J,** then this distinction between orientation and alignment can also be understood easily from the point of view of quantum mechanics. Let m_J be the projection of **J** onto an axis fixed in space. For either orientation or alignment the distribution of m_J values is not uniform. For alignment the distribution is symmetric but not uniform; the probability of finding m_J is the same as the probability of finding $-m_J$, but the value of, for example, m_0 is not equal to that of m_1. For orientation, on the other hand, the distribution is not symmetric, so that the probability of finding positive values of m_J is different than the probability of finding negative ones.

Orientation and alignment are useful for understanding molecular collisions. For example, if we could arrange to orient the C–I bond of CH_3I with respect to the relative velocity vector between CH_3I and an attacking Rb atom, we might then see how the reaction probability depended on angle.[n] Alternatively, we might be able to align the rotation vector of a reactant with respect to an incoming reactant. For example, in the reaction of vibrationally excited HF with Sr, it is possible to align the **J** vector describing HF rotation with respect to the incoming Sr.[o] In such an experiment, we might then see how the reaction probability depends on whether the Sr atom approaches the HF in the plane of its rotation or perpendicular to that plane.

---

[n]D. H. Parker, K. K. Chakravorty, and R. B. Bernstein, *J. Phys. Chem.* **85,** 466 (1981); see also P. R. Brooks and E. M. Jones, *J. Chem. Phys.* **45,** 3499 (1966) for a discussion of methyliodide-alkali atom systems.

[o]Z. Karny and R. N. Zare, *J. Chem. Phys.* **68,** 3360 (1978); see also J. G. Pruett and R. N. Zare, *J. Chem. Phys.* **64,** 1774 (1977) for a discussion of Ba + HF.

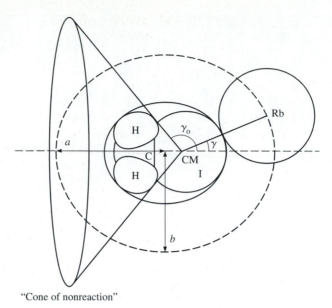

"Cone of nonreaction"

■ **Figure 8.37**

Pictorial representation of the steric aspects of the CH_3I + Rb reaction.

From S. E. Choi and R. B. Bernstein, *J. Chem. Phys.* **83,** 4463 (1985). Reprinted with permission from the *Journal of Chemical Physics*. Copyright, American Institute of Physics, 1985.

Let us examine in more detail results from an orientation experiment. By using a hexapole focusing field Parker, Chakravorty, and Bernstein were able to orient the methyl iodide, i.e., to choose the value of $< \cos \gamma > = KM/(J^2 + J)$, where γ is the angle of the C–I bond with respect to the relative velocity between Rb and CH_3I and J, K, and M are the quantum numbers specifying the rotation of the CH_3I. They found that the reaction, $Rb + CH_3I \rightarrow RbI + CH_3$, was much more probable if the CH_3I were in the orientation with the I end pointed toward the Rb than if it were in the orientation with the methyl end toward the Rb. The results are summarized in **Figure 8.37,** which shows a "cone of nonreaction" centered on the methyl group with a width of about 53°.

8.7.2   Reactive Collisions: Bond-Selective Chemistry

It has long been a goal in the field of reaction dynamics to use selective excitation to control the course of a reaction. A naive concept of how such control might be achieved assumes that we might be able to use, for example, a laser to excite vibrationally and then perhaps break a particular bond in a molecule. A large number of efforts have shown that, in general, this dream cannot be realized. In accord with the equilibrium assumptions of the RRKM theory (Section 7.5.4), energy put into a particular bond is redistributed among the other vibrational degrees before the molecule has a chance to react or dissociate. The excitation of hydrogen or deuterium stretching motions, however, provides an exception to this general rule. Because of the light mass of H or D, these stretching vibrations are of very high frequency and are not coupled well to the other, lower-frequency vibrations in most molecules. Sinha, Crim, and coworkers have recently used this observation to produce selective reaction at the OH or OD bond of HOD.

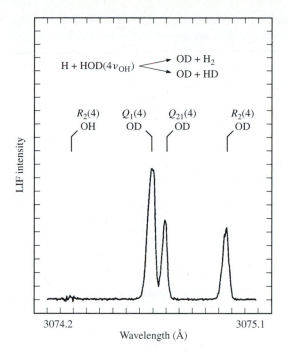

■ **Figure 8.38**

Laser-induced fluorescence (LIF) excitation spectrum of OH and OD resulting from the HOD $(4\nu_{OH})$ + H reaction.

From A. Sinha, M. C. Hsiao, and F. F. Crim, *J. Chem. Phys.* **94,** 4928 (1991). Reprinted with permission from the *Journal of Chemical Physics.* Copyright, American Institute of Physics, 1991.

Figure 8.38 shows the reaction scheme. HOD was selectively excited by a laser to a level with four quanta of the O–H stretch. These excited HOD molecules were then reacted with H atoms to form either H_2 + OD or HD + OH. The resulting OD and OH were probed using a second laser to excite fluorescence from the products. As the spectrum shows, several lines corresponding to the OD product are detected, while very little intensity is observed on lines corresponding to the OH product. Evidently, reaction occurred at the O–H bond, the bond excited by the laser, in preference to reaction at the O–D bond. The enhancement in reactivity at the O–H bond was found to be at least a factor of 100, far more than would be expected by a simple chemical isotope effect. The experiment shows that, while difficult, it is possible to direct the reaction to one of two otherwise nearly equivalent channels by carefully selecting the initial state of the reactants.

8.7.3   Potential Energy Surfaces from Spectroscopic Information: van der Waals[p] Complexes

Although much information about potential energy surfaces can be gained from scattering experiments or from state-to-state cross sections, the most accurate determination of the potential comes from spectroscopic measurements. Unfortunately,

---

[p]Professor J. D. van der Waals of Amsterdam won the Nobel Prize in Physics in 1910 for his work on the equation of state for gases and liquids.

spectroscopic measurements are rarely possible for a wide range of internuclear configurations, but where they are possible, the information they provide is usually the best available. An example is the recent determination of the potential energy surface for an argon atom interacting with an HF molecule.

The attractive well between closed-shell species like Ar and HF is, of course, extremely shallow, but supersonic expansion techniques (see p. 264) have enabled their observation. Because most of the internal energies for the molecules in an expansion have been converted into directed translational motion, the local temperature in the beam is very cold. A "thermometer" riding along with the molecules might register a temperature of only a few degrees Kelvin. At these low temperatures, collisions of even molecules that are only weakly bound can result in clustering.

Figure 8.39 shows the spectrum of the ArHF complex obtained using direct absorption of infrared light by the molecular beam. The lines in the spectrum correspond to different rotational transitions superimposed on vibrational excitation of a combination mode consisting of one quantum of H–F stretch and one quantum of Ar–HF stretch.

From spectra such as these it has been possible to construct the potential energy surface for the ArHF system shown in **Figure 8.40.** The figure depicts the potential that would be felt if an argon were placed at a specific angle and distance from the HF for a fixed HF bond distance. The "pillar" rising out of the figure is the increase in potential energy for argon positions very close to the HF, with the H in the right rear of the figure and the F in the left foreground. For larger distances, we see that as the argon approaches the HF, the potential is attractive for all angles, as shown by the "trough" around the HF. The depth of the trough varies with angle.

There are several points to note from the figure. First, the bonding between Ar and HF is extremely weak, only 216 cm^{-1} or 0.15 kJ mole^{-1}. Second, although the Ar is more tightly bound to the H end of HF, there is also a potential energy well at the F end. Finally, note that the barrier to bending is small, so that the ArHF complex is quite floppy.

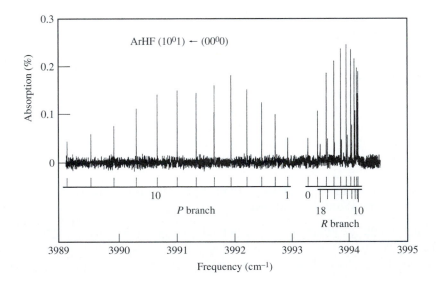

Figure 8.39

Spectrum of the ArHF ($10^{0}1$) $-$ ($00^{0}0$) van der Waals stretch combination band.

From C. M. Lovejoy and D. J. Nesbitt, *J. Chem. Phys.* **91,** 2790 (1989). Reprinted with permission from the *Journal of Chemical Physics.* Copyright, American Institute of Physics, 1989.

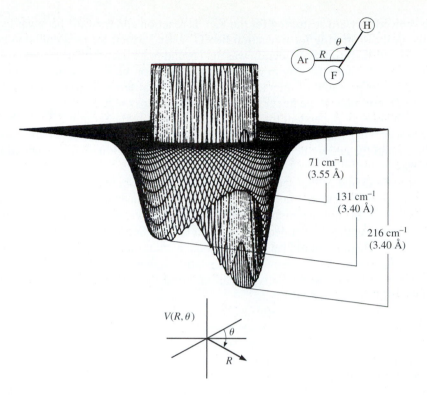

71 cm^{-1}
(3.55 Å)

131 cm^{-1}
(3.40 Å)

216 cm^{-1}
(3.40 Å)

$V(R, \theta)$

■ **Figure 8.40**

Two-dimensional bend-stretch potential energy surface for Ar + HF($v = 1$) determined from infrared data.

From D. J. Nesbitt, M. S. Child, and D. C. Clary, *J. Chem. Phys.* **90,** 4855 (1989). Reprinted with permission from the *Journal of Chemical Physics.* Copyright, American Institute of Physics, 1989.

8.8   SUMMARY

We started this section with the simple reaction F + D$_2$ → DF + D and asked how the overall rate constant for the reaction was related to the "state-to-state" rate constants. The answer we obtained was that

> The rate constant at a particular energy is the average of the state-to-state rate constants over initial states and their sum over final states.

Specifically,

$$k(T) = \int_{\epsilon=0}^{\infty} \sum_{i=1}^{\infty} \sum_{f=1}^{f_{max}} P(i) v_r G(\epsilon_r) \sigma(\epsilon_r, i, f)\, d\epsilon_r, \qquad \textbf{(8.8)}$$

where $v\sigma(\epsilon_r, i, f) = k(\epsilon_r, i, f)$, the state resolved rate constant at a particular energy.

Much of our knowledge of reaction dynamics has come from scattering experiments using crossed molecular beams. A key to understanding these studies is the *Newton diagram,* which relates the velocities in the laboratory frame to the velocities in the center-of-mass frame. By measuring the angular distribution of products in the laboratory and using the Newton diagram to convert the results to the center-of-mass frame, one can obtain the differential cross section. An example for the F + D$_2$ reaction was given in **Figure 8.13,** which showed that the DF product is scattered backward with respect to the incoming F atom. Other examples were discussed that showed forward scattering for the K + I$_2$ reaction and forward-backward symmetry in the scattering for a reaction like O + Br$_2$, which forms a collision complex. The differential cross section was also found to be useful in elastic and inelastic processes, where the way in which the cross section varies with angle is closely related to the potential of interaction. The differential cross section is also related to the rate constant. If we integrate the differential cross section over all angles, we obtain the energy-dependent total cross section. We saw in Section 3.3 that the rate constant for the reaction was obtained by averaging the product of the relative velocity and the energy-dependent cross section over the Boltzmann distribution of energies.

A main theme of this section has been the relation between the dynamics of a reaction and the potential energy surface. This relationship can be explored in an intuitive way by the technique of classical trajectory calculations. For example, simple considerations of trajectories lead to the prediction that an early barrier in the reaction channel should lead to vibrational excitation in the products and should require that the reactant energy be in translation rather than vibration. By contrast, a late barrier leads preferentially to translation in the products but favors the consumption of vibrational energy.

Of course, classical trajectories, although often accurate, do miss quantum mechanical features. Yet for all but the simplest systems, a full quantum mechanical treatment is often impractical. Semiclassical approaches, which typically use a classical trajectory to describe the relative motion while treating internal motion quantum mechanically, are often a good compromise. Using such an approach, we found that the probability for production of final state f starting in initial state i under the perturbation of a time-dependent potential is given in the Born approximation by

$$|a_f(\infty)|^2 = \left| \frac{1}{i\hbar} \int_{-\infty}^{\infty} V_{fi}(t)e^{-i\omega t}\,dt \right|^2. \qquad (8.29)$$

The transfer of energy between electronic, vibrational, rotational, and translational degrees of freedom is important in a variety of chemical processes and is related to the potential of interaction. For example, studies of translational energy transfer can give reasonable estimates for interatomic potential for collisions of most heavy particles even if classical mechanics is assumed. The potential also appears in equations describing other types of transfer, but semiclassical methods are needed because the internal degrees of freedom are quantized. By assuming a straight-line trajectory we saw that **equation 8.29** leads to the conclusion that the probability for vibrational energy transfer is exponentially dependent on the adiabaticity parameter, $\xi = 2\pi\omega\rho/v$, proportional to the ratio of the collision time over the period of vibration. When this parameter is much larger than 1, the perturbation is slow compared to the resonant frequency of the internal motion, and the transfer is inefficient. As this parameter approaches 1, the perturbation will produce Fourier components at the frequency of the internal motion, and transfer becomes more efficient. Rotational and electronic transfer are also dependent on the adiabaticity parameter. In the latter case, we found that the probability of curve crossing is given by the Landau-Zener formula:

$$P_{12} = \exp\left(-\frac{\pi^2 \Delta E_o^2}{h\Delta F v} \right). \qquad (8.46)$$

The field of molecular reaction dynamics is rapidly expanding, and it is impossible in a short description to give an idea of its breadth. We considered a few new directions: the use of techniques to orient or align reactants so as to examine the dependence of chemical processes on the angle of approach; the use of laser excitation to control the outcome of reactions; and the spectroscopy of van der Waals clusters to learn directly about the potential energy surface.

More generally, *Chemical Kinetics and Reaction Dynamics* has endeavored to explore the microscopic basis for chemical change. We began by considering collisions between gas molecules and saw that the ideas of Bernoulli led naturally to the Maxwell-Boltzmann distribution of molecular velocities. This distribution enabled us to deduce the rate of molecular collisions and the mean free path. We saw that thermal conductivity, viscosity, and diffusion had their basis in the properties carried by molecules: energy, momentum, and density. Chapter 3 provided an introduction to empirical kinetics and taught us how to characterize reaction rates by their order. We also learned how elementary reactions could be put together into simple mechanisms, for example, opposing reactions, parallel reactions, and consecutive reactions. This last mechanism led us to the steady-state principle and a number of applications: unimolecular reactions, catalysis, free radical reactions, and reactions in solution. We returned to the theory of collisional events and saw that reaction rate constants could be predicted by collision theory. Alternatively, activated complex theory, based in statistical mechanics, also provided an estimation of the rate constant. Either of these theories concentrated on but a few features of the potential energy surface governing the forces between approaching reactants and separating products. After digressing into some special aspects concerning reactions in liquid solutions, we examined in more detail the potential surface and entered the field of molecular reaction dynamics, with applications to surface kinetics, photochemistry, and gas-phase reactions. The excitement in this field comes from being able to deduce through more and more sophisticated measurements exactly how and why reactions occur. Deducing the "how and why" of reactions led to a better understanding of many practical areas: catalysis, atmospheric chemistry, and chemical lasers, for example. However, like those who have gone before us, our primary reason for seeking a better understanding has been simply to satisfy our desire to know.

suggested readings

P. B. Armentrout and T. Baer, "Gas-Phase Ion Dynamics and Chemistry," *J. Phys. Chem.* **100,** 12866 (1996).

Z. Bačić and R. E. Miller, "Molecular Clusters: Structure and Dynamics of Weakly Bound Systems," *J. Phys. Chem.* **100,** 12945 (1996).

M. Baer, ed., *The Theory of Chemical Reaction Dynamics* (CRC Press, Boca Raton, FL, 1985).

D. R. Bates, *Quantum Theory I. Elements* (Academic Press, New York, 1961).

R. B. Bernstein, ed., *Atom-Molecular Collision Theory: A Guide for the Experimentalist* (Plenum Press, New York, 1979).

A. W. Castleman, Jr. and K. H. Bowen, Jr., "Clusters: Structure, Energetics, and Dynamics of Intermediate States of Matter," *J. Phys. Chem.* **100,** 12911 (1996).

M. S. Child, *Molecular Collision Theory* (Academic Press, New York, 1974).

D. C. Clary, ed., *The Theory of Chemical Reaction Dynamics* (D. Reidel, Boston, 1986).

F. F. Crim, "Bond-Selected Chemistry: Vibrational State Control of Photodissociation and Bimolecular Reaction," *J. Phys. Chem.* **100,** 12725 (1996).

G. W. Flynn, C. S. Parmenter, and A. M. Wodtke, "Vibrational Energy Transfer," *J. Phys. Chem.* **100,** 12817 (1996).

P. L. Houston, "New Laser-Based and Imaging Methods for Studying the Dynamics of Molecular Collisions," *J. Phys. Chem.* **100,** 12757 (1996).

J. D. Lambert, *Vibrational and Rotational Relaxation in Gases* (Oxford University Press, Oxford, 1977).

R. Levine and J. Jortner, eds., *Molecular Energy Transfer* (Wiley, New York, 1976).

R. D. Levine, *Quantum Mechanics of Molecular Rate Processes* (Oxford University Press, London, 1969).

R. D. Levine and R. B. Bernstein, *Molecular Reaction Dynamics and Chemical Reactivity* (Oxford University Press, New York, 1987).

H. S. W. Massey, *Atomic and Molecular Collisions* (Taylor and Francis, London, 1979).

W. H. Miller, ed., *Dynamics of Molecular Collisions* (Plenum Press, New York, 1976).

C. B. Moore and I. W. M. Smith, "State-Resolved Studies of Reactions in the Gas Phase," *J. Phys. Chem.* **100,** 12848 (1996).

J. N. Murrell and S. D. Bosanac, *Introduction to the Theory of Atomic and Molecular Collisions* (Wiley, New York, 1989).

E. E. Nikitin, *Theory of Elementary Atomic and Molecular Processes in Gases* (Clarendon Press, Oxford, 1974).

G. C. Schatz, "Scattering Theory and Dynamics: Time-Dependent and Time-Independent Methods," *J. Phys. Chem.* **100,** 12839 (1996).

G. C. Schatz and M. A. Ratner, *Quantum Mechanics in Chemistry* (Prentice-Hall, Englewood Cliffs, NJ, 1993).

R. Schinke and J. M. Bowman, eds., *Molecular Collision Dynamics* (Springer-Verlag, Berlin, 1983).

G. Scoles, ed., *Atomic and Molecular Beam Methods,* Vols. 1 and 2 (Oxford University Press, London, 1988–1992).

A. G. Suits and R. E. Continetti, eds., *Imaging in Chemical Dynamics* (Am. Chem. Soc., Washington, 2001).

D. G. Truhlar, ed., *Potential Energy Surfaces and Dynamics Calculations* (Plenum Press, New York, 1981).

problems

8.1 The rate constant at a particular energy is which of the following: (a) an average of state-selected rate constants over the initial states, (b) an average of state-selected rate constants over the final states, (c) a sum of state-selected rate constants over the initial states, (d) a sum of state-selected rate constants over final states?

8.2 Collisions that form complexes are characterized by (a) scattering that is independent of final center-of-mass angle, (b) forward scattering, (c) backward scattering, or (d) symmetric forward and backward scattering.

8.3 The *rainbow* impact parameter is the one for which (a) there is no deflection angle, (b) the deflection angle is most negative, (c) the deflection angle is a maximum, or (d) the deflection angle is infinite.

8.4 Maxima in the scattering intensity $I(\Theta)$ occur when the change of deflection angle with impact parameter is (a) infinite, (b) zero, (c) a constant, or (d) a large value.

8.5 Translation-to-vibrational energy transfer generally becomes more efficient with which of the following: (a) an increase in T, (b) a decrease in T, (c) an increase in the duration of the collision, (d) a decrease in the duration of the collision?

8.6 The potential in **Figure 8.41** for the BrHI system depicts the collinear approach of the reactants [from M. Broida and A. Persky, *Chem. Phys.* **133,** 405 (1989)]. Contours are given in kcal/mole, and the dark point represents the position of the barrier. For the Br + HI → HBr + I reaction, is vibration or translation of the reagent more likely to promote the reaction? For the same reaction, would you expect the products to be translationally or vibrationally excited?

8.7 In Section 2.4.1 we discussed the principle of detailed balance: in a system at equilibrium, any process and its reverse proceed at the same rate. This principle can often be used to check the internal consistency of experimental data. The table on page 311 gives the detailed state-to-state rate constants [in "experimental" units of 10^{-4} (torr n.s.)$^{-1}$] for rotational energy transfer in CN from

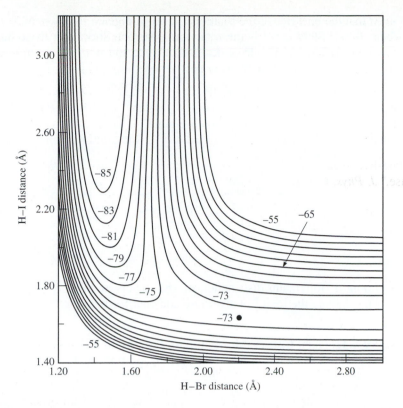

■ **Figure 8.41**

Potential energy surface for the collinear BrHI system.

From M. Broida and A. Persky, *Chem. Phys.* **133,** 405 (1989). Reprinted from *Chemical Physics,* copyright 1989, with permission from Elsevier Science.

rotational level N_i to rotational level N_f as well as the rate constant for the reverse process (some measurements were repeated, as shown by double entries).

N_i, N_f	$k_{i \to f}$	$k_{f \to i}$
1, 3	36.66 ± 2.76	15.05 ± 0.87
1, 5	15.84 ± 1.98	4.89 ± 0.76
1, 7	11.18 ± 1.40	3.99 ± 0.37
2, 4	23.82 ± 2.54	19.44 ± 1.49
2, 6	15.48 ± 2.43	8.44 ± 1.25
3, 5	26.35 ± 1.61	17.94 ± 1.22
11, 12	7.85 ± 1.77	8.12 ± 1.92
11, 14	3.36 ± 0.38	7.19 ± 0.83
	4.70 ± 0.60	5.93 ± 0.90
11, 15	5.67 ± 0.91	9.76 ± 1.40
11, 17	1.81 ± 0.26	5.60 ± 0.61
	1.43 ± 0.33	4.19 ± 0.83
12, 15	2.92 ± 0.93	6.42 ± 1.20
14, 17	3.74 ± 0.55	5.56 ± 1.03

The data are taken from R. Fei, H. M. Lambert, T. Carrington, S. V. Filseth, C. M. Sadowski, and C. H. Dugan, *J. Chem. Phys.* **100,** 1190(1994). Reprinted with permission from the *Journal of Chemical Physics.* Copyright, American Institute of Physics, 1994.

Given that the energies of the rotational levels are given by $E_N = N(N + 1)B$, where $B = 1.8997$ cm^{-1} is the rotational constant, show that these data are consistent with the principle of detailed balance and with the room temperature at which the experiments were performed.

8.8   For the reaction F + HCl(v) → HF(v') + Cl, the relative rate constants have been measured by J. L. Kirsch and J. C. Polanyi [*J. Chem. Phys.* **57**, 4498 (1972)] and are shown in the following table, where the rates are relative to the specific rate constant for $v = 0$ and $v' = 2$ and are for a translational temperature of 1700 K.

v	v'	$k_{relative}$	v	v'	$k_{relative}$
0	1	0.47	1	1	0
0	2	1.0	1	2	2.1
0	3	0.09	1	3	4.9
			1	4	0.3

Assuming a Boltzmann distribution of HCl levels at 1700 K ($\nu_{HCl} = 2990$ cm^{-1}), calculate the total (relative) rate constant for the reaction F + HCl → HF + Cl.

8.9   **Figure 8.42,** taken from A. M. Moutinho, J. A. Aten, and J. Los, *Physica* **53**, 471 (1971), shows data on the cross section, proportional to F plotted logarithmically on the ordinate, as a function of velocity, plotted on the abscissa, for the reactions Na + I → Na$^+$ + I$^-$ (open circles) and Li + I → Li$^+$ + I$^-$ (closed circles). Explain qualitatively why the cross section first rises and then falls as the velocity is increased.

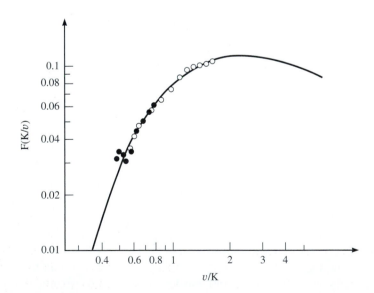

Figure 8.42

Cross section as a function of velocity for Na + I → Na$^+$ + I$^-$ (open circles) and Li + I → Li$^+$ + I$^-$ (closed circles).

From A. M. Moutinho, J. A. Aten, and J. Los, *Physica* **53**, 471 (1971). Reprinted from *Physica*, copyright 1971, with permission from Elsevier Science.

8.10 Potential energy surfaces such as those given in Section 3.2 and **Figure 8.24** are useful representations, but they are somewhat misleading. What we would really like to have is an analog representation of the collision, i.e., a potential representation on which a ball of unit mass rolling along the surface would represent the trajectory of the collision. The surfaces we have used so far contain no information about the mass of the atoms and so are not useful in this regard. However, a simple transformation can produce what we would like. Consider the potential for a collinear reaction $A + BC \rightarrow AB + C$ in terms of the bond lengths R_{AB} and R_{BC}, so that $V = V(R_{AB}, R_{BC})$. Now introduce two new coordinates Q_1 and Q_2 such that $Q_1 = aR_{AB} + bR_{BC}\cos\beta$ and $Q_2 = bR_{BC}\sin\beta$, where $a = [m_A(m_B + m_C)/M]^{1/2}$, $b = [m_C(m_B + m_A)/M]^{1/2}$, and $\cos^2\beta = m_A m_C/[(m_B + m_C)(m_A + m_B)]$, with $M = m_A + m_B + m_C$.

a. Show that in the new coordinate system the total kinetic energy is given by $T = \frac{1}{2}(\dot{Q}_1^2 + \dot{Q}_2^2)$. Note that this form is that for the energy of a unit mass particle moving on coordinates Q_1 and Q_2. *Hint:* You may wish to start by writing the total kinetic energy as the sum of the relative motion of A with respect to B and of C with respect to the AB diatom:

$$T = \frac{1}{2}\mu_{A-B}(\dot{R}_{AB})^2 + \frac{1}{2}\mu_{AB-C}\left(\dot{R}_{BC} + \frac{m_A}{m_A + m_B}\dot{R}_{AB}\right)^2.$$

b. The geometrical effect of this transformation is to *skew* and *mass-weight* the two bond distances at an angle to one another rather than at the normal angle of 90°. Show that a point $V(R_{AB}, R_{BC})$ when plotted axes separated by angle β and mass weighted so that the ordinate is bR_{BC} and the abscissa is aR_{AB} gives the point $V(Q_1, Q_2)$.

c. What is the skew angle for the $F + H_2$ surface?

d. How would the dynamics of a ball rolling on the surface of **Figure 8.24** differ from those on an appropriately skewed surface?

8.11 Given the definitions of $\mu' = m_3 m_4/M$, $M = m_3 + m_4$, and $\mathbf{v}_r' = \mathbf{u}_4 - \mathbf{u}_3$, show that **equation 8.11** simultaneously satisfies **equations 8.10** and **8.9**.

8.12 A beam of fluorine atoms traveling with a speed of 800 m/s intersects a beam of deuterium molecules traveling at 1500 m/s at right angles. Reaction produces DF in various vibrational levels. A detector is located at the laboratory angle corresponding to the center-of-mass direction and 50 cm from the interaction zone; it detects the DF product. A chopper located between the reaction center and the detector is used to time the flight of the products.

a. Given that the exothermicity of the reaction is 142 kJ/mol, explain why two peaks in the arrival time-of-flight distribution are observed for the $HF(v = 3)$ product of the reaction (for which the exothermicity is 39.8 kJ/mol) whereas only one peak is observed for the $HF(v = 0)$ product. (*Hint:* This observation is due purely to the kinematics of the reaction and not to any feature of the differential cross section.)

b. What is the arrival time difference between the two peaks corresponding to $HF(v = 3)$?

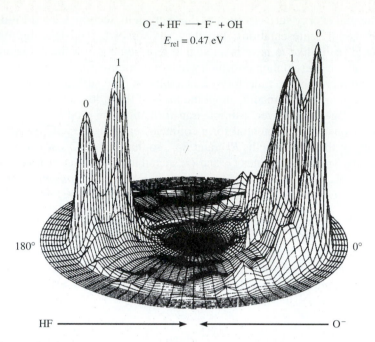

$O^- + HF \longrightarrow F^- + OH$

$E_{rel} = 0.47$ eV

Figure 8.43

Differential cross section for $O^- + HF \rightarrow F^- + OH$.

From J. M. Farrar, *Ann. Rev. Phys. Chem.* **46,** 525–554 (1995). With permission, from *The Annual Review of Physical Chemistry,* Volume 46 © 1995 by Annual Reviews www.AnnualReviews.org.

8.13   The differential cross section for F^- product of the reaction $O^- + HF \rightarrow F^- + OH$ at a collision energy of 45 kJ mol^{-1} is shown in **Figure 8.43.** We are given that ΔH for the reaction is -46 kJ mol^{-1}, that the distance between the peaks marked 1 in the cross section corresponds to 3000 m s^{-1}, and that the OH vibrational quantum is 3568 cm^{-1} (= 43 kJ mol^{-1}).

   a.   Show that the peaks are correctly assigned as belonging to OH($v = 1$) and OH($v = 0$).
   b.   What is the mechanism of the reaction? [The data and figure are from J. M. Farrar, *Ann. Rev. Phys. Chem.* **46,** 525–554 (1995).]

Answers and Solutions to Selected Problems

Chapter 1

1.1 (a)

1.3 Zero.

1.5 7.

1.7 $x_{mp} = L/2; <x> = L/2$.

1.9 The average molecular energy, $<\epsilon>$, will be given by

$$<\epsilon> = \int_0^\infty \epsilon G(\epsilon)\, d\epsilon,$$

where

$$G(\epsilon)\, d\epsilon = 2\pi \left(\frac{1}{\pi kT}\right)^{3/2} \sqrt{\epsilon} \exp\left(-\frac{\epsilon}{kT}\right) d\epsilon.$$

We can simplify the integral by letting $a \equiv \sqrt{\epsilon}$, so that

$$<\epsilon> = 2\pi \left(\frac{1}{\pi kT}\right)^{3/2} \int_0^\infty \epsilon \sqrt{\epsilon} \exp\left(-\frac{\epsilon}{kT}\right) d\epsilon$$

$$= 2\pi \left(\frac{1}{\pi kT}\right)^{3/2} \int_0^\infty a^3 \exp\left(-\frac{a^2}{kT}\right) d(a^2)$$

$$= 2\pi \left(\frac{1}{\pi kT}\right)^{3/2} 2 \int_0^\infty a^4 \exp\left(-\frac{a^2}{kT}\right) da$$

$$= 2\pi \left(\frac{1}{\pi kT}\right)^{3/2} 2 \frac{1}{2} \sqrt{\pi} \frac{3}{4} (kT)^{5/2} = \frac{3}{2}kT,$$

where the integral was evaluated using **Table 1.1**.

1.11 $f(\epsilon^*) \approx 10^{-100,000,000,000}$.

1.13 2.95×10^3 independent of T.

1.15 $0.673[kT/m]^{1/2}$.

1.17 \$1/molecule, independent of T.

Chapter 2

2.1 (c)

2.3 The reaction is first order.

2.5 The Arrhenius form of the rate law is $k = A \exp(-E_a/kT)$. A can be evaluated by (a) measurement of the rate at infinite temperature, because as $T \to \infty$, $-E_a/RT \to 0$ and the exponential $\to 1$, or (d) or (e) measurement of the intercept of a plot of $\ln k$ versus $1/T$: $\ln k = \ln A - (E_a/RT)$. Thus, the correct answer is (f).

2.7 (c)

2.9 (b)

2.11 The substrate concentration must be small.

2.13 (a) and (c) are both true, so the correct answer is (d).

2.22 $B = E_0 = 1.13 \times 10^{-5}\ M; A = 1.27 \times 10^{-9}\ M/s; b = 1.05 \times 10^{-2}\ s^{-1}; k_3 = 1.12 \times 10^{-4}\ s^{-1}. k_2 = 0.304\ s^{-1}$.

2.26 $E_a = 48.1$ kJ/mol.

2.27 $E_a = 112.21$ kJ/mol.

2.28 $k = 5.04 \times 10^{-12}\ cm^3\ molec^{-1}\ s^{-1}$. The activation energy is 1250 times R, or $(1250)(8.314\ J\ mol^{-1}\ K^{-1}) = 10.4$ kJ $mole^{-1}$, while the preexponential factor is $A = 5.41 \times 10^{-11}\ cm^3\ molec^{-1}\ s^{-1}$.

2.32 $k_v = 0.014\ \mu s^{-1}\ torr^{-1} = 1.4 \times 10^4\ s^{-1}\ torr^{-1}; k_1 = 714\ s^{-1}\ torr^{-1}; k_2 = 61\ s^{-1}\ torr^{-1}$.

2.34 The rate constant is $0.46 \times 10^{-3}\ s^{-1}$.

Chapter 3

3.1 Nine coordinates.

3.3 (d)

3.5 The activated complex for elimination of HBr would be more "ordered" than that for elimination of Br, so the elimination of HBr has a more negative entropy of activation and thus a smaller rate constant.

3.6 (b) is true.

3.10 (b) 8.8%; 4.6%; 25%.

3.11 (a) $A = 1.28 \times 10^{11}$ L mol^{-1} s^{-1}. (b) $p = 7.82 \times 10^{-4}$.

3.13 (a) $k = 2.4 \times 10^{11}$ L mol^{-1} s^{-1} exp$(-2.67) = 1.7 \times 10^{10}$ L mol^{-1} s^{-1}. The preexponential factor from collision theory, 2.4×10^{11} L mol^{-1} s^{-1}, agrees well with the experimentally observed result, 2×10^{11} L mol^{-1} s^{-1}. (b) $A = 0.67 \times 10^{11}$ L mol^{-1} s^{-1}, as compared to the experimental value of 2×10^{11} L mol^{-1} s^{-1}.

3.14 (a) $A(T) \propto T^{3/2}$. (b) $A(T) \propto T^2$. (c) $A(T) \propto T^3$.

3.16 (b) The values of k_H/k_D calculated for the given vibrational frequencies are CH, 7.78; OH, 10.3; NH, 8.96; SiH, 6.29.

Chapter 4

4.1   The viscosity coefficient, like the thermal conductivity coefficient, is independent of the molecular density because the mean free path depends inversely on density, whereas the amount of the quantity carried (momentum in this case and energy in the case of thermal conductivity) depends linearly on density. Thus, the density cancels in the final expression, provided, of course, that the density is high enough so that the mean free path is small compared to the macroscopic dimensions of the system.

4.3   The correct answers are (b), (d), and (e).

4.5   The coefficient κ is proportional to the average molecular speed, and helium is faster than argon because it is lighter. Although N$_2$ is lighter than argon, it has rotational degrees of freedom, so that its heat capacity is larger. Recall that κ is proportional to C_v.

4.7   (b) increases; (c) decreases.

4.9   The correct answer is (d).

4.11 800 W.

4.13 (a) 5.56×10^{-2} W m^{-1} K^{-1}. (b) $J_z A = (44.4$ W m$^{-2})$ $(1$ m$^2) = 44.4$ W. (c) 6.34×10^{-3} torr.

4.15 0.635 J.

Chapter 5

5.1 Independent of size.

5.3 The differential equation for the system is

$$\frac{d[C]}{dt} = k_f[A][B] - k_r[C].$$

Let $x = x(t)$ be $A(t) - A_e$, or equivalently $B(t) - B_e$ or $C_e - C(t)$. Then

$$-\frac{dx}{dt} = k_f(x + A_e)(x + B_e) - k_r(C_e - x)$$

$$= k_f A_e B_e - k_r C_e + k_f x(A_e + B_e) + k_r x + k_f x^2.$$

In the last equation, the first line is zero by the definition of the equilibrium and the last line can be neglected if we assume that the perturbation is small. The resulting equation is

$$\frac{dx}{dt} = -[k_f(A_e + B_e) + k_r]x,$$

$$x = x(0)\exp\{-[k_f(A_e + B_e) + k_r]t\}.$$

Substitution of this equation into $A(t) = A_e + x$ gives the desired result.

5.5 $k = 9.08 \times 10^9$ L mol^{-1} s^{-1}.

Chapter 6

6.1 (a) increases monotonically.

6.3 Only (b) or (d) could possibly be correct. Usually the barrier to diffusion is about 25% of that for desorption, so (d) is most nearly the correct answer.

6.5 Answer (c) is correct.

6.7 This is a case where one of the products of the reaction occupies sites needed by the reactant. In the presence of both NH$_3$ and H$_2$ in the gas phase and at temperatures where both adsorb to any appreciable extent, the surface coverage of NH$_3$ is given by **equation 6.5**, where "A" here refers to ammonia and "B" to hydrogen. In the limit when $K_B[B] >> K_A[A] >> 1$, the equation gives the surface coverage of ammonia as $\theta_A \approx K_A[A]/K_B[B]$. The rate for the unimolecular reaction is simply $k\theta_A$, which is first order in ammonia pressure but inversely proportional to hydrogen pressure.

6.9 The activation energy is assumed to be zero and we have already noted in the text that the partition function for the unoccupied site, q_s, can be taken as unity. Thus the rate constant is given simply by $k = (kT/h)\, q^{\ddagger}/q_g$. For atomic adsorption, q_g contains only translational motion: $q_g = (2\pi mkT/h^2)^{3/2}$. For the activated complex, by assumption there are (in addition to the vibrational coordinate perpendicular to the surface—that is, the reaction coordinate) two degrees of translational motion. Thus $q^{\ddagger\prime} = (2\pi mkT/h^2)^{2/2}$. The rate constant is thus

$$k = \frac{kT}{h}\frac{1}{(2\pi mkT/h^2)^{1/2}}$$

$$= \frac{(kT)^{1/2}}{(2\pi m)^{1/2}}$$

$$= \frac{(1/4)(16kT)^{1/2}}{(2\pi m)^{1/2}} = \frac{1}{4}\sqrt{\frac{8kT}{\pi m}}$$

$$= \frac{1}{4}<v>.$$

Since this equation gives the rate constant for adsorption, the rate of adsorption is simply $\frac{1}{4}<v>$ times the concentration of the gaseous reactants, n^*; or rate $= \frac{1}{4}n^*<v>$. This is also the rate at which atoms hit the surface, as we have seen in Sections 6.1 and 4.3.2.

6.10 $D = 4.0 \times 10^{-15}$ m^2 s^{-1}.

Chapter 7

7.1   (a), (c), (d), and possibly (e) are correct.

7.3   The basic assumption is that the electrons move much more rapidly than the nuclei, so that the molecule can change its electronic configuration on a time scale short compared to any changes in nuclear arrangement. Note that the assumption here is the same as that used in the Born-Oppenheimer approximation.

7.5   All of these processes depend on the density of states.

7.7   The differential equations for the three species are

$$\frac{dn_1}{dt} = 2j_a n_2 - k_b n_1 n_2 n_M + j_c n_3 - k_d n_1 n_3,$$

$$\frac{dn_2}{dt} = -j_a n_2 - k_b n_1 n_2 n_M + j_c n_3 + 2k_d n_1 n_3,$$

$$\frac{dn_3}{dt} = k_b n_1 n_2 n_M - j_c n_3 - k_d n_1 n_3.$$

Summing the first and third of these equations and applying the steady-state principle we obtain

$$\frac{dn_1}{dt} + \frac{dn_3}{dt} = 2j_a n_2 - 2k_d n_1 n_3 \approx 0.$$

Neglecting all but the second and third terms in the differential equation for n_1 and application of the steady-state equation to this intermediate yields

$$\frac{dn_1}{dt} \approx 0 \approx -k_b n_1 n_2 n_M + j_c n_3,$$

$$n_1 = \frac{j_c n_3}{k_b n_2 n_M}.$$

Substitution of this last equation into the equation for dn_1/dt + dn_2/dt gives

$$j_a n_2 - k_d n_1 n_3 \approx 0,$$

$$j_a n_2 - k_d \frac{j_c n_3}{k_b n_2 n_M} n_3 \approx 0,$$

$$n_3 = n_2 \sqrt{\frac{j_a k_b n_M}{j_c k_d}}.$$

7.10 $\mathbf{v}_{O_2} = (2714) \times [\sin(141.5)]/\sin(30) = 3379$ m/s, for an arrival time of 101 μs. The difference in arrival times is 90 μs.

7.11   (a) Estimating the spacing between the peaks in the upper plot in **Figure 7.35** gives about 1300 fs; $\nu = 1/t = 1/(1300 \times 10^{-15}$ s$) = 7.7 \times 10^{11}$ Hz. (b) To find a rate, assume an exponential rise and estimate the time, τ, at which a concentration of (maximum conc) $\times [1 - (1/e)]$ is reached. Take the inverse of this time to get the rate constant, $k = 1/\tau$. The rise time is about 580 fs, so $k = 1/(580 \times 10^{-15}$ s$) = 1.7 \times 10^{12}$ s^{-1}.

7.13   One method is to solve **equation 7.5** for $1/\rho(\nu)$, substitute in the blackbody result for $\rho(\nu)$, and equate terms with the same temperature dependence. Let $BF \equiv (g_2/g_1) \times \exp(-h\nu/kT)$. Then

$$B_{12}\rho(\nu) = B_{21}\rho(\nu)BF + A_{21}BF,$$

$$\rho(\nu) = \frac{A_{21}BF}{(B_{12} - B_{21}BF)},$$

$$\frac{1}{\rho(\nu)} = \frac{B_{12}}{A_{21}BF} - \frac{B_{21}}{A_{21}},$$

$$\left[\exp\left(\frac{h\nu}{kT}\right) - 1\right]\frac{c^3}{8\pi h^3} = \frac{B_{12}}{A_{21}BF} - \frac{B_{21}}{A_{21}}.$$

The first term on the left-hand side and the first term on the right-hand side of the last equation depend on temperature, while the second terms do not. Equating the temperature independent terms we obtain the second equation of **equation 7.6**. Equating the temperature dependent terms we find that

$$B_{12} = A_{21}\frac{g_2}{g_1}\exp\left(-\frac{h\nu}{kT}\right)\exp\left(\frac{h\nu}{kT}\right)\frac{c^3}{8\pi h\nu^3}$$

$$= \frac{g_2}{g_1}B_{21},$$

where in obtaining the last line we have used the second equation of **equation 7.6** obtained above.

7.15   Let the two fragments have center-of-mass recoil velocities of u_1 and u_2 and internal energies of ϵ_1 and ϵ_2. Let the energy of the photon be $E = h\nu$ and the bond dissociation energy be D_0^0. Then the total energy available to the photofragments following dissociation is $\epsilon = h\nu - D_0^0$ and is known. By conservation of energy, $\epsilon = \epsilon_1 + \epsilon_2 + \frac{1}{2}\mu u_r^2$, where u_r is the relative velocity of recoil and is related to u_1 and u_2 by the equation $\frac{1}{2}\mu u_r^2 = \frac{1}{2}m_1 u_1^2 + \frac{1}{2}m_2 u_2^2$. Putting these equations together gives $h\nu - D_0^0 = \epsilon_1 + \epsilon_2 + \frac{1}{2}m_1 u_1^2 +$

$\frac{1}{2}m_2u_2^2$. By conservation of linear momentum, $m_1u_1 = m_2u_2$, so that $u_2 = (m_1/m_2)u_1$. Substitution yields $h\nu - D_0^0 = \epsilon_1 + \epsilon_2 + \frac{1}{2}m_1u_1^2 + \frac{1}{2}m_2(m_1/m_2)^2u_1^2$. Solving for ϵ_2, we find that $\epsilon_2 = h\nu - D_0^0 - \epsilon_1 - \frac{1}{2}m_1u_1^2 - \frac{1}{2}m_2(m_1/m_2)^2u_1^2 = h\nu - D_0^0 - \epsilon_1 - \frac{1}{2}m_1[1 + (m_1/m_2)]u_1^2$. Thus, if all the quantities on the right-hand side of this equation are known, ϵ_2 can be calculated.

7.20 The square root dependence on light concentration suggests that there is an initiation step in which light dissociated Br_2 into

$$Br_2 + h\nu \rightarrow 2\ Br.$$

It is then likely that some sort of chain reaction takes place:

$$Br + CH_4 \rightarrow HBr + CH_3,$$

$$CH_3 + Br_2 \rightarrow CH_3Br + Br,$$

with a likely termination step as

$$Br + Br\,(+ M) \rightarrow Br_2\,(+ M).$$

Note that this is the same mechanism as the Br_2/H_2 reaction, except for the initiation step, so that we would expect the same result: linear dependence on the CH_4 concentration and square root dependence on the initiation step; that is, the rate should be proportional to the square root of light intensity and the square root of Br_2 concentration.

Chapter 8

8.1 The answers (a) and (d) are correct.

8.3 The correct answer is (b).

8.5 The answers (a) and (d) are correct.

8.9 The cross section will be proportional to the total probability given in **equation 8.47:** $P = 2\,P_{12}(1 - P_{12})$, where P_{12} is given in **equation 8.46.** Note that as P_{12} approaches either zero or unity, the total probability P will approach zero. This is because there are two possibilities for crossings, one on the way in and one on the way out. For the process to occur, one and only one surface hop must occur. Thus, P will be a maximum when $P_{12} = 0.5$. Because P_{12} increases as the velocity increases, P will at first increase as P_{12} increases, but will then decrease as P_{12} becomes greater than 0.5. A more quantitative treatment is discussed by A. P. M. Baede, *Adv. Chem. Phys.* **30,** (1975), see in particular section IA3 and Figure 5.

8.11 Substituting **equation 8.11** into the left-hand side of the first equation of **equation 8.10,** we find that

$$\frac{1}{2}m_3\left(\frac{m_4}{M}\mathbf{v}_r'\right)^2 + \frac{1}{2}m_4\left(\frac{m_3}{M}\mathbf{v}_r'\right)^2$$

$$= \frac{1}{2}\frac{m_4}{M}\left(\frac{m_3m_4}{M}\mathbf{v}_r'^2\right) + \frac{1}{2}\frac{m_3}{M}\left(\frac{m_3m_4}{M}\mathbf{v}_r'^2\right)$$

$$= \frac{1}{2}\left(\frac{m_3m_4}{M}\mathbf{v}_r'^2\right)$$

$$= \frac{1}{2}\mu'\mathbf{v}_r'^2.$$

Thus, **equation 8.10** is satisfied. **Equation 8.11** also satisfies **equation 8.9:**

$$m_3\mathbf{u}_3 + m_4\mathbf{u}_4 = -m_3\frac{m_4}{M}\mathbf{v}_r' + m_4\frac{m_3}{M}\mathbf{v}_r' = 0.$$

8.13 (a) From what we are given, the distance from the center of mass to the peak marked 1 is 3000/2 = 1500 m/s. By measuring with a ruler, we find that the distance between peaks 0 is 1.3 times the distance between peaks 1, so the distance from the center of mass to peak 0 is (1.35)(1500 m/s) = 2025 m/s. These then are radii of the Newton spheres that we are asked to show belong to $OH(v = 1)$ and $OH(v = 0)$. We need to show that these velocities make sense given the energetics of the reaction. If the collision energy is 45 kJ mol^{-1} and the exothermicity is 46 kJ mol^{-1}, then there are 91 kJ mol^{-1} available to the OH and F$^-$ products. For $OH(v = 0, J = 0)$, all of this energy would be in translation, so the relative velocity of the products, u, would be given by $\frac{1}{2}\mu'u^2 = 91$ kJ/mole. With $\mu' = (17)(19)/(17 + 19) = 8.97$ g/mole, we find that the relative velocity is $u = [(2 * 91\ \text{kJ/mole})(1000\ \text{g/kg})/(8.97\ \text{g/mole})]^{1/2} = 4504$ m s^{-1}. The diagram shows the F$^-$ products, and this velocity is 17/(17 + 19) of u, or 2126 m s^{-1}, somewhat larger than the 2025 m/s calculated for the $v = 0$ peak above. On the other hand, there may be some rotational excitation of the OH, and this would reduce the size of the product sphere. Note that a measurement (with a ruler) of total width of the diagram shows that all the scattering is within a sphere corresponding to 4416 m s^{-1}, in rough agreement with the total calculated value of 4054 m s^{-1}. For $OH(v = 1)$, there is less energy available by 43 kJ mol^{-1}, so that the total available is $91 - 43 = 48$ kJ mol^{-1}. The relative velocity is calculated as $u = (2 * 48\ \text{kJ/mole})(1000\ \text{g/kg})/(8.97\ \text{g/mole})]^{1/2} = 3271$. The F$^-$ velocity would then be 1545. Since this is again a bit larger than the measured value of 1500, there is probably some rotational excitation for the $v = 1$ products. In general, however, the assignment seems realistic. (b) Note that the reaction is peaked in both the forward and backward directions, suggesting that a long-lived complex is likely formed.

Index

Page numbers followed by t, f, and n indicate tables, figures, and notes, respectively.